T0296268

PHOTOSYNTHESIS IN ACTION

PHOTOSYNTHESIS IN ACTION

Harvesting Light, Generating Electrons, Fixing Carbon

Edited by

ALEXANDER RUBAN
School of Biological and Behavioural Sciences, Queen Mary University of London, United Kingdom

CHRISTINE H. FOYER
School of Biosciences, College of Life and Environmental Sciences, University of Birmingham, Edgbaston, United Kingdom

ERIK H. MURCHIE
Division of Plant and Crop Sciences, School of Biosciences, University of Nottingham, Sutton Bonington, Loughborough, United Kingdom

Academic Press is an imprint of Elsevier
125 London Wall, London EC2Y 5AS, United Kingdom
525 B Street, Suite 1650, San Diego, CA 92101, United States
50 Hampshire Street, 5th Floor, Cambridge, MA 02139, United States
The Boulevard, Langford Lane, Kidlington, Oxford OX5 1GB, United Kingdom

Library of Congress Cataloging-in-Publication Data
A catalog record for this book is available from the Library of Congress

British Library Cataloguing-in-Publication Data
A catalogue record for this book is available from the British Library

ISBN: 978-0-12-823781-6

For information on all Academic Press publications
visit our website at https://www.elsevier.com/books-and-journals

Publisher: Nikki P. Levy
Acquisitions Editor: Nancy Maragioglio
Editorial Project Manager: Pat Gonzalez
Production Project Manager: Omer Mukthar
Cover Designer: Matthew Limbert

Typeset by STRAIVE, India

Contents

Contributors

Martin Battle School of Life Sciences, University of Essex, Colchester, Essex, United Kingdom

Roel Brienen School of Geography, Faculty of Earth and Environment, University of Leeds, Leeds, United Kingdom

Alexandra J. Burgess School of Biosciences, University of Nottingham, Sutton Bonington, Loughborough, United Kingdom

Amanda P. Cavanagh School of Life Sciences, University of Essex, Colchester, Essex, United Kingdom

Luca Dall'Osto Department of Biotechnology, University of Verona, Verona, Italy

Robyn Emmerson School of Life Sciences, University of Essex, Colchester, Essex, United Kingdom

Christine H. Foyer School of Biosciences, College of Life and Environmental Sciences, University of Birmingham, Edgbaston, United Kingdom

David Galbraith School of Geography, Faculty of Earth and Environment, University of Leeds, Leeds, United Kingdom

Emanuel Gloor School of Geography, Faculty of Earth and Environment, University of Leeds, Leeds, United Kingdom

Rodrigo L. Gomez Department of Biotechnology, University of Verona, Verona, Italy

Zeno Guardini Department of Biotechnology, University of Verona, Verona, Italy

Guy Hanke School of Biological and Chemical Sciences, Queen Mary University of London, London, United Kingdom

Jeremy Harbinson Laboratory of Biophysics, Helix Building, Wageningen University, Wageningen, The Netherlands

Tanja A. Hofmann Suffolk One Sixth Form College, Ipswich, Suffolk, United Kingdom

Elias Kaiser Horticulture and Product Physiology Group, Radix Building, Wageningen University, Wageningen, The Netherlands

Diana Kirilovsky Institute for Integrative Biology of the Cell (I2BC), CEA, CNRS, Paris-Saclay University, Gif-sur-Yvette Cedex, France

Anja Krieger-Liszkay Institute for Integrative Biology of the Cell (I2BC), CEA, CNRS, Paris-Saclay University, Gif-sur-Yvette Cedex, France

Tracy Lawson School of Life Sciences, University of Essex, Colchester, Essex, United Kingdom

Lorna McAusland School of Biosciences, University of Nottingham, Sutton Bonington, Loughborough, United Kingdom

Jun Minagawa National Institute for Basic Biology, Division of Environmental Photobiology, Okazaki, Japan

Alejandro Sierra Morales Center for Crop Systems Analysis, Radix Building, Wageningen University, Wageningen, The Netherlands

Conrad W. Mullineaux School of Biological and Behavioural Sciences, Queen Mary University of London, London, United Kingdom

Erik H. Murchie School of Biosciences, University of Nottingham, Sutton Bonington, Loughborough, United Kingdom

Jacob Pullin School of Life Sciences, University of Essex, Colchester, Essex, United Kingdom

Christine A. Raines School of Life Sciences, University of Essex, Colchester, Essex, United Kingdom

Andrew J. Simkin School of Biosciences, University of Kent, Canterbury, Kent, United Kingdom

Herbert van Amerongen Laboratory of Biophysics, Wageningen University, Wageningen, The Netherlands

Shellie Wall School of Life Sciences, University of Essex, Colchester, Essex, United Kingdom

Emilie Wientjes Laboratory of Biophysics, Wageningen University, Wageningen, The Netherlands

Foreword

In the rapidly changing world, profound knowledge of the processes that enable its existence is crucial for humankind. Our planet is entering a new geological era—the Anthropocene. Throughout the history of humanity, we have changed Earth's biosphere: first, its nature—plants, animals, microbes—then our nature with advances in medicine, and finally, its climate. We began shaping the world armed with science and technology, and driven by the ambition of economic growth. As a result, we came to gradually realise that our engagement in the evolution (or deterioration) of the biosphere is here to stay. The recent violent manifestations in climate patterns, such as heat waves, floods, forest fires, and hurricanes, prompt conscientious people to think about how they can protect and preserve life on our planet.

The driving force behind life on Earth is the Sun that gives us energy in the form of light. This energy feeds the biosphere's 'wheel of life' consisting of the two opposing chemical processes: the oxidation and reduction of oxygen. The former is photosynthesis, the process that gives us molecular oxygen to breathe, food to eat, and many materials that we rely upon. The latter process is respiration. Most lifeforms, even those that possess no lungs or gills, need to respire—to use oxygen for slowly and intelligently 'burning' organic matter in order to build the structures of life. Photosynthesis, however, starts it all. It captures the Sun's energy and converts it into the energy and matter of life. That is why photosynthetic organisms are called autotrophs; they are feeding themselves. They do so on a diet of carbon dioxide, light, and water—a tough diet, but it has been working for several billions of years without any detrimental impact on the biosphere. In reality, almost all energy we use today came and keeps coming via photosynthetic organisms, including fossil fuels.

So, what is the purpose of this book, and why does it claim to be different from the numerous photosynthesis books published in the past? The following are the objectives of the book: (i) to reveal to the scholars and readers the fundamental knowledge of the stages (and the integration) of energy and matter transformation in photosynthesis, that supports all life; (ii) to show how flexible and adaptable to the environment these stages have evolved to be; and (iii) to explain how we can alter these stages, their principles, mechanisms, and efficiency in order to improve the photosynthesis for our own benefits and the benefit of the biosphere. Increasing photosynthetic adaptability, productivity, and survival in the changing environment is necessary to meet the food demands of the Anthropocene. This is photosynthesis in action as we currently know it. This is up-to-date knowledge, and it is our duty to address the demands of this new geological era. The book is written by a group of excellent experts working in the all-encompassing spectrum of the topics of photosynthesis research. It took me little effort to convince them to commit to this project, since all of them believe in the beauty and strength of the photosynthesis research for the sake of humankind. Have we succeeded? The answer, I assume, is here for the reader to learn, judge, and, hopefully, act. I, therefore, believe that the book has the potential to teach a broad audience of students, researchers, engineers, and policymakers.

Alexander Ruban

Preface

Food production is facing one of the most critical challenges of our times. As argued in the recently approved Green Deal, "our food system is under threat and must become more sustainable and resilient"—not only at European level but also on a global scale (European Commission, 2020). It comes as no surprise that the 2020 Nobel Prize for peace was awarded to the 'World Food Program', which has addressed the economic, social, and ecological emergencies that threaten food security worldwide. The main objective of the food security program is to increase the production yield of cereals around the globe and thereby ensure the incremental demand for food, animal feed, and biofuels (Miraglia et al., 2019; Prosekov and Ivanova, 2018). The Green Deal rules out the extensive use of fertilisers due to their high pollution potential. Genetic improvement is an essential part of the strategy, as is the increasing the efficiency of plant use of resources (Bailey-Serres et al., 2019). Photosynthesis is the major driver of carbon gain, biomass production, and crop yield. It is hence a central target for the improvement and sustainable production of crops in a changing climate. Climate change is considered the most important factor limiting the production and yield of cereals. Understanding the fundamental mechanisms of photosynthesis, its functional operation, and how the photosynthetic processes are regulated in response to changing environmental inputs and signals, as well as the central role of photosynthesis in plant growth, development, and survival, is a key driver of strategies for environmental-friendly agriculture. The information provided in this book will enable and support readers in (1) understanding the mechanisms of photosynthesis at a functional level, (2) quantitatively determining the relationships between each individual mechanism with the context of the operation of photosynthesis as a whole, and (3) highlighting possible interventions where the improvement of efficiency and function might be achieved. The 12 chapters that comprise this book are from outstanding scientists, who are internationally recognised experts in their respective fields. Each author has not only provided essential information but also included personal perspectives and insights that allow a deeper understanding of the topic and its importance.

Chapters 1, 4, and 7 introduce the readers to the principles of light harvesting (1) adaptations of the processes of light capture to abiotic stress (4) and possible ways to manipulate the process in order to improve the photosynthetic efficiency and protection against environmental factors (7).

Chapters 2, 5, and 8 provide a comprehensive introduction to the photosynthetic electron transport system in plants (2), the differences between the photosynthetic electron transport systems in chloroplasts and in cyanobacteria (8), and the systems that limit the production of reactive oxygen species (ROS) and hence chloroplast signalling.

Chapters 3, 6, and 9 introduce the mechanisms and limitations of CO_2 uptake and assimilation in leaves and chloroplasts, the barriers to diffusion and the metabolic processes underlying conversion of this gas into carbohydrates. Environmental 'stress' places restrictions on photosynthesis in leaves that can commonly restrict productivity and yield (6). Improving carbon fixation in crops to underpin yield improvement will be a substantial contribution to global food security, and routes for achieving this via the Calvin Benson pathway are described (9).

Chapters 10 and 11 provide an integration of the multiple scales on which photosynthesis operates. The role of photosynthesis in shaping Earth's climate and biosphere is described. Regulation of photosynthesis can be perceived at the molecular, cellular canopy, and ecosystem level, and this is integrated using tools such as mathematical modelling.

We recommend this book to undergraduate students, postgraduate students, and researchers who are interested in gaining a deeper knowledge of photosynthesis and its role in carbon gain and crop yield, as well as its responses to a changing climate. The content of the book will also be of interest to specialists who are working and studying various aspects of photosynthesis and its dynamic regulation, as well as nonspecialists, who are simply interested in knowing more about this environment-friendly and evolving process that sustains all life on Earth.

A.V. Ruban, E. Murchie, and C.H. Foyer

References

Bailey-Serres, J., Parker, J.E., Ainsworth, E.A., Oldroyd, G.E. D., Schroeder, J.I., 2019. Genetic strategies for improving crop yields. Nature 575, 109–118.

European Commission, 2020. A Farm to Fork strategy for a fair, healthy and environmentally-friendly food system.

Miraglia, M., Marvin, H.J.P., Kleter, G.A., et al., 2019. Climate change and food safety: an emerging issue with special focus on Europe. Food Chem. Toxicol. 47 (5), 1009–1021.

Prosekov, A.Y., Ivanova, S.A., 2018. Food security: the challenge of the present. Geoforum 91, 73–77.

PART I

Principles

CHAPTER

1

Harvesting light

Herbert van Amerongen and Emilie Wientjes

Laboratory of Biophysics, Wageningen University, Wageningen, The Netherlands

1 Introduction, what is light harvesting and why is it needed?

As already mentioned in the foreword of this book, the driving force behind life on earth is the sun that provides energy in the form of light. Photosynthesis is the process that captures this energy and converts it into the energy of life. In this chapter, we address how the energy is captured and delivered to the reaction centres (RCs), entities that are located in the thylakoid membrane and where the conversion process starts by a charge separation. The combination of capturing and delivery is usually called light harvesting and is performed by the light-harvesting complexes, also called antennae.

Electron transfer or charge separation: The transfer of an electron from a donor molecule in an excited state to a neighbouring cofactor. In photosynthesis this process occurs in a photosynthetic RC.

Excitation-energy transfer: The nonradiative transfer of excitation energy from an excited pigment, the donor, to a neighbouring pigment, the acceptor. Upon transfer the donor falls back to its ground state and the acceptor goes to the excited state. This process allows to deliver the excitations to the RCs.

The RCs will be discussed in more detail in Chapter 2; here, we summarise a few essential aspects. In organisms performing oxygenic photosynthesis, two types of RCs are present, type I and II. The type II RC complex, for instance, is a large pigment–protein complex consisting of several closely associated proteins with a number of cofactors, including several chlorophyll *a* (Chl *a*) pigments. If one of them, the primary electron donor, is in the (first) excited state, it can donate an electron to a neighbouring cofactor, which forms the start of a cascade of electron transfer steps, ultimately leading to the generation of a chemical potential that is fuelling ATP formation and the generation of reductive power.

The pigment–protein ratio in the RCs is rather low and a type II RC contains only 6 Chls *a* (less than 6 kDa), whereas the total molecular mass is hundreds of kDa. One Chl *a* in full sunlight can typically absorb ~10 photons per second and in isolation the RC would be inactive most of the time. To use their 'expensive' RCs more efficiently, photosynthetic organisms usually have 'cheap' light-harvesting complexes (LHCs) that surround the RCs to provide them with excitations. In that case, the excited-state energy is transferred from pigment to pigment towards the RC where it can be used for charge separation. The highly abundant light-harvesting

Photosynthesis in Action
https://doi.org/10.1016/B978-0-12-823781-6.00007-1

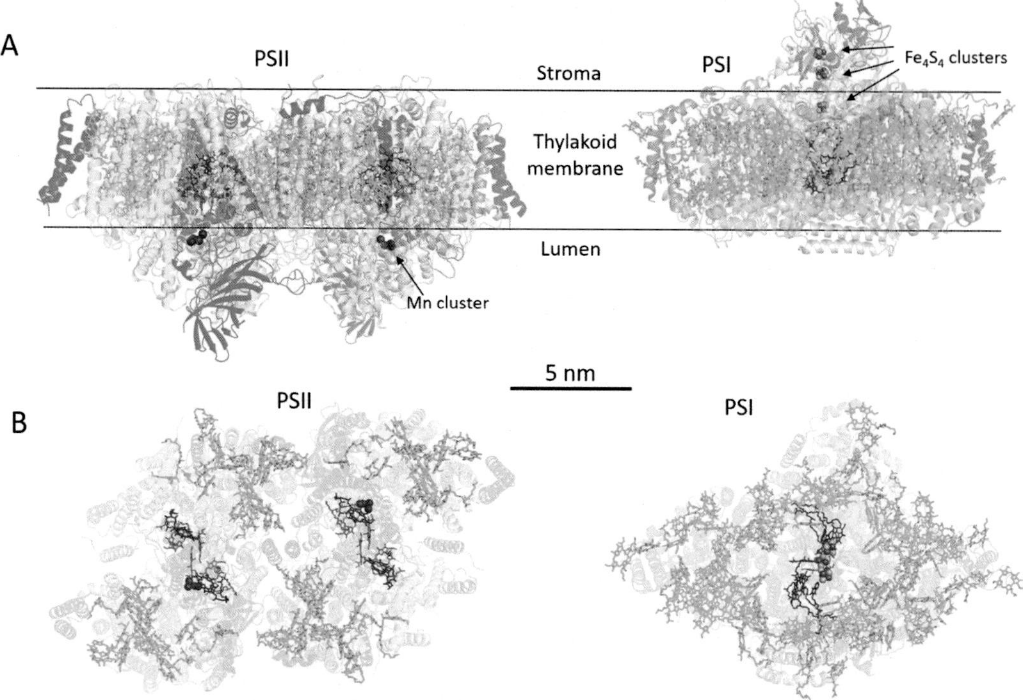

FIG. 1 (A) PSI and PSII are embedded in the thylakoid membrane. Here, the PSI core (Caspy et al., 2020, pdb 6YAC) and the PSII core (Su et al., 2017, pdf 5XNL) from higher plants are shown. The polypeptides are displayed as transparent coloured cartoons, the chlorophylls in *green*, the RC Chl in *red*, pheophytin (PSII) in *orange*, the manganese cluster in *purple*, and the iron sulphur clusters (PSI) in *yellow/orange*. (B) Top view of PSII and PSI core.

complex II (LHCII) in plants, for instance, has a far higher pigment density (15 kDa of pigments per 25 kDa of protein) than the RC, like all members of the LHC family. The antenna systems usually contain hundreds of pigments per RC. The combination of a type II RC in organisms performing oxygenic photosynthesis and its associated LHCs is called photosystem II (PSII) whereas photosystem I (PSI) contains a type I RC. Fig. 1 shows the PSI and PSII core complexes, which are composed of the RC and tightly associated LHCs. This minimal antenna system is usually enlarged by peripheral LHCs. The high number of antenna pigments in PSII is also required because the conversion process initiated by the RCs involves multielectron redox processes. If the rate of electron generation is not high enough, this causes a substantial loss of energy because of charge recombination processes. A light-harvesting system is therefore a cheap way to increase the turn-over rate of the RC by typically two orders of magnitude, at the same time facilitating multielectron redox processes.

2 The concepts of light-harvesting capacity and efficiency

The absorption capacity or absorption cross-section σ_{abs} of an antenna system associated with an RC can be increased by augmenting the number of LHCs. This may be desirable in low-light conditions when the number of

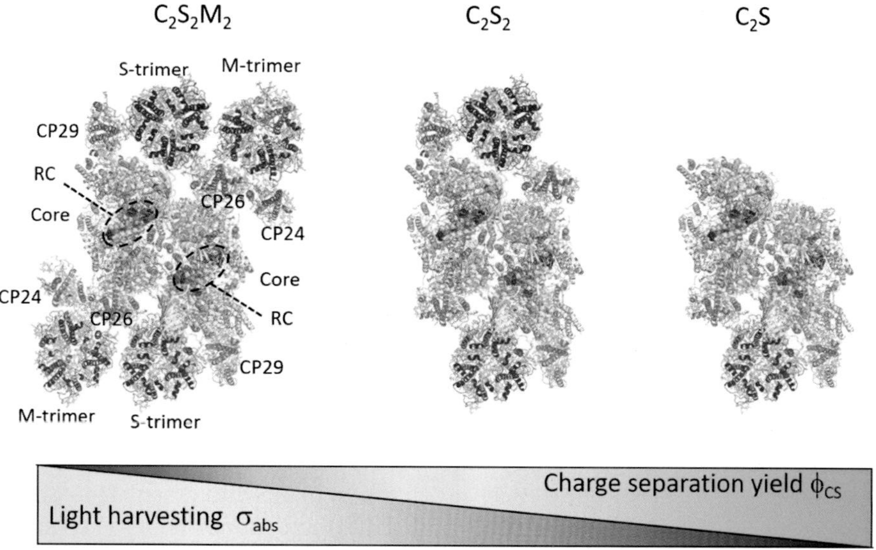

FIG. 2 PSII supercomplex with different antenna sizes: C2S2M2, C2S2, and C2S supercomplexes. PSII core proteins *yellow*, LHCII S-trimer *blue*, LHCII M-trimer *red*, minor antenna (CP24, CP26, CP29) *orange*, chlorophylls *green* sticks, carotenoids *orange* sticks, Mn cluster *purple* spheres, pheophytin *red* spheres, and RC Chls 405–406 *orange* spheres. The bottom schematic diagram reflects the fact that a photosystem with a small antenna has a relatively low σ_{abs} and high ϕ_{CS}, whereas a larger antenna corresponds to a higher value of σ_{abs} and a lower one of ϕ_{CS}.

excitations delivered to the RC is below the maximum photosynthetic capacity. The term σ_{abs} is used here in an intuitive way to denote the capacity to absorb available photons; a doubling of σ_{abs} implies that the number of absorbed photons doubles when the available light intensity remains the same. However, an increase of σ_{abs} does not necessarily mean that also the light-harvesting capacity increases. The latter also depends on the quantum efficiency of charge separation ϕ_{CS} after absorption. The value of ϕ_{CS} is simply the fraction of absorbed photons that leads to charge separation in the RC. The total light-harvesting capacity scales with the product σ_{abs} ϕ_{CS}. Whereas the value of σ_{abs} goes up when the number of LHCs increases, the value of ϕ_{CS} typically goes down because on average excitations need to travel longer to reach the RC and the probability to get lost via competing deexcitation processes increases (Fig. 2). These counteracting effects lead to an estimated maximum value of the light-harvesting capacity for PSII when the antenna contains 200–250 Chls *a* (Wientjes et al., 2013). For PSI, this number is significantly higher (Croce and Van Amerongen, 2020). The maximum capacity can be useful in low-light conditions, but when the organisms are exposed to high-light intensities, saturation of the photosynthesis process may occur, the surplus of excitations may lead to photodamage and the light-harvesting capacity should decrease rapidly to minimise deleterious effects. This can be done by a decrease of σ_{abs} and/or ϕ_{CS}, and both processes occur in nature as a quick response to increasing light intensities. σ_{abs} can be lowered on a short time scale by detaching part of the antenna system, whereas ϕ_{CS} can be decreased by reducing the excited-state lifetime of the photosystem by introducing excitation quenchers. When organisms grow in high-light conditions, the maximum light-harvesting capacity is usually not needed and the antenna size is smaller.

Below we present the solar spectrum and introduce various pigments, their structures, and absorption spectra and illustrate that different pigments are used to deal with the available light and optimise the absorption cross-section. Various examples will be given of light-harvesting complexes occurring in nature. After that, we discuss the properties and mechanisms that underlie the quantum efficiency ϕ_{CS} and discuss the variability of the photosystems and some regulatory aspects.

3 Solar spectrum and its coverage by photosynthetic pigments

In Fig. 3A, the solar irradiance spectrum at the surface of the earth is given. Also, the absorption spectrum of Chl *a* is shown, the most important pigment in oxygenic photosynthesis, which shows prominent peaks around 670 and 430 nm. Both fall within the visible region ranging from 400 to 700 nm, often called photosynthetically active radiation (PAR). Photons absorbed in this wavelength region can drive oxygenic photosynthesis, although also photons that are slightly outside this region can contribute to some extent (Fig. 3A). There are some bacteria that can use light between 700 and 1000 nm for anoxygenic photosynthesis, but here, we will focus on the oxygenic process (Fig. 3A).

The primary electron donors in PSI and PSII are both Chls *a* and they are often indicated as P700 and P680, referring to the wavelengths of their respective absorption maxima. In Fig. 3A, the absorption spectrum of β-carotene is also given, which is partly complementary to that of Chl *a*. Both pigments can absorb visible light due to their extended conjugated π-electron systems with (exceptionally) high extinction coefficients of the order of $\sim 1 \times 10^5 M^{-1} cm^{-1}$ (Fig. 3).

Together, Chl *a* and β-carotene are responsible for all the absorption of solar photons in the cores of PSI and PSII. The carotene molecules transfer their excitation energy within 1 ps to a neighbouring Chl *a* molecule, after which excitation energy transfer (EET) proceeds between Chl *a* molecules until one of the primary donors becomes excited and charge separation can take place.

Additional light-harvesting complexes, also called outer antenna complexes, can increase the absorption cross-section of the photosystems. In plants and green algae, additional major (LHCII) and minor complexes are present that show a large homology to each other and they will be discussed in more detail in the next paragraph. In addition to Chl *a*, they also contain Chl *b* (Fig. 3) and a number of different xanthophylls (oxygenated carotenoids) like lutein, neoxanthin, and violaxanthin, together broadening/increasing the absorption cross-section (Fig. 4).

Another class of outer antenna complexes can be found in diatoms, which form a very diverse phytoplanktonic group that is responsible for 20% of the global primary productivity. Besides Chl *a*, they contain Chl *c* molecules (Fig. 3) as pigments together with a number of carotenoids. These Chl *a*/*c* complexes show a large homology with the Chl *a*/*b* complexes of plants and green algae. However, the absorption spectrum of Chl *c* shows more absorption in the blue part of the spectrum, which is useful in the aquatic environments in which diatoms live and where substantially less red light is present. These Chls are structurally peculiar because they do not contain a long hydrophobic phytol chain that is common to the other Chls and BChls, leaving space in the light-harvesting complexes for additional carotenoids, which

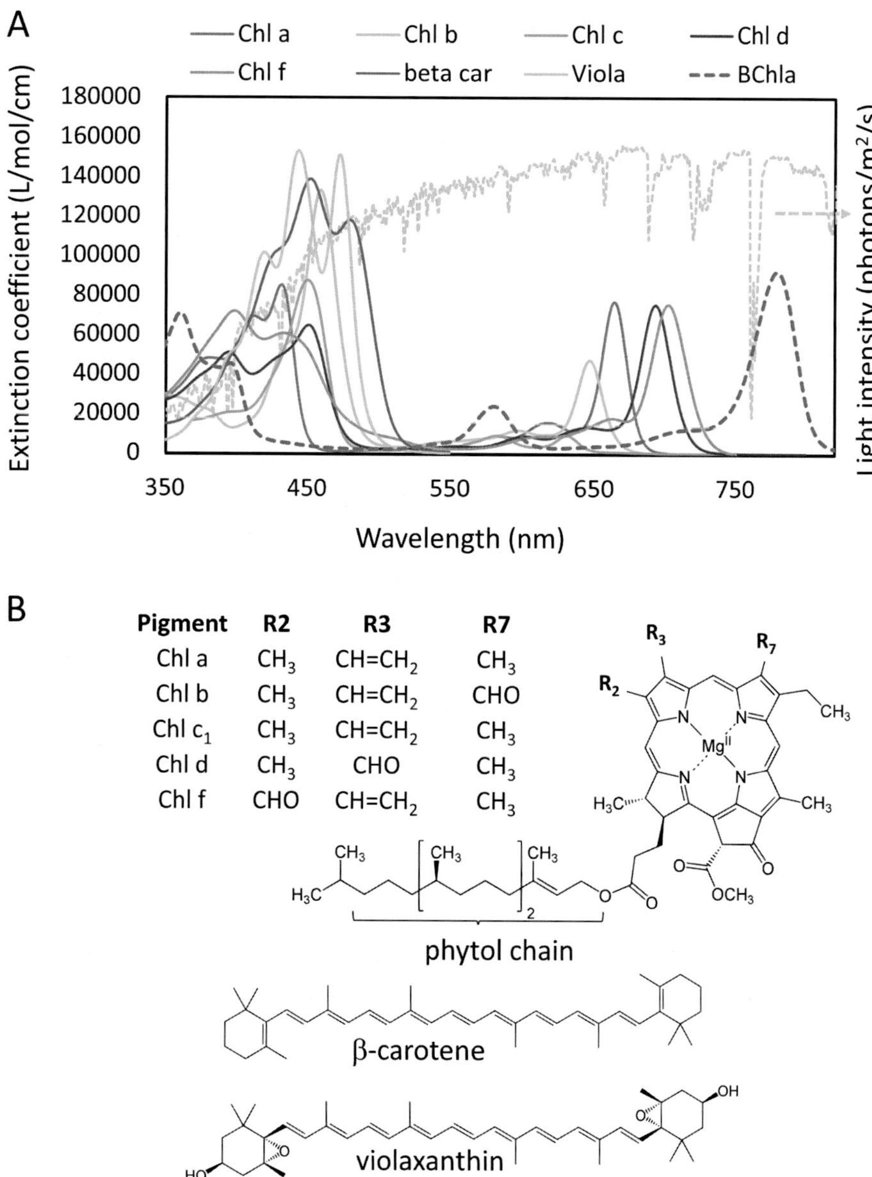

FIG. 3 (A) Spectrum of photosynthetic pigments and sunlight. The sun spectrum is based on the American Society of Testing and Materials G-173 spectrum converted to photons m^{-2} s^{-1} scale (U.S. Department of Energy (DOE)/NREL/ALLIANCE). The spectra of the pigments are on a molecular extinction coefficient scale, the pigment spectra are reported in 80% acetone, only that of bacteriochlorophyll *a* (BChl *a*) is in toluene. BChl *a* is only found in a number of organisms that perform nonoxygenic photosynthesis. (B) Structures of the pigments of (A). Note that Chl *c* lacks the phytol chain, which is present in all other chlorophylls.

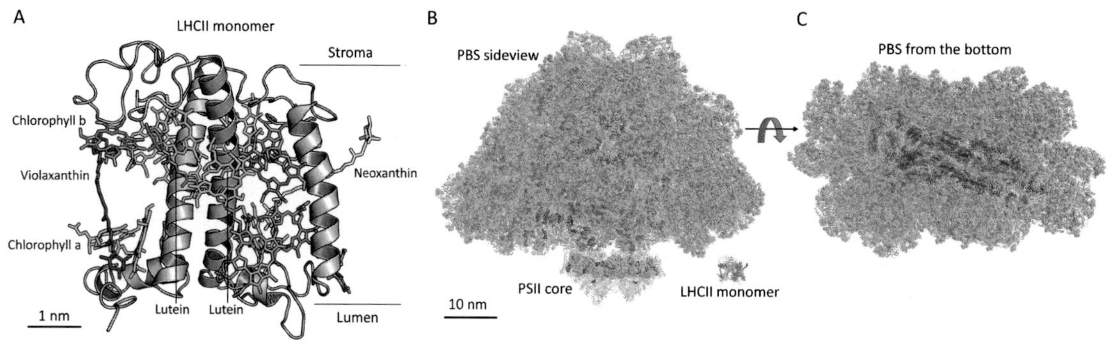

FIG. 4 (A) Structure of LHCII monomer from plants (Su et al., 2017). The protein is shown in *grey*, Chl *a* in *light green*, Chl *b* in *dark green*, violaxantin in *red*, neoxanthin in *yellow*, and lutein in *orange*. For clarity, the phytol chains of the chlorophylls are not shown. (B) Phycobilisome of the red alga *Porphyridium purpureum* (Ma et al., 2020). Polypeptides are displayed as cartoons: phycocyanins are displayed in *orange*, allophycocyanins in *red*, and all others in *grey*. Pigments are displayed as spheres: phycourobilin in *cyan* (abs max 498nm), phycoerythrobilin in *green* (abs max 540 and 565nm), and phycocyanobilins in *orange* (abs max 620nm in phycocyanin and 650nm in allophycocyanins). For comparison, the PSII core and LHCII monomer are displayed on the same scale. Polypeptides in *grey cartoons*, pigments in spheres: chlorophylls in *green*, carotenoids in *orange*, and the manganese cluster in *purple*. (C) Bottom view of PBS displayed in B showing that the pigments that absorb energy at lower energy are located at the bottom where PBS interacts with the photosystem.

can further increase the absorption cross-section in the blue/green.

Cyanobacteria form another important class of photosynthetic organisms. Most of them live in a water environment, and they are the only bacteria performing oxygenic photosynthesis. They exist nearly everywhere on our planet, as long as there is some light available. The outer antenna complexes of cyanobacteria and some red algae, called phycobilisomes, are completely different from those of plants, algae, and diatoms. They are huge, consisting of many water-soluble pigment–protein complexes, and they are appressed to the membrane-embedded photosystems (Fig. 4). Their pigments are bilins, open-chain tetrapyrroles that are covalently bound to the proteins. They typically absorb between 550 and 650 with an absorption maximum that is strongly dependent on the protein environment and pigment configuration.

Finally, while most cyanobacteria contain Chl *a* as their only type of chlorophyll, there are a few species that contain Chl *d* or Chl *f* (Fig. 3), which allows them to absorb light in the near-infrared. These cyanobacteria also contain a number of bilins, attached to phycobiliproteins that absorb above 700nm.

4 Light-harvesting complexes, a few examples

The most abundant light-harvesting complex on earth is LHCII (Fig. 4A) that usually occurs as a trimer in plants and green algae. It is mainly a light-harvesting complex for PSII, but a variable fraction (dependent on light conditions) can also bind to PSI. It is composed of pigments, which are coordinated by a protein scaffold. The protein is folded in three transmembrane helices plus some parts that are located outside the membrane. One monomer typically harbours eight Chls *a* and six Chls *b* together with four carotenoids: two luteins that form a central cross, one neoxanthin, and one violaxanthin (Fig. 4A). A crucial aspect of this structure, which is determined by the protein, is the dense pigment packing that guarantees short distances between pigments, which is advantageous for efficient EET. The dense packing also minimises protein synthesis, thereby saving energy. The presence of some protein seems to be essential in oxygenic photosynthesis because similar Chl *a* concentrations in solution lead to efficient excitation quenching (converting the excited-state energy into heat), also called concentration quenching, which would be detrimental for efficient photosynthesis. Other important aspects to notice are the close contacts between the carotenoids and chlorophylls and the short distances between Chls *b* and Chls *a*, which will be addressed in the next paragraph.

LHCII is, in fact, a combination of three proteins in different compositions: lhcb1, lhcb2, and lhcb3, which are slightly different. In addition, most plants also have three additional minor light-harvesting complexes, lhcb4, lhcb5, and lhcb6 (see Fig. 2) as part of PSII, also called CP29, CP26, and CP24, respectively. These occur as monomers only and share a large sequence homology with LHCII. On the other hand, lhca1, lhca2, lhca3, and lhca4 bind to PSI but also show a large sequence homology. Noticeably, green algae do not contain lhcb4, whereas they have in addition to lhca1–4 also lhca5–9. All these complexes contain Chl *a* and *b* and a variety of different carotenoids.

The outer light-harvesting complexes of diatoms are called fucoxanthin-chlorophyll proteins or FCPs, and they are also members of the large LHC family, but instead of Chl *b* they bind Chl c. They show substantial sequence homology with LHCII, and the pigment organisation has many common motifs. It is of interest to mention that again all carotenoids are in van der Waals contact with Chl molecules and that each Chl *c* molecule is in the close vicinity of a Chl *a* molecule.

In Fig. 4, the structure of a phycobilisome is also shown. A striking difference, as compared to the LHC family, is the far lower pigment/protein ratio in phycobilisomes for these evolutionary old light-harvesting complexes, which is costly from a protein-synthesis point of view. Depending on the type of subcomplex, the protein to pigment weight ratio can vary from 7.5 to 16. It also implies that EET is less efficient because of the larger pigment-to-pigment distances, leading to slower EET steps. However, these antenna systems have an efficient funnel-like organisation, meaning that there is a gradient of high-energy pigments to low-energy pigments going from the outside of the radial rods to the central allophycocyanin rods, with some red-shifted bilins, which are in contact with the photosystems. The advantage of such a funnel is that the excitation energy is directed towards the RC. This is not the case for the antenna systems of higher plants and algae in which the peripheral antenna are almost isoenergetic with the RC.

5 Pigment properties in more detail: Absorption shifts and broadening

In Fig. 5, an energy level diagram is given for Chl *a*, Chl *b*, and a typical carotenoid. A transition from the ground state to the Q_y state of either Chl *a* or Chl *b* corresponds to the lowest-energy absorption band in the red part of the solar spectrum. The transitions to the Soret bands of these Chls correspond to absorption of blue light. The transition from the ground state to the triplet state T of Chl is forbidden and will not be observed in the absorption spectrum. For carotenoids, the transition from the ground state to the T state and the S_1 state are both forbidden, whereas absorption to S_2 is allowed.

Although the energy level diagram suggests that the optical transitions correspond to sharp absorption lines, it is obvious from the absorption spectra above that this is not the case. In fact, there are many ways in which the (protein) environment of the pigments can shift the absorption lines of individual pigments. In a dynamic and somewhat heterogeneous protein environment, this leads to inhomogeneous broadening of the absorption spectrum. Moreover, the electronic transitions couple to vibrations of the pigments

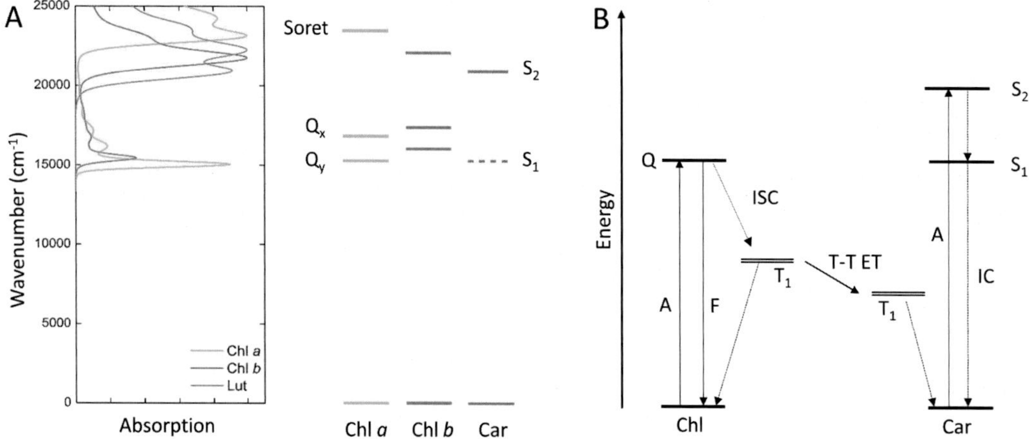

FIG. 5 (A) Absorption spectra and energy level diagram of ground and singlet excited states of Chl *a*, Chl *b*, and carotenoids. The optical forbidden S_1 state of Cars is indicated as a *dashed line*. Absorption spectra are of pigments in 80% acetone. (B) Triplet excited state energy levels of Chl and Car relative to the singlet excited state levels. Absorption (A), fluorescence (F), internal conversion (IC), intersystem crossing (ISC), and triplet–triplet energy transfer (T-T ET) are indicated in the figure.

themselves and of the protein environment (phonons), which leads to additional homogeneous broadening of the absorption spectrum. The electrostatic environment of the pigments is a well-known reason for absorption shifts to occur, but also H-bonding and puckering of the Chl rings contribute. Whereas the structural deformation of Chls and, thus, the shifts in absorption are somewhat limited, the effects can be larger for carotenoids and, in particular, for bilin molecules, leading to substantial shifts in absorption. Also, the occurrence of excitonic states and charge-transfer states can substantially contribute, but before addressing those phenomena, we need to introduce the transition dipole moment.

5.1 Transition dipole moment

The optical transition from the ground state of a pigment to one of its excited states is the result of the interaction between the electric field component of the incident light $\vec{E}$ and the transition dipole moment $\vec{\mu}$ of the pigment. $\vec{\mu}$ is a quantum mechanical parameter defined as

$$\vec{\mu} = \int \Psi_f^* \left(-e\vec{r}\right) \Psi_i \, dxdydz$$

where Ψ_f and Ψ_i are the excited-state and ground-state wavefunctions of the pigment and $\left(-e\vec{r}\right)$ represents the 3-D dipole moment operator. $\vec{\mu}$ has a vectorial character, and the transition probability is proportional to

$$\left(\vec{E} \cdot \vec{\mu}\right)^2 = E^2 \mu^2 \cos^2\theta$$

Here, E^2 is a measure of the incident light intensity, μ^2 is also called the dipole strength of a particular transition, which scales with the molar extinction coefficient ε_A for absorption (see equation below), and θ is the angle between the vectors $\vec{E}$ and $\vec{\mu}$.

The relation between ε_A and μ^2 is given by

$$\varepsilon_A(\tilde{\nu}) = \frac{8\pi^3}{3} \frac{(N'\tilde{\nu})}{(\ln(10)hcn)} \mu^2(\tilde{\nu})$$

where $\tilde{\nu}$ is the wavenumber in cm^{-1}, N' is Avogadro's constant divided by 1000, h is Planck's constant, c is the speed of light in vacuum in cm/s, and n is the refractive index of the solvent/environment. Integration over the entire absorption band leads to the well-known relation

$$\mu^2 = 9.186 \times 10^{-3} n \int [\varepsilon(\tilde{\nu})/\tilde{\nu}] d\tilde{\nu}$$

When two Chls are located close together (~1 nm), they will usually strongly interact with a coupling strength V_{12}. This interaction leads to two excited states (exciton states) at different energies that are shared between both pigments; each exciton state is, in fact, a linear combination of the excited states of the individual pigments. One could say that both molecules form a new supermolecule with delocalised excitations. The coupling strength is often approximated by the dipole–dipole interaction leading to the following expression for V_{12}:

$$V_{12} = \left(\frac{1}{4\pi\varepsilon\varepsilon_0}\right)\left(\frac{\vec{\mu}_1 \cdot \vec{\mu}_2 - 3\left(\widehat{R_{12}} \cdot \vec{\mu}_1\right)\left(\widehat{R_{12}} \cdot \vec{\mu}_2\right)}{R_{12}^3}\right) = \left(\frac{1}{4\pi\varepsilon\varepsilon_0}\right)\left(|\vec{\mu}_1||\vec{\mu}_2|\right)\left(\frac{\kappa}{R_{12}^3}\right)$$

in which ε is the dielectric constant, ε_0 is the vacuum permittivity, $\vec{\mu}_1$ and $\vec{\mu}_2$ are the transition dipole moments of molecule 1 and 2, R_{12} is the centre-to-centre distance between the pigments, and $\widehat{R_{12}}$ is the corresponding normalised vector. The parameter κ is given by

$$\kappa = \widehat{\mu_1} \cdot \widehat{\mu_2} - 3\left(\widehat{R_{12}} \cdot \widehat{\mu_1}\right)\left(\widehat{R_{12}} \cdot \widehat{\mu_2}\right)$$

in which $\widehat{\mu_1}$ and $\widehat{\mu_2}$ are the normalised transition dipole moment vectors.

In case of two isoenergetic pigments (having equal site energies), the two exciton states are separated by $2V_{12}$ as shown in Fig. 6. In addition to the exciton splitting by $2V_{12}$, a shift of the average energy level, denoted as displacement energy, may also occur. The dipole strength of the absorption transitions to each exciton level depends on the relative position and orientation of both pigments and two extreme examples are given in Fig. 6. In the head-to-tail orientation of the transition dipole, the transition to the lowest exciton level is strongly allowed and the dipole strength is approximately equal to twice the dipole strength for a monomer. In the parallel staggered configuration, the situation is reversed.

The splitting in energy between the two exciton states is $2|V_{12}|$, when both monomers are isoenergetic, i.e., their site energies are equal. When the site energies of both monomers differ, the excitonic coupling increases the separation

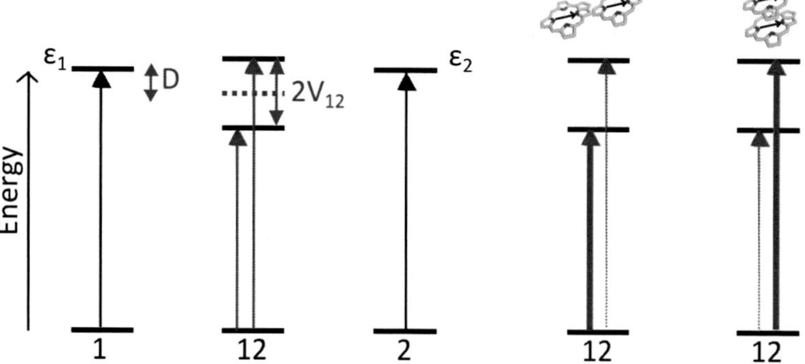

FIG. 6 Excitonically coupled dimer. If two monomeric pigments of similar energy have a strong excitonic coupling ($V_{1,2}$), they form an excitonically coupled dimer. The average energy level is lowered by the displacement energy (D), and the energy levels are split by $2V_{12}$. On the right side, the allowed electronic transitions for two specific orientations of the transition dipole moments (represented by the *arrows* in the plane of the Chls are shown).

between the energy levels. This effect becomes smaller the larger the difference in site energies is, and for large differences in site energies, the excitonic effect becomes negligible. A direct consequence is that only a slight broadening of the absorption bands will occur and no visible splitting if the excitonic coupling is small compared to the absorption bandwidth of the monomers. In case there are more than two interacting pigments, the number of exciton states will go up proportionally although this may not be noticeable in the absorption spectrum because of overlap due to broadening.

Depending on the protein environment Chl *a* can also function as an electron donor or an electron acceptor, which is what is happening in the RCs where Chl *a* molecules function both as primary electron donor and acceptor. But there may also be partial electron transfer from one Chl to another in the excited state, leading to a so-called charge-transfer (CT) state. CT states are characterised by increased spectral broadening and bathochromic shifts to lower energies. Transitions to CT states are usually only weakly allowed, but if these states couple to excitonic states, they may borrow dipole strength and become visible in absorption. An example of this situation are the red-forms in the antenna of PSI. The red-forms correspond to Chl a dimers that due to excitonic interaction and mixing with a CT state absorb around 710nm, which is more than 30nm red-shifted from the absorption of the bulk Chl a molecules. This red-shifted absorption is thought to be important under canopies, where the blue and red light is absorbed by upper leaves and the available light is strongly enriched in the far-red region.

6 Pigment properties in more detail: Decay of the excited states

After an individual Chl or Car has reached one of its excited states by light absorption, it falls back to the first excited electronic state within several hundreds of femtoseconds. Whereas a Car stays in this state for typically 10ps, the excited-state lifetime of an individual Chl is several ns. The Chl molecule can spontaneously fall back from the excited state to the ground state by emitting a photon, a process called fluorescence. The corresponding decay rate (radiative rate or rate of fluorescence k_F) is directly proportional to μ^2, which is the dipole strength that also corresponds to absorption from the ground state to the first excited state. The relation between this rate and the dipole strength is given by the following equations:

$$A(\tilde{\nu}) = \frac{\left(64\pi^4 n(\tilde{\nu})^3\right)}{3h}\mu^2(\tilde{\nu})$$

$$\int A(\tilde{\nu})\,d\tilde{\nu} = k_F$$

For Chl *a*, a typical radiative rate is $\sim$0.05 ns^{-1}, but the exact value depends on the environment/solvent. It can also fall back via internal conversion in which case the excited-state energy is transformed into heat with a typical rate constant $k_{IC} = 0.02\,\text{ns}^{-1}$ for Chl *a* in a solvent. In the thylakoid membranes, this rate may be at least a factor 10 larger. In addition, the excited state may relax to a lower-lying triplet state, a process called intersystem crossing (see Fig. 5B), and for isolated Chls, this is the most likely decay pathway with a typical rate constant of $k_{ISC} = 0.13\,\text{ns}^{-1}$. In the absence of excitation energy transfer or electron transfer, these three processes determine the excited-state lifetime $\tau = (k_d)^{-1} = (k_F + k_{IC} + k_{ISC})^{-1}$ of individual Chls, which is around 4–5ns in solution, whereas it is presumably around 2ns for Chl *a* pigments in the antenna of PSII. The processes of excitation energy transfer and electron transfer are crucial in photosynthesis, they substantially lower the excited-state lifetime of individual Chls. It is important to mention that triplet formation on Chls is potentially dangerous for the organism because Chl triplets can react with oxygen to produce singlet oxygen, which is very reactive and damages lipids, amino acids, and nucleic acids. To prevent this, every Chl *a* in the antenna is in van der Waals contact with a carotenoid molecule and Chl triplets are efficiently transferred to these Cars. Because Car triplets are substantially lower in energy, they do not lead to singlet oxygen formation and Cars can even scavenge singlet oxygen in case it is produced.

7 Excitation energy transfer, the Förster equation

The Chl excitations created in the antenna need to be transported to the primary donor of one of the RCs, and to obtain a high quantum efficiency, this process of excitation-energy transfer (EET) should be much faster than the competing decay processes introduced above. How is this achieved? Excitation energy transfer is based on Coulombic interactions between the pigments and this leads to radiationless EET between the various pigments. In first approximation, many of the transfer steps can be described by the Förster equation. The latter is an expression for the rate of EET from molecule 1 to molecule 2 and it is a manifestation of Fermi's Golden rule and scales with V_{12}^2, where the coupling strength V_{12} between molecules 1 and 2 is approximated by the dipole–dipole coupling, that was introduced above. This is represented by the following equation:

$$k_{DA} = \frac{4\pi^2\kappa^2}{h^2 c n^4 R_{DA}^6}\int \mu_D^2(\tilde{\nu})\mu_A^2(\tilde{\nu})\,d\tilde{\nu}$$

in which the subscripts 1 and 2 have been replaced by D (donor) and A (acceptor). The fact that the coupling strength is squared leads to the R_{DA}^{-6} dependence of the transfer rate k_{DA}. κ^2 is an orientational factor depending on the relative positioning and orientation of both pigment (κ is defined above, see exciton coupling) and the refractive index n refers to the (protein) medium between the two pigments. Combining this expression with the expressions for the (fluorescence) dipole strength for the donor and the (absorption) dipole strength of the acceptor leads to the Förster equation

$$k_{DA} = k_F\left(\frac{R_0}{R_{DA}}\right)^6$$

Here, k_F is the radiative rate or rate of fluorescence that was introduced above and the Förster radius R_0 is given below:

$$R_0^6 \equiv \frac{9\ \ln(10)}{128\pi^5 n^4 N'}\kappa^2 \int \frac{f_D(\tilde{\nu})\varepsilon_A(\tilde{\nu})}{\tilde{\nu}^4}\,d\tilde{\nu}$$

It is important to note that the overlap integral should be determined from the normalised fluorescence spectrum f_D of the donor (the area of which should be normalised to 1 in this equation) and the absorption spectrum ε_A of the acceptor, both in situ. Although the appearance of the Förster equation might suggest that the donor molecules emit a fluorescence photon that is absorbed by the acceptor molecule, this is not a correct view. As already stated earlier, the transfer is a radiationless process, and the size of the overlap integral reflects how well the energy levels of donor and acceptor match with each other. The orientation factor κ^2 and the distance R_{DA} can be obtained from structural data about the pigment–protein complexes, if available.

Strictly speaking, the Förster equation gives only a good approximation if the coupling between the pigments is relatively weak, meaning that the coupling strength is substantially smaller than the absorption and fluorescence bandwidths of the pigments involved. Moreover, the pigments should be far enough apart for the dipole–dipole approximation to be correct. In practice, this requirement is satisfied when the centre-to-centre distances of the pigments are 1.5 nm or more, but even with a distance of around 1.0 nm, the approximation still works rather well (Van Amerongen et al., 2000; Croce and Van Amerongen, 2020). However, for a better description, more sophisticated theories are required (see later).

The spectral overlap integral is larger for downhill EET, when the acceptor molecule has a lower excited-state energy than the donor than for uphill EET, but the reverse process can also take place. The ratio of downhill and uphill transfer rates is ruled by Boltzmann statistics and described by the detailed balance equation $k_{12}/k_{21} = e^{-\Delta E/kT}$. Here, k_{12} is the rate of EET from molecule 1 to 2 and k_{21} is the rate for reverse EET, ΔE is the excited-state energy difference of both pigments, k is the Boltzmann constant, and T is the absolute temperature.

8 Excitation energy transfer, beyond the Förster equation

As was pointed out earlier, the Förster equation is strictly speaking only valid if the coupling between the pigments is weak and the dipole–dipole approximation is sufficiently accurate. When, on the other hand, the coupling is very strong (larger than the bandwidth), exciton splitting will take place and in that case relaxation between delocalised exciton levels will occur that can be described by Redfield theory. Excitations are differently distributed over the pigments contributing to the different exciton states and the energy corresponding to the difference in energy levels is either dissipated as thermal energy to the direct environment of the pigments or reversely taken up from this environment. However, such strong coupling does typically not occur in photosynthetic systems, and we end up with an intermediate situation in which coupling energies are of the same order of magnitude as the bandwidths and neither the Förster equation nor the Redfield theory can be applied and more sophisticated theories are needed, like generalised Förster theory, modified Redfield theory, combined Redfield–Förster approach, Lindblad theory, and hierarchical equations of motion (HEOM) theory. However, those theories go beyond the level of the current chapter and more details can be found in the following reviews (Valkunas et al., 2018; Mirkovic et al., 2017; Novoderezhkin and Van Grondelle, 2018).

9 Overall trapping in photosynthetic units

The average time between the absorption of a photon somewhere in a photosystem and charge separation in the RC is called the trapping time τ_{trap}, and the corresponding rate $k_{phot} = \tau_{trap}^{-1}$ is sometimes called the rate of photosynthesis, which is a bit of a misnomer. The trapping time τ_{trap} can be considered as the sum of two contributions, the migration time τ_{mig} and the overall charge separation time τ_{cs}, i.e., $\tau_{trap} = \tau_{mig} + \tau_{cs}$. The migration time τ_{mig}, also called first passage time, reflects the average time for an excitation to reach the primary electron donor. The overall rate of charge separation (k_{cs}) is the inverse of the overall charge separation time τ_{cs}, which is determined both by the intrinsic rate of charge separation k_{ics} between primary donor and acceptor and the probability p_{pd} that the excitation resides on the primary donor according to the Boltzmann distribution, i.e., $k_{cs} = (\tau_{cs})^{-1} = k_{ics}/p_{pd}$. If k_{ics} would, for instance, be $1\,\mathrm{ps}^{-1}$ and the entire photosystem would contain 100 isoenergetic Chls, τ_{cs} would be 100 ps. More generally, p_{pd} is given by

$$\frac{e^{-E_{pd}/kT}}{\sum_i e^{-E_i/kT}}$$

where E_{pd} is the excited-state energy of the primary donor and E_i reflects the excited-state energy of all the pigments connected to the primary donor, including the primary donor itself. T is the absolute temperature, and k is the Boltzmann constant.

When the migration is infinitely fast and τ_{mig} can be ignored, the overall trapping is called trap-limited, whereas in the opposite case with infinitely fast charge separation, the process is termed migration-limited. In reality, the truth lies probably somewhere in between, but in PSII, there is a substantial contribution of the migration time to the overall trapping time (Croce and Van Amerongen, 2020). Whatever the dominating term in the expression for the overall trapping time, it is imperative that τ_{trap} is substantially smaller than the normal average excited-state lifetime of the pigments involved (i.e., the lifetime τ in the absence of trapping) to avoid that too many excitations get lost. For this reason, antenna systems cannot exist as 1-D systems, because the excitations would 'diffuse' far too slowly. However, as we know, the photosystems are organised in 2-D thylakoid membranes and the pigment–protein complexes even have some 3-D characteristics with pigments on the stromal and luminal sides. The relation between trapping, excited-state lifetime, and quantum efficiency of excitation trapping ϕ_{phot} can be expressed by the following equations: $\phi_{phot} = k_{phot}/(k_{phot} + k_d)$. In the case of PSII, this is equal to $1 - \tau_{open}/\tau_{closed} = 1 - F_0/F_m$. Here, τ_{open} and τ_{closed} are the excited-state lifetimes with the RC in the open and closed state, respectively, and F_0 and F_m are the corresponding steady-state fluorescence levels in the presence of open and closed RCs, respectively. When applying the above equations, it is implicitly assumed that we are dealing with separate photosynthetic units, consisting of a reaction centre and accompanying antenna system. In reality, different photosystems can be connected to each other, and if one RC is closed, the excitation might migrate to another RC. Moreover, there may be some heterogeneity in the composition of the various photosystems and some care should be taken while interpreting the data that are usually collected with time-resolved or steady-state fluorescence measurements.

In general, the quantum efficiency ϕ_{phot} of PSI is very high, even when the antenna size is very large. For instance, the PSI-LHCI complex of the green alga *Chlamydomonas reinhardtii* has around 250 Chl *a* molecules, but its excited-state lifetime is as short as ~50 ps, corresponding to a value of $\phi_{phot} = 0.97–0.98$ for the quantum efficiency, using a value of $0.5\,ns^{-1}$ for k_d. In contrast, the largest PSII-LHCII complexes of plants can contain around 130 Chl *a* molecules per RC, corresponding to a lifetime of ~300 ps and a value of $\phi_{phot} = 0.85$.

As was already mentioned earlier, the photosystems can adapt to changing light conditions, which can either be an increase/decrease of light intensity or a change in spectral composition. The response can be a change of antenna size of PSII and/or PSI but also the reversible induction of excitation quencher. This will be further discussed in Chapter 4, 'Abiotic stress and adaptation in light harvesting', by Jun Minagawa.

10 Summary: The ideal antenna system—The role of the protein

An ideal light-harvesting antenna has to fulfil a list of requirements. First, it needs to be able to strongly absorb the light spectrum that is available. This is achieved by selecting the right pigments, but also by the protein that tunes the absorption wavelengths of the pigments. The absorbed light energy needs to be transferred to the RC and trapped by the RC before the excitation energy is lost by other decay processes. This means that the excited state lifetime of the antenna complexes in the absence of photosynthesis should be long, e.g., in ns range. The other side of the coin is that the excitation energy transfer must be fast. As such, the protein has to keep the pigments at a proper orientation and distance from each other that allows for efficient energy transfer. Furthermore, the Chls *a* should be in close proximity to Cars to avoid the production of reactive singlet oxygen by the reaction of Chls *a* in the triplet state with molecular oxygen. Finally, the antenna should be adaptive. In low light, its absorption cross section should be large, and its decay rate should be low. On the other hand, when the light intensity is high, and more energy is absorbed than can be used for photosynthesis, the antenna system should become smaller, and/or the decay rate of the antenna should increase. The on and off switch of the antenna is regulated at the protein level.

References

Caspy, I., Borovikova-Sheinker, A., Klaiman, D., et al., 2020. The structure of a triple complex of plant photosystem I with ferredoxin and plastocyanin. Nat. Plants 6, 1300–1305.

Croce, R., Van Amerongen, H., 2020. Light harvesting in oxygenic photosynthesis: structural biology meets spectroscopy. Science 369, 6505eaay2058.

Ma, J., You, X., Sun, S., et al., 2020. Structural basis of energy transfer in *Porphyridium purpureum* phycobilisome. Nature 579, 146–151.

Mirkovic, T., Ostroumov, E.R., Anna, J.M., Van Grondelle, R., Scholes, G.D., 2017. Light absorption and energy transfer in the antenna complexes of photosynthetic organisms. Chem. Rev. 117, 249–293.

Novoderezhkin, V.I., Van Grondelle, R., 2018. Modeling of energy transfer in photosynthetic light harvesting. In: Croce, R., van Grondelle, R., van Amerongen, H., van Stokkum, I. (Eds.), Light Harvesting in Photosynthesis. CRC Press, pp. 269–303. Chapter 13.

Su, X., Ma, J., Wei, X., et al., 2017. Structure and assembly mechanism of plant C2S2M2-type PSII-LHCII supercomplex. Science 357, 815–820.

Valkunas, L., Chmeliov, J., Van Amerongen, H., 2018. The exciton concept. In: Croce, R., van Grondelle, R., van Amerongen, H., van Stokkum, I. (Eds.), Light Harvesting in Photosynthesis. CRC Press, pp. 249–268. Chapter 12.

Van Amerongen, H., Valkunas, L., Van Grondelle, R., 2000. Photosynthetic Excitons. World Scientific, Singapore.

Wientjes, E., Van Amerongen, H., Croce, R., 2013. Quantum yield of charge separation in photosystem II: functional effect of changes in the antenna size upon light acclimation. J. Phys. Chem. B 117, 11200–11208.

Further reading

Blankenship, R.E., 2014. Molecular Mechanisms of Photosynthesis, second ed. Wiley Blackwell, Hoboken, NJ.

CHAPTER

2

Transport of electrons

Anja Krieger-Liszkay and Diana Kirilovsky

Institute for Integrative Biology of the Cell (I2BC), CEA, CNRS, Paris-Saclay University, Gif-sur-Yvette Cedex, France

1 General principles of photosynthetic conversion of light energy

Chlorophyll (Chl) is the major light-absorbing pigment in most photosynthetic organisms. In cyanobacteria, in addition to chlorophyll, the phycobilins, open tetrapyrroles that are covalently bound to phycobiliproteins, forming the phycobilisome, the cyanobacterial extramembrane antenna, are very important light harvesting pigments. Each time a photon is absorbed by a Chl molecule, an electron in the ground state is excited and it populates one unoccupied excited state. In the chlorophyll absorbance spectrum, two principal bands are observed, one in the red and the other one in the blue region of the spectra. The first one corresponds to the transition to the first excited state (S1) and the second one to the second excited state (S2). This state decays very rapidly by internal conversion to the more stable S1 state. From the S1 state, the excited Chl molecule can dissipate the excitation energy in three different ways: (1) internal conversion where the excitation energy is converted into heat whilst the electron returns to the ground state; this is the most rapid process; (2) the excited electron returns to the ground state by emitting a photon, fluorescence emission; (3) the exciton can be transferred to other pigment (chlorophyll or carotenoid); and finally (4) it can be used for photochemistry. In the last case, the excited Chl (Chl*), which is a strong reducing agent, transfers an electron to an acceptor molecule, which will be reduced whilst the Chl is oxidised, creating a hole-charge pair (Chl^+/$Acceptor^-$). This is possible since the electron in its excited state is on a higher energy level and is easier to be transferred to another molecule than in its ground state. The cationic Chl^+ will come back to its initial state by oxidising another molecule or by charge recombination (Chl^+ + $Acceptor^-$ = Chl + Acceptor) since the couple Chl^+/$Acceptor^-$ is an unstable system. In addition to these four ways of energy dissipation at the level of the S1 state, there exists also the possibility for the forbidden transition to a triplet state of excited Chl. This triplet state can either be directly quenched by carotenoids in its neighbourhood or react with oxygen, $^3O_2$, a triplet in its fundamental state, to the highly reactive and oxidising singlet oxygen, $^1O_2$.

The Chls in the antenna complexes, which play a light-absorption role, principally transfer the energy to an adjacent Chl molecule. The

Photosynthesis in Action
https://doi.org/10.1016/B978-0-12-823781-6.00003-4

energy can be transferred to a large number of molecules (in the same or different proteins) during the lifetime of the excited state. When the excitation energy arrives at a particular Chl *a* or a pair of Chl *a* molecules (primary electron donor) in the reaction centre, charge separation takes place and the electron is transferred to the primary electron acceptor in a few picoseconds. This is the reaction in which light energy is converted into chemical energy and it occurs in all reaction centres: Type I reaction centres including those of green sulphur bacteria and reaction centre I (RCI) (cyanobacteria, algae, and plants) and Type II reactions centres including those of purple bacteria, Chloroflexi, and reaction centre II (RCII) (cyanobacteria, algae, and plants).

The nature of the primary acceptors and the characteristics of the special Chl pair (primary donor), at the start of the electron transport process, are different in each reaction centre (see the sections on PSII and PSI). Upon charge separation, the positive charge can be shared by the two Chl of the dimer as it is the case in RCI or be centred on one Chl (like in photosystem II (PSII)). In each reaction centre, the primary donor is designated P (for pigment) followed by a number that indicates its wavelength of absorbance (bleaching when the system absorbs light): P_{680} in PSII, P_{700} in PSI. The primary acceptor in PSII is a pheophytin (Pheo) molecule, whilst in PSI, it is a Chl molecule (Ao).

As mentioned earlier, the couple P^+/ $Acceptor^-$ is unstable since the small distance between them promotes a rapid charge recombination: the acceptor molecule gives back the electron to the donor. In the recombination reaction, all the energy is lost (mostly as heat) and cannot be stored. To avoid this, fast reactions (much faster than recombination reactions) occur involving more spaced secondary acceptors and donors that will separate (in distance) and stabilise the charges in both sides of the membrane. The charges are stabilised not only by increases in distance but also by energy losses in each reaction step, leading to a further stabilisation.

2 Linear electron transport

In linear photosynthetic electron transport chain, the two photosystems are arranged one after each other with H_2O as substrate for photosystem II (PSII) at the beginning of the chain and with photosystem I (PSI) and the final acceptor $NADP^+$ at the end. The order of events was discovered by Emerson in 1957. When chloroplasts were illuminated with light at wavelengths >690 nm, the relatively constant quantum yield of O_2 evolution dropped dramatically, whilst it reached back its maximum by illuminating at 650 nm and 700 nm. The far-red light absorbed by PSI can only be efficiently used for O_2 evolution when light of a shorter wavelength is present, which is absorbed by PSII. The Emerson effect can be used to determine the action spectra of the two photosystems. Compared to linear electron transport, cyclic electron flow (see below) does not show an Emerson effect.

Fig. 1 shows the 'Z-scheme' of the electron transport system first proposed by Hill and Bendall in 1960. In PSII, charge separation takes place after excitation of a chlorophyll *a* molecule, P_{680}, the primary acceptor is a pheophytin (a chlorophyll molecule without the Mg^{2+} ion), and then the electron is transferred to the primary plastoquinone acceptor Q_A, a one electron acceptor, and from there to the secondary quinone acceptor Q_B, a two electron acceptor. After a second charge separation event and electron transport to Q_B^-, it becomes double reduced and protonated. The plastoquinol (QH_2) leaves its binding pocket (Q_B-site) and diffuses to the plastoquinone pool. A plastoquinone binds to the Q_B-site and becomes Q_B. At the donor side of PSII, the water-splitting complex, a Mn_4O_5Ca cluster, is bound. After four excitations of PSII, the Mn_4O_5Ca cluster accumulates four positive charges, 4 electrons are taken from 2 H_2O molecules, and 1 molecule O_2 is generated.

After excitation of the chlorophyll *a* dimer, P_{700}, in the reaction centre of PSI, an electron is transferred to the acceptor A_0, a monomeric

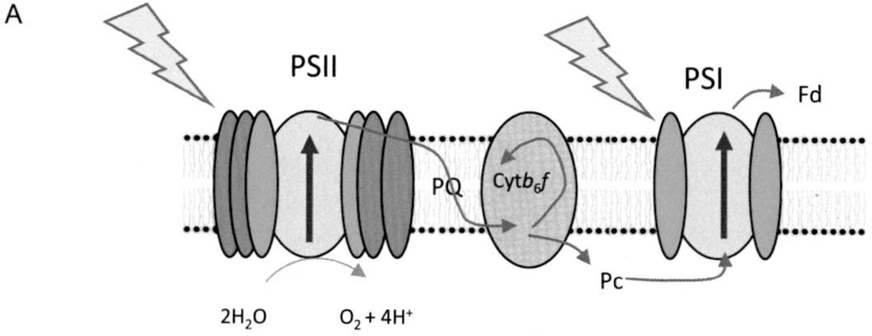

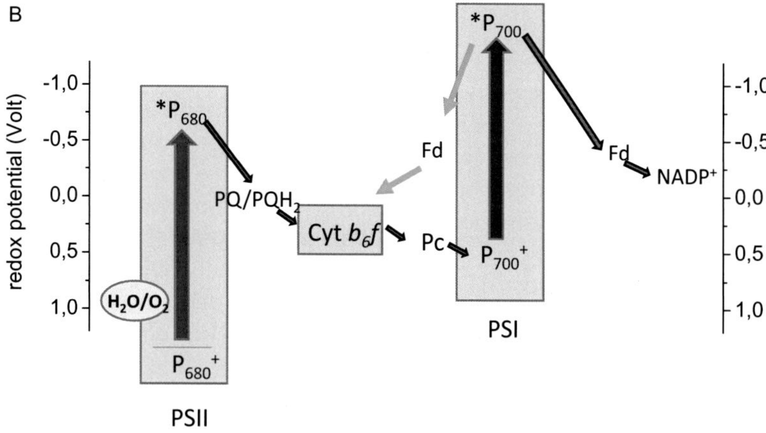

FIG. 1 Photosynthetic electron transport.(A) Simplified scheme of linear photosynthetic electron transport. *Yellow flashes*, excitation by light; *red flashes*, charge separation, and electron transport within the reaction centres; *blue flashes*, electron transport from water to ferredoxin (Fd), including the Q-cycle; *Cyt b₆f*, cytochrome b_6f complex; *Pc*, plastocyanin; *PQ*, plastoquinone; *PS*, photosystem. (B) Z-scheme of the photosynthetic electron transport. The *red arrows* indicate the potential difference between the excited and the ground states of the central chlorophylls of the reaction centres, P680 in photosystem II (PSII) and P700 in photosystem I (PSI). The linear electron transport is shown by *dark blue flashes*, cyclic electron flow by *light blue*, and the Q-cycle by *orange flashes*. *OEC*, oxygen evolving complex; *Tyr_Z*, redox-active tyrosin residue of the D1 protein in PSII; *Pheo*, pheophytin; *Q_A*, Q_B plastoquinone acceptors in PSII; *PQ*, mobile plastoquinone; *cyt b_l and cyt b_h*, hemes of the cytochrome b_6 with a low redox potential (b_l) or a high redox potential (b_h); *2Fe–2S*, iron–sulphur cluster of the Rieske protein; *cyt f*, cytochrome *f*; *PC*, plastocyanin; *A_0, A_1*, electron acceptors in PSI; *Fx, F_A, F_B*, 4Fe–4S clusters in PSI; *Fd*, ferredoxin. This scheme shows the functional relationship and not the distribution of the complexes within the thylakoid membrane. *Credit: Made by authors.*

chlorophyll *a*, then to the acceptor A_1 (phylloquinone) and afterwards to three 4Fe–4S clusters (see below for more details on PSI).

An intermediate redox chain connects the two photosystems. Q_B delivers electrons to the mobile plastoquinone pool that diffuses within the membrane. The cytochrome b_6f complex oxidises plastoquinol by reduction of its b-type cytochromes cyt b_l and cyt b_h. Upon the oxidation of a PQH_2 molecule, one electron is donated from the cyt b_l (low potential) to the cyt b_h (high potential) and the second one reduces the 2Fe–2S cluster of the Rieske protein, which reduces cyt *f*. Plastocyanin, a copper-containing soluble protein in the thylakoid lumen, is reduced by cyt *f* and delivers an electron to P_{700}^+.

At the acceptor side of PSI (stromal side), the terminal acceptor F_AF_B reduces the soluble ferredoxin (Fd) that contains a 2Fe–2S cluster. The flavin-containing ferredoxin-$NADP^+$-oxidoreductase (FNR) finally transfers the electrons to the acceptor $NADP^+$.

Experimental evidence for the Z-scheme was provided by the spectra of the oxidation or reduction of the cofactors. Measurements of the difference spectra of cyt *f* in intact chloroplasts, for example, have shown that preferential excitation of PSI induces an oxidation of this redox component whilst excitation of PSII leads to a reduction of cyt *f*. Similar results have been found for plastoquinone and P_{700}, to name just a few of the components of the linear electron transport chain. Another approach to identify the components of the electron transport chain is the use of specific inhibitors. A large number of substances is available that interrupt electron transport at well-defined sites (see Fig. 2).

The electron transport from H_2O to $NADP^+$ is accompanied with the generation of a proton gradient (ΔpH) across the thylakoid membrane. Three reactions contribute to the formation of the ΔpH:

1. At the level of PSII, four H^+ are taken from the stroma upon reduction of 2 PQ to PQH_2 and four protons are released in the lumen upon oxidation of two water molecules to form one oxygen molecule.
2. At the level of cyt b_6f complex, oxidation of two PQH_2 in the Q_0 site releases 4 protons to the lumen whilst oxidation of a PQH_2 in the Q_i site releases 2 protons to the stroma.
3. The reduction of $NADP^+$ at the stoma side consuming one H^+ per e^-.

As a consequence, per e^-, which are moved from H_2O to $NADP^+$, two H^+ are transferred to the thylakoid lumen: one at the PSII, one at the cyt b_6f, and one H^+ is consumed in the stroma upon reduction of $NADP^+$.

During electron transport, in parallel with the ΔpH, an electrochemical gradient $\Delta\Psi$ is generated across the thylakoid membrane. The ΔpH together with $\Delta\Psi$ forms the proton motive force (pmf), which is the energetic basis for the photosynthetic generation of ATP by the ATP synthase of the thylakoid membrane.

To achieve oxidation of water, positive potentials are required (+0.82 V) whilst the negative midpoint potential of ferredoxin (−430 mV) permits the reduction of $NADP^+$ to NADPH (Em −320 mV). The energy for these chemically highly demanding reactions comes from the absorption of sunlight (680 nm in the case of PSII and 700 nm in the case of PSI), followed by charge separation. The oxidation potential of $P680^+$ is estimated to be +1.2 to +1.3 V, the highest known in biology. The midpoint potentials of the electron carriers are shown in the Z-scheme (Fig. 1). The electron transfer reactions following the charge separation events are exergonic, leading to a stabilisation of the radical pairs.

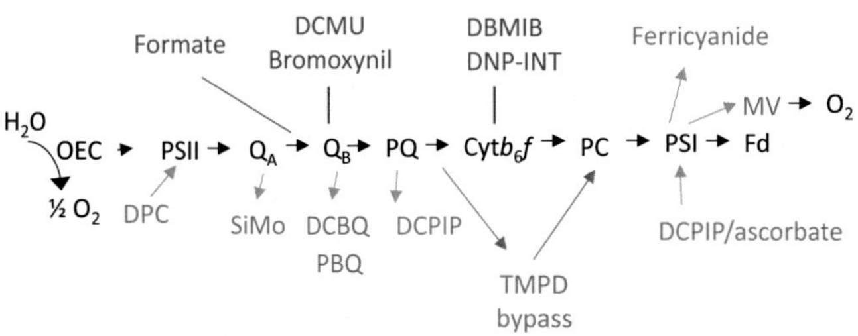

FIG. 2 Action of inhibitors *(red)*, artificial electron donors *(orange)*, and acceptors on photosynthetic electron transport. *DBMIB*, 2,5-dibromo-6-isopropyl-3-methyl-1,4-benzoquinone; *DCBQ*, 2,6-dichloro-1,4-benzoquinone; *DCMU*, 3-(3,4-dichlorophenyl)-1,1-dimethylurea; *DCPIP*, dichlorphenopindophenol; *DNP-INT*, 2,4-dinitrophenylether of iodonitrothymol; *DPC*, diphenylcarbazide; *MV*, methylviologen or paraquat; *PBQ*, phenyl-*p*-benzoquinone; *SiMo*, silicomolybdate; *TEMPD*, *N,N,N′,N′*-tetramethyl-*p*-phenylenediamine. *Credit: Made by authors.*

3 Cyclic electron transport

Cyclic electron transport (CET) involves the transfer of electrons from the PSI acceptor side via ferredoxin to the PQ pool and their subsequent return to PSI via the cyt b_6f complex and plastocyanin. CET results in no net formation of NADPH, but it contributes protons to the ΔpH and leads thereby to the production of extra ATP. It is important for well regulating the ATP/NADPH ratio needed for the variable demands of the metabolism. Together with other regulatory mechanisms, CET plays an important role in protecting the photosynthetic apparatus against photo-oxidative damage when more light is absorbed than needed to fulfil the metabolic demands. Two Fd-PQ dependent CET pathways do exist: (1) the NDH-dependent pathway involving the NDH complex, a homologue of mitochondrial complex I that uses Fd as a substrate and couples electron transfer to PQ with proton pumping to the lumen and (2) the PGR5-PGRL1-dependent antimycin-A sensitive pathway. The exact pathway of this second type of CET is still unclear. PGR5-PGRL1 may act directly as a Fd-PQ reductase or alternatively somehow regulate PQ reduction at the Q_i site of cyt b_6f by Fd. Heme c_i (see below) of cyt b_6f may play a role as a redox module facilitating the reduction of PQ by cyt b_h at the Q_i site. Cyclic electron transport around PSI may involve membrane reorganisation and supercomplex formation between PSI and cyt b_6f as has been evidenced in the green alga *Chlamydomonas reinhardtii*.

4 The protein complexes involved in electron transport and ATP synthesis

4.1 Photosystem II

In the chloroplast of plants and green algae, photosystem II (PSII) with its light harvesting complex II (LHCII), the external antenna of PSII, is located in the grana region of the thylakoid membrane. In the membranes, the PSII is a dimer formed by two identical complexes that function independently. Each complex, containing 36 chlorophyll and 9 carotenoids and around 20 subunits, is largely conserved in all oxygenic photosynthetic organisms (Fig. 3). Whilst the proteins involved in charge separation and charge stabilisation are almost identical, differences appear in secondary proteins, especially in those involved in the stabilisation of the complex. The photochemical reaction centre is formed by two homologue proteins, D1 (encoded by the *psbA* gene) and D2 (encoded by the *psbD* gene), forming a symmetric heterodimer that binds all the cofactors needed for charge separation and stabilisation. The presence of the cytochrome b_{559} (formed by two subunits PsbE and PsbF) is also essential for PSII activity since the minimum PSII complex able to perform charge separation is composed of D1/D2/cyt b_{559}. D1 and D2 comprise five α-helices exposing the N-termini to the stromal side of the thylakoid membrane. An inorganic complex, Mn_4CaO_5, called oxygen evolving complex (OEC), involved in the oxidation of water and production of oxygen, is attached to the luminal side of D1. Specific luminal extra-membrane proteins (OE33 (or PsbO), OE23 (PsbP), and OE16 (PsbQ) in plants and PsbO, PsbV (cyt c_{550}), and PsbU in cyanobacteria) stabilise the binding and activity of the OEC complex. In cyanobacteria, there are several isoforms of D1 with slight differences rendering the protein more or less resistant to high light stress. As a consequence, these isoforms are differentially expressed depending on the light conditions. D1 is the first component of PSII that is damaged under high light intensities and must be replaced very often. To increase the light absorption, two antenna subunits are located surrounding the D1/D2 core. CP43 (CP for Chl binding protein) is composed by five α helices and contains 14 chlorophylls whereas CP47 contains 16 chlorophylls and is composed

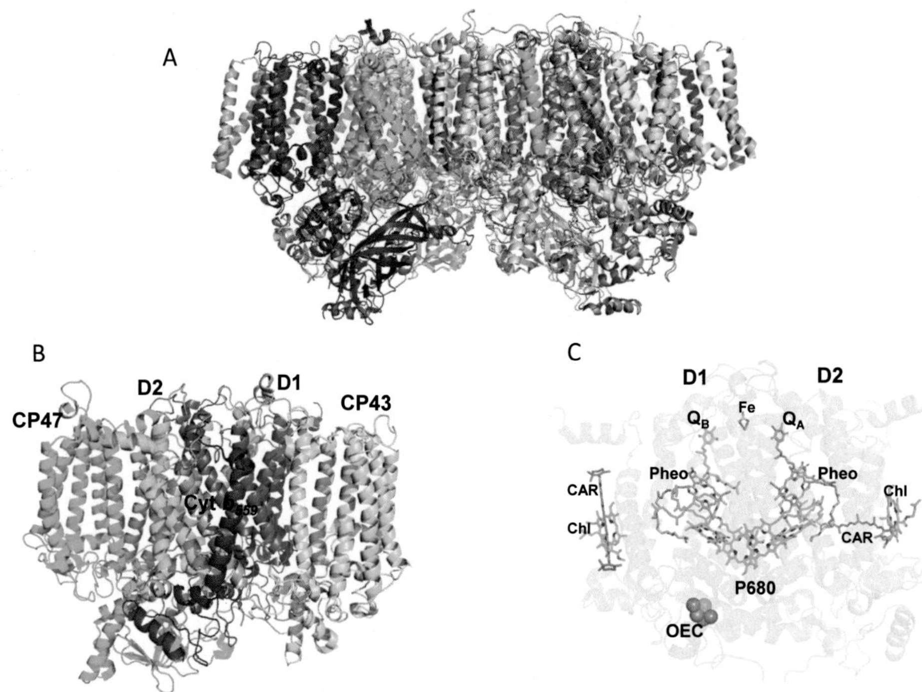

FIG. 3 Photosystem II structure from the thermophilic cyanobacterium *Thermosynechococcus elongatus*. (A) Photosystem II dimer. The PSII is shown from the side, with the luminal side of the membrane at the bottom and the stromal side at the top of the figure. (B) The reaction centre II (D1/D2/cyt b_{559}) and the inner antennae (CP43 and CP47). (C) The cofactors and pigments attached to the heterodimer D1/D2. *Chl*, chlorophyll; *CAR*, carotenoid; *Pheo*, pheophytin; Q_A *and* Q_B, primary and secondary quinone acceptors; *OEC*, oxygen evolving complex. *Credit: Figure generated from PDB file 2axt using Pymol.*

by six α helices and three β-sheets. The other subunits are small, and they are generally composed by only one transmembrane α helix. They stabilise the dimer D1/D2 structure and improve electron transport activity. In plants and green algae, trimeric and monomeric LHCII proteins are attached to the dimeric PSII complex forming supercomplexes containing different quantities of these LHC proteins. In cyanobacteria and red alga, these membrane LHCII complexes do not exist, and the principal antenna of the PSII is a huge extramembrane complex, named phycobilisome.

As mentioned earlier, the PSII belongs to the type II reaction centres. The first and best characterised type II reaction centre is that of purple bacteria. Our understanding of photosystem II structure was first based on the knowledge from the purple bacteria reaction centre. Nevertheless, comparative studies led to the discovery of several differences between PSII and the bacterial reaction centre. Here is described what we knows nowadays about electron transport in PSII.

The heterodimer D1/D2 contains all the cofactors involved in electron transport in PSII (Fig. 3C): 4 Chls (PD1, PD2, ChlD1, ChlD2), 2 pheophytins (PhD1 and PhD2; only PhD1 is involved), two quinones (Q_A covalently bound to D2 and Q_B attached to D1), an Fe atom, two redox-active tyrosine residues (TyrZ and TyrD), and finally the manganese cluster (Mn_4CaO_5), formed by 4 Mn, 5 O, and 1 Ca, principally attached to D1 (Fig. 4).

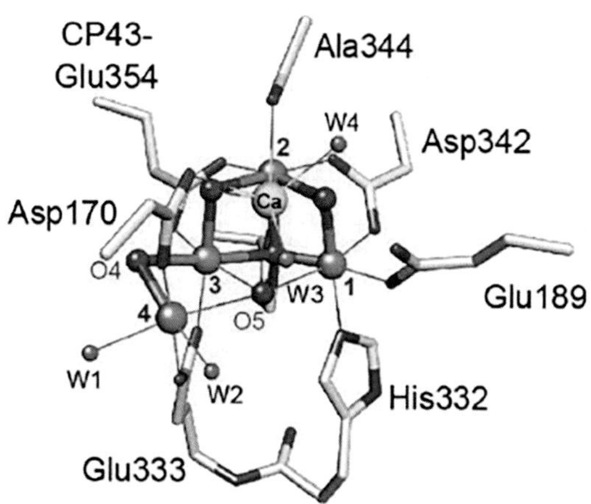

FIG. 4 Structure of the water oxidising Mn_4O_5Ca cluster in PS II with four Mn ions (Mn1 to Mn4, *purple*) and one Ca *(yellow)*, bridged by oxygen ligands *(red)*. Three Mn ions (1 to 3) and the Ca form a distorted cube bridged by oxygen ligands, the fourth Mn (Mn4) is dangling. Mn4 and the Ca carry two water molecules each (W1 to W4, orange). The coordination of the metal ions by amino acid ligands from D1 and CP43 is also shown. *Credit: From Lubitz, W., Chrysina, M., Cox, N., 2019. Water oxidation in photosystem II. Photosynth Res. 142, 105–125. https://doi.org/10.1007/s11120-019-00648.*

In the purple bacteria reaction centre when the energy arrives to the special Chl pair P_LP_M, which are strongly coupled, the Chl gives an electron to the Chl $BChl_{L1}$ (with BChl being bacteriochlorophyll that absorbs at much longer wavelengths than Chl *a*) forming the redox pair $P^+BChl_L^-$. This pair that is formed just three picoseconds after the formation of excited state P* is stabilised by passing the electron to the bacterial pheophytin (BPh) in 1 ps generating $P^+BPh^-BChl_L$.

In PSII, the special pair is less strongly coupled, and the energy arriving at the reaction centre is shared by the four Chls ($P_{D1}P_{D2}Chl_{D1}Chl_{D2}$). The energy is not localised on a specific Chl, leading to different charge separation reactions between the Chls. A few tens of picoseconds after the first charge separation, the redox pair $P_{D1}^+Ph_{D1}^-$ (equivalent to $P^+BPh^-Chl_L$) is formed (Fig. 5). The positive charge is localised at P_{D1}. Then, the electron is transferred from Pheo to Q_A in 400 ps with loss of energy forming the more stabilised pair $P_{D1}^+Q_A^-$. P_{D1}^+ is a strong oxidant (1.2–1.4 V) and takes an electron from the Tyr160 of D1 (Tyr_Z) in several tens of nanoseconds (50 to 250 ns) depending on the oxidation state of the Mn cluster (see Fig. 6). Finally, the neutral tyrosyl radical Tyr_Z. oxidises the Mn cluster in around 55 μs (in S1) to 1 ms (in S3; for an explanation of the S-states, see Fig. 6). Four charges separations are needed to accumulate 4 positive charges at the Mn_4O_5Ca cluster and to oxidise two water molecules. At the acceptor side, Q_A^- gives the electron to Q_B, forming Q_B^- in 0.2–0.4 ms. Q_B^- has a strong affinity for its binding pocket in D1 and waits for a second electron. Once double reduced and taking two protons (QBH_2), it leaves its binding pocket at D1 and it is replaced by a new oxidised quinone molecule from the plastoquinone pool of the membrane. In darkness, depending on the redox state of the plastoquinone (PQ) pool, not only Q_B but Q_B^- can also be present in the Q_B pocket. In this case, Q_A^- gives the electron to Q_B^- in 0.8 ms forming directly QH_2. The plastoquinol molecule is reoxidized by the cytochrome b_6f complex (see the next section).

Nowadays, the structure of the oxygen-evolving complex (Mn_4CaO_5) and the mechanism of charge accumulation and water oxidation are the most active research subjects on PSII, the catalytic mechanism is still a matter of debate, and the picture is changing very often. Joliot and Kok were the first who observed that the quantity of oxygen formed by each flash in a series of flashes varies. When a dark adapted sample is illuminated by a series of flashes, no or very little O_2 is formed in the first two flashes; then a maximum of O_2 is produced in the third flash and finally less in the fourth flash. Thus, the production of oxygen oscillates with a period of four (Fig. 6A). This is due to the fact

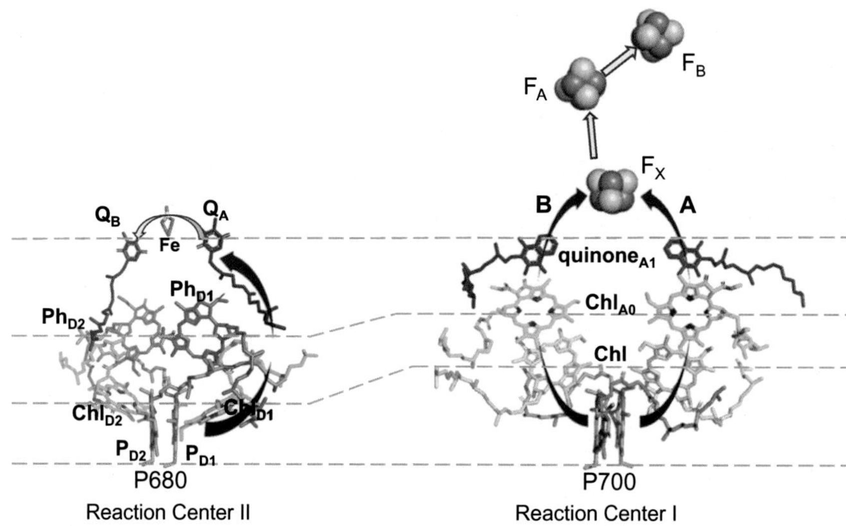

FIG. 5 Electron transport in photosystem II and photosystem I reaction centres (RC). In photosystem II, 4 chlorophyll *a* molecules (P_{D1}, P_{D2}, Chl_{D1}, Chl_{D2}), 2 pheophytins (Ph_{D1} and Ph_{D2}; only Ph_{D1} is involved in electron transport), two plastoquinones (primary quinone Q_A covalently bound to D2 and secondary quinone Q_B attached to D1), and a Fe atom with bicarbonate as ligand between Q_A and Q_B are shown. The energy arriving at RCII is shared by the 4 Chls; charge separation reaction occurs, and finally, the redox pair $P^{+}_{D1}Ph^{-}_{D1}$ is formed in a few tens of ps. Then, the electron is transferred to Q_A and then to Q_B. A second photon and charge separation is needed to double reduce Q_B and form PQH_2 that leaves the RCII. In reaction centre I, the two symmetric branches from P700 to Fx are active. P700 donates very fast an electron to the chl A^{+}_{A} forming the $P^{+}_{700}A^{-}_{0A}$ (A^{-}_{0B}) redox couple. Then, the electron is transferred to A_{1A} (A_{1B}) (a phylloquinone) creating the more stable $P^{+}_{700}A^{-}_{1A}$ (A^{-}_{1B}) redox couple and finally to F_X. Then, the electron is transferred to F_A or F_B Fe–S clusters in PsaC.

that the Mn cluster must accumulate 4 positives charges in order to oxidise two water molecules and to produce one oxygen molecule. Four charge separation reactions are needed to accumulate these 4 positives charges in the (Mn_4CaO_5) complex, which goes through 5 different redox states in which the oxidation of Mn increases: S0, S1, S2, S3, and S4 (Fig. 6B). In darkness, most of the centres are in S1. This explains why the maximum of oxygen is formed in the third flash. Each flash advances the complex to the next S state (Fig. 6A).

4.2 Photosystem I

Photosystem I (PSI), the other transmembrane pigment-protein complex of the photosynthetic apparatus in which light-induced charge separation occurs, is present in the stroma lamellae (unstacked thylakoids) in the chloroplasts. Plant PSI is monomeric whereas cyanobacterial PSI can be monomeric or trimeric depending on the growth conditions. Most of the time it is isolated as a trimer. In some cyanobacteria strains, even the presence of tetrameric PSI was observed. The PSI monomer is formed by 12 subunits, including the heterodimer PsaA–PsaB (82–83kDa) containing all the cofactors needed for charge separation and stabilisation: the special chlorophyll pair (P_{700}), four additional chlorophylls (pair A and pair A_0), two phylloquinones molecules, and an iron–sulphur (4Fe–4S) cluster (Fx) (Fig. 7). Three extrinsic proteins at the stroma side of the thylakoid membrane, PsaC, PsaD, and PsaE, are also essential for PSI activity. PsaC binds two other 4Fe–4S clusters (F_A and F_B) that are located next to F_X at the stromal side. The Fe–S clusters are ligated by cysteine residues. PsaF, a

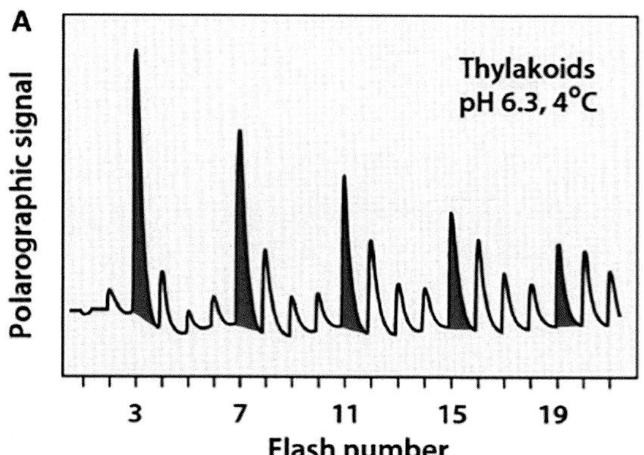

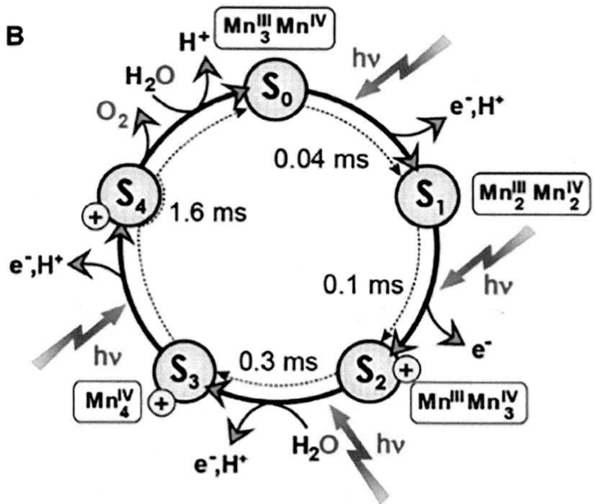

FIG. 6 (A): Release pattern of O_2 measured polarographically after illumination with successive light flashes of spinach thylakoids at 4°C. O_2 release follows a 4-flash pattern (the starting dark stable state is S1). The original experiment was performed by Pierre Joliot as early as 1969 . (B) Water oxidation cycle (Kok cycle) detailing the five basic S states (S0 to S4), the light-induced $1e^-$ oxidation steps and the proton release pattern, the uptake of the two substrate waters, and the Mn oxidation states. The reaction times for the single electron oxidation steps are also indicated. Note that here the 'S' stands for 'state'. *Credit: From Lubitz, W., Chrysina, M., Cox, N., 2019. Water oxidation in photosystem II. Photosynth Res. 142, 105–125. https://doi.org/10.1007/s11120-019-00648.*

transmembrane protein subunit, forms the docking site of plastocyanin at the donor side of PSI. In addition to the cofactors needed for electron transport, the PsaA–PsaB dimer attaches around 79 chlorophyll molecules that serve as internal antenna. Other 11 chlorophylls and 22 carotenoids are coordinated to the small subunits PsaJ, PsaK, PsaL, PsaM, and PsaX. In plants, LHCI antenna complexes are attached to the monomeric PSI.

In PSI, there are two symmetric branches for electron transport from P_{700} to F_x(Fig. 5). Both branches are active. The PSI photoactive core is formed by 4 excitonically coupled chlorophylls: Chl *a* (PsaB) and Chla' (PsaA) forming P_{700} and two accessory chlorophylls (A_A (PsaA) and A_B (PsaB)). Although it was generally assumed that P_{700} gives the electron to A_0 (another Chl) to form the first pair radical, more recently it was proposed that A_A (or A_B) gives the electron to A_0 and then very fast P_{700} donates an electron to A_A^+ forming the $P_{700}^+A_{0A}^-$ (A_{0B}^-) redox couple in around 3.7 ps. Then the electron is transferred in 20–30 ps to A_{1A} (A_{1B}) (a phylloquinone) creating the more stable $P_{700}^+A_{1A}^-$ (A_{1B}^-) redox couple and finally to F_X in about 200 ns. Then, the electron is transferred to F_A or F_B 4Fe–4S clusters in PsaC. Together, PsaC, PsaD, and PsaE form the docking site of ferredoxin, the soluble electron acceptor containing a 2 Fe–2S cluster. Ferredoxin has a very negative

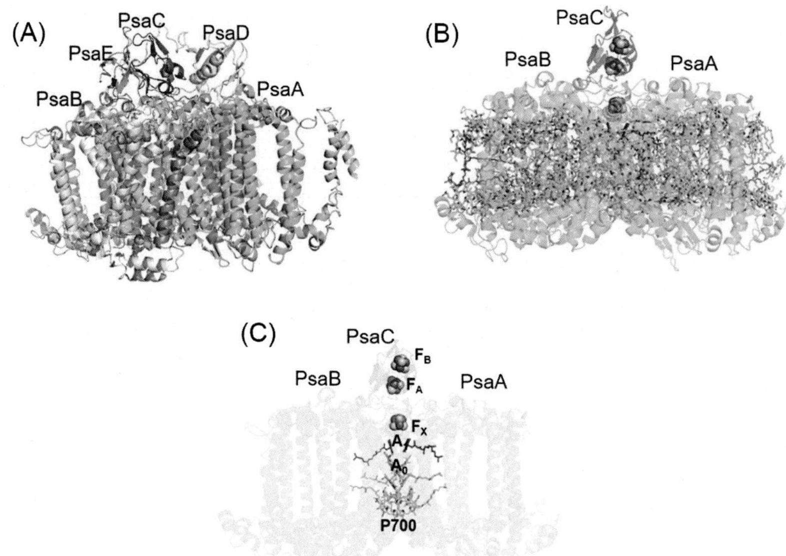

FIG. 7 Photosystem I structure from the thermophilic cyanobacterium *Thermosynechococcus elongatus*. (A) Photosystem I monomer. The PSI is shown from the side, with the luminal side of the membrane at the bottom and the stromal side at the top of the figure. The following subunits are visible in this figure: PsaA *(green)*, PsaB *(cyan)*, PcaC *(magenta)*, PsaD *(lemon)*, psaD *(pink)*, psaJ *(orange)*, psaX *(yellow)*, and psaF *(grey)*. Subunits psaI, psaK, psaL, and psaM are not visible. PsaC, PsaD, and PsaE are extra membrane proteins. (B) The three proteins carrying all the pigments and cofactors are as follows: PsaA, PsaB, and PsaC. All the chlorophylls *(green)* and carotenoids *(orange)* are shown in addition to the cofactors involved in electron transport. (C) The cofactors involved in electron transport. *Credit: Figure generated from PDB file 1jb0.*

reduction potential (−430 mV), and it is easily reduced by F_A or F_B clusters (−550 mV). It accepts only one electron, which is shared by the 2 Fe molecules.

The final step of linear electron transport is the reduction of the $NADP^+$ by Fd via the ferredoxin-NADP reductase (FNR), a 35–45 kDa soluble protein. FNR accepts two electrons, one at a time, from Fd. The electrons are stored at the Flavin adenine dinucleotide (FAD) cofactor, which can exist in three forms: fully oxidised, semireduced, and completely reduced going from FAD to FADH to $FADH_2$. Then, it realises the two electron reduction of $NADP^+$ into NADPH.

P_{700}^+ is re-reduced by the luminal soluble protein plastocyanin that binds a Cu atom. The plastocyanin attaches to the cytochrome b_6f and receives an electron from the cytochrome *f*. Then, the plastocyanin dissociates from the cytochrome and diffuses in the lumen till it reaches PSI. The binding to PSI is facilitated by its PsaF subunit. In cyanobacteria and green algae, in addition to, or instead of plastocyanin, cytochrome c_6, a c-type cytochrome, is present. Under low concentrations of copper in the medium, the amount of cytochrome c_6 increases.

4.3 Cytochrome b_6f

The cytochrome b_6f complex (cyt b_6f) is the third essential membrane complex involved in photosynthetic electron transport and is equally distributed between stacked and unstacked regions of the thylakoid membrane. It is the link between PSII and PSI. Cyt b_6f oxidises the membrane diffusing plastoquinol molecule formed in

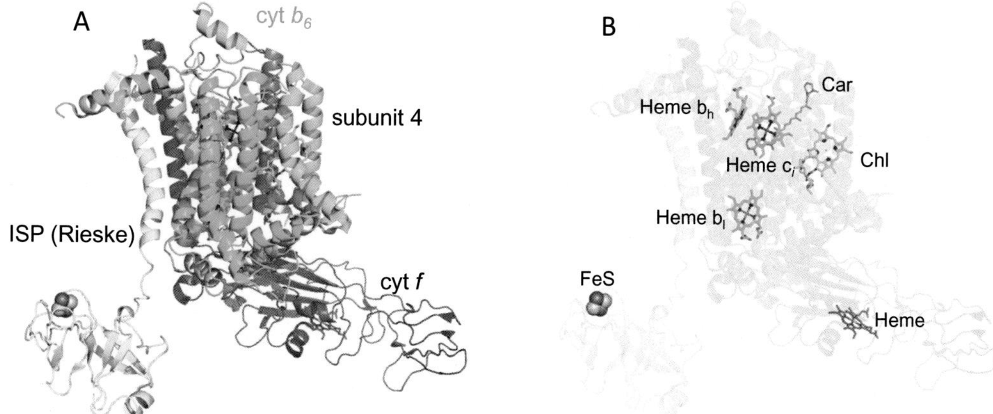

FIG. 8 Structure of the cytochrome b_6f complex from the cyanobacterium *Nostoc* PCC 7120. (A) Cytochrome b_6f monomer. The cyt b_6f is shown from the side, with the luminal side of the membrane at the bottom and the stromal side at the top of the figure. In the figure, the following subunits are visible: cyt b_6 *(green)*, cyt f *(magenta)*, Rieske (ISP) *(yellow)*, subunit IV *(cyan)*, PetG (subunit 5, *violet*), and PetN (subunit 8, *orange*). PetL (subunit 6) and PetM (subunit 7) are not visible. (B) The cofactors of the cyt b_6f monomer. *Credit: Figure generated from PDB file 2zt9.*

PSII and reduces the luminal plastocyanin (or cytochrome c_6) that reduce P_{700}^+ in PSI. Plastocyanin diffusion is a crucial regulatory element of plant photosynthetic electron transport. The cyt b_6f, like PSII, is a dimer. Each monomer contains four large polypeptide subunits, including two cytochromes (cytochrome b_6 and cytochrome f), a 2Fe–2S protein (the Rieske protein) and protein IV, and four small subunits consisting of one transmembrane α-helix (PetG, PetL, PetM, and PetN) with a structural role (Fig. 8). The large subunits contain the redox cofactors. The cyt b_6 possesses four transmembrane helices and coordinates two heme groups (b_L and b_H) and, an additional heme group, c_i that is electronically coupled with b_H. Cytochrome f is formed by a single transmembrane helix and a large soluble luminal domain, which binds a heme group. The Rieske Fe–S protein consists also in a single transmembrane helix and a soluble luminal domain, which coordinates a 2Fe–2S cofactor (instead of a heme). A chlorophyll molecule and a carotenoid molecule of unknown roles are also bound to the complex between subunit IV and cyt b_6. In each monomer, there are two quinone binding pockets: one is at the luminal side (Q_0), binds a plastoquinol molecule, and is involved in its oxidation, and the other one is in the stromal side (Q_i) binding an oxidised plastoquinone.

The electron transfer within the complex is known as the 'Q cycle'. One plastoquinol (PQH_2) binds to the Q_0 site at cytochrome b_6, its oxidation leads first to the transfer of one electron to the Rieske Fe–S centre and the release of one proton to the lumen. The Rieske protein moves away (around 20 Å) from the cyt b_6 towards the cyt f. Once the cyt f is reduced, the Rieske protein comes back to its original position and the oxidised PQ molecule dissociates from the Q_0 site. The cyt f gives the electron to the plastocyanin (or cytochrome c_6). The second electron is transferred to the low redox potential heme (b_L) and then to the high redox potential heme (b_H), both in the cyt b_6, and a second proton is released to the lumen. A plastoquinone molecule attached to the stromal Q_i site is semi-reduced by the electron coming from b_H.

A proton is taken up from the stroma. The semiquinol remains in the site till a second electron coming from the reoxidation of a second PQH_2 in the Q_0 site. The full Q cycle drives the oxidation of one PQH_2, the reduction of two plastocyanins and the accumulation of two extra protons in the lumen.

The cyt b_6f is the point of control and regulation of the photosynthetic electron transport chain. In most of environmental conditions, the rate of the electron transport depends on the rate of reoxidation of the PQ pool by the cyt b_6f. In addition, cyt b_6f is inhibited when the lumen pH is very low, a situation occurring at high light intensities and other stress conditions. Finally, cyt b_6f is an essential element of a photoprotective mechanism called 'States transitions', which regulates the relative activities of the photosystems by partial migration of LHCII complexes from PSII to PSI and vice versa. This mechanism involves a LHCII specific kinase that is activated by the cyt b_6f when the PQ pool is largely reduced.

4.4 ATP synthesis

The formation of an anhydride binding between adenosine diphosphate, ADP, and phosphate, Pi, generates adenosine triphosphate, ATP. ATP is often referred to as the 'molecular unit of currency' of the cell. This generation of ATP requires energy that is provided by the proton motive force (pmf), with $\text{pmf} \approx -60\ \text{mv} \cdot \Delta\text{pH} + \Delta\Psi$ at room temperature. The most widely accepted values is 4 for the H^+/ATP ratio in plant ATP synthase. Plants produce ATP by the chloroplast cFocF1-ATP synthase, which is very similar to the ATP synthase in the mitochondria. It consists of a hydrophobic, membrane-spanning part (Fo) and a hydrophilic, catalytically active head (F1) (Fig. 9). The total molecular weight of the

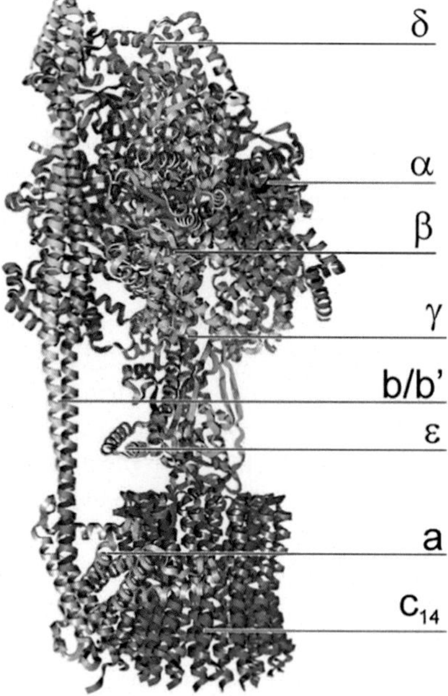

FIG. 9 Cryo-electron microscopy structure of chloroplast ATP synthase from spinach (PDB ID 6fkh). The differently coloured subunits are labelled in Greek (CF1) and Latin letters (CFo). The redox domain in the γ-subunit, exclusively found in chloroplasts of the green lineage, is coloured in *light green*. *Credit: From Buchert, F., 2020. Chloroplast ATP synthase from green microalgae. Adv. Bot. Res. 96, 75–118. https://doi.org/10.1016/bs.abr.2020.07.001 (Elsevier).*

ATP synthase is about 500 kDa. The ATP synthase consists of 26 protein subunits, 17 of them wholly or partly membrane-embedded. ATP synthesis in the hydrophilic α3β3 head (cF1) is powered by the cFo rotary motor in the membrane. cFo contains a rotor ring of 14 c subunits, each with a conserved protonatable glutamate. Subunit a conducts the protons to and from the c-ring protonation sites. The central stalk of subunits γ and ε transmits the torque from the Fo motor to the catalytic cF1 head, resulting in the synthesis of three ATP per complete turn. The peripheral stalk subunits b, b′, and δ act as a stator to prevent unproductive rotation of cF1 with cFo. The head is composed of 3 αβ-subunits, the nucleotide, and phosphate binding sites, which are connected to the F0 via the subunits δ, ε, and γ. The activity of the enzyme depends on the pmf and additionally on the thiol-modification of the regulatory cysteine residues of the γ-subunit. Reduction of the disulphide bridge by the thioredoxin system activates the ATP synthase. This regulation is important to limit the reverse reaction, i.e., ATP-hydrolysis, in the dark.

CHAPTER

3

Carbon fixation

Tracy Lawson[a], Robyn Emmerson[a], Martin Battle[a], Jacob Pullin[a], Shellie Wall[a], and Tanja A. Hofmann[b]

[a]School of Life Sciences, University of Essex, Colchester, Essex, United Kingdom [b]Suffolk One Sixth Form College, Ipswich, Suffolk, United Kingdom

1 Introduction

Carbon fixation is the process by which plants and algae convert the carbon found in inorganic molecules in the atmosphere into organic matter to produce biological building blocks and fuel for cellular respiration. Whilst heterotrophs rely on breaking down existing organic materials via food and digestion, for both energy and growth, photoautotrophy allows plants to use the light energy captured in the light dependent reaction (see Chapter 2) to drive the fixation of carbon from carbon dioxide, with the help of the most abundant protein on Earth, the enzyme rubisco.

The light reactions convert energy from the sun into chemical energy in the form of ATP and reductant (NADPH) needed for carbon fixation, often referred to as the light independent reaction, Calvin–Benson–Bassham Cycle (CBBC) or the C3 cycle. The CBBC is an autocatalytic cycle regenerating the original acceptor molecules as part of the process and can be divided into three phases: carboxylation, reduction, and regeneration. In this chapter, we will describe each of these phases, as well as variation that exists in photosynthetic pathways. Carbon fixation in C3 plants is usually limited by the concentration of CO_2 ($[CO_2]$) at the carboxylation sites (C_C) inside the chloroplasts (Evans et al., 2009). To reach the site of carboxylation, CO_2 must travel from the atmosphere to rubisco and there are many limitations to this diffusion pathway (Ouyang et al., 2017) that can greatly influence the photosynthetic rate, and we will start by describing the various resistance components along this route.

2 Diffusion of CO_2 from the atmosphere into the leaf

CO_2 encounters a number of barriers on the journey to the site of carboxylation, which are often described as a series of resistances (Lundgren and Fleming, 2020). As the leaf cuticle is virtually impermeable to gaseous diffusion, the majority of gas exchange takes place via stomata, small pores found on the aerial parts of leaves and other organs (Lawson et al., 2010; Simkin et al., 2020). The rate of diffusion of gases into and out of the leaf depends on the concentration gradient and the resistance to

Photosynthesis in Action
https://doi.org/10.1016/B978-0-12-823781-6.00008-3

diffusion along the pathway (Weyers and Meidner, 1990). CO_2 must first diffuse through the unstirred layer of air surrounding the leaf, known as the boundary layer, before entering the plant through these stomatal pores. Inside the leaf the general structure consists of different layers of photosynthetic mesophyll cells, interspersed amongst air spaces of different sizes and shapes. The path of CO_2 from the external atmosphere into these intercellular air spaces represents the gaseous phase of diffusion. As a consequence of stomatal opening to facilitate CO_2 uptake, water is lost via the same pathway but in the reverse direction. Water evaporates from the extensive surface area of the mesophyll cell walls (which create the necessary negative pressure required for mass flow of water from the soil through the plant, essential for photosynthesis) into the intercellular air spaces before being lost from the plant through stomata (Fig. 1). The evaporative loss of water from the leaf is also an important component of thermal dissipation and leaf cooling to maintain the appropriate temperature for metabolic processes (including carbon fixation), as well as being vital for nutrient uptake via mass flow (Vialet-Chabrand and Lawson, 2020). Once in the air spaces, CO_2 then has to travel to the chloroplast stroma, the site of carbon fixation, and in order to do so must cross the plasma membrane of mesophyll cells and enter the liquid phase. The liquid phase consists of resistances imposed by the cell wall, plasmalemma, cytosol, chloroplast envelope, and chloroplast stroma (Fig. 1; Cousins et al., 2020; Evans et al., 2009; Tholen et al., 2012). Even though the path length of the liquid phase is much smaller, the diffusion coefficient for CO_2 in water is about 1000 times lower than in air (Rumble, 2020); therefore, both liquid and gaseous phases are thought to contribute more or less equally to the diffusional constraint of CO_2 on photosynthesis (Flexas et al., 2013). Water loss from the plant is an order of magnitude greater than CO_2 uptake due to differences in the path lengths. Water evaporates directly from the cell wall into the air spaces and does not have the resistance imposed by the liquid phase that CO_2 must travel through. Additionally, the diffusion gradients are substantially different, with the inside of the leaf considered to be 100% water saturated whilst CO_2 concentration outside the leaf is only ca. ⅓ higher than inside (Jones, 1992). Researchers often refer to the stomatal or mesophyll conductance, which is the reciprocal of resistance when describing or quantifying these contributions. In this section, we will describe each phase in more detail and discuss current approaches to reducing diffusional limitations of CO_2 on photosynthesis.

2.1 The gaseous phase

The boundary layer, the unstirred region of air surrounding the leaf, represents the first resistance pathway to CO_2 fluxes. The thickness of this layer is determined by the shape of the leaf (Stokes et al., 2006) and the presence of hairs (trichomes), as well as environmental factors including wind speed (Jones, 1992). The majority of plants have trichomes on their leaves (Haworth and McElwain, 2008) as well as other parts of the plant (Zhou et al., 2020), and the presence of these increases the thickness of the boundary layer and therefore resistance to gaseous diffusion (Schuepp, 1993; Amada et al., 2017). Furthermore, trichomes, which are found in different densities in different species (Joel et al., 1994), also play a role in reflecting incoming irradiance, helping to cool the leaf and reducing water loss (Amada et al., 2017) and are thought to contribute greatly to drought tolerance, as plants with densely pubescent leaves are often located in dry environments (Johnson, 1975; Ehleringer and Mooney, 1978). Boundary layer conductance/resistance is further complicated by the air movement around the leaf influenced by wind direction, speed, and leaf flapping (Weyers and Meidner, 1990).

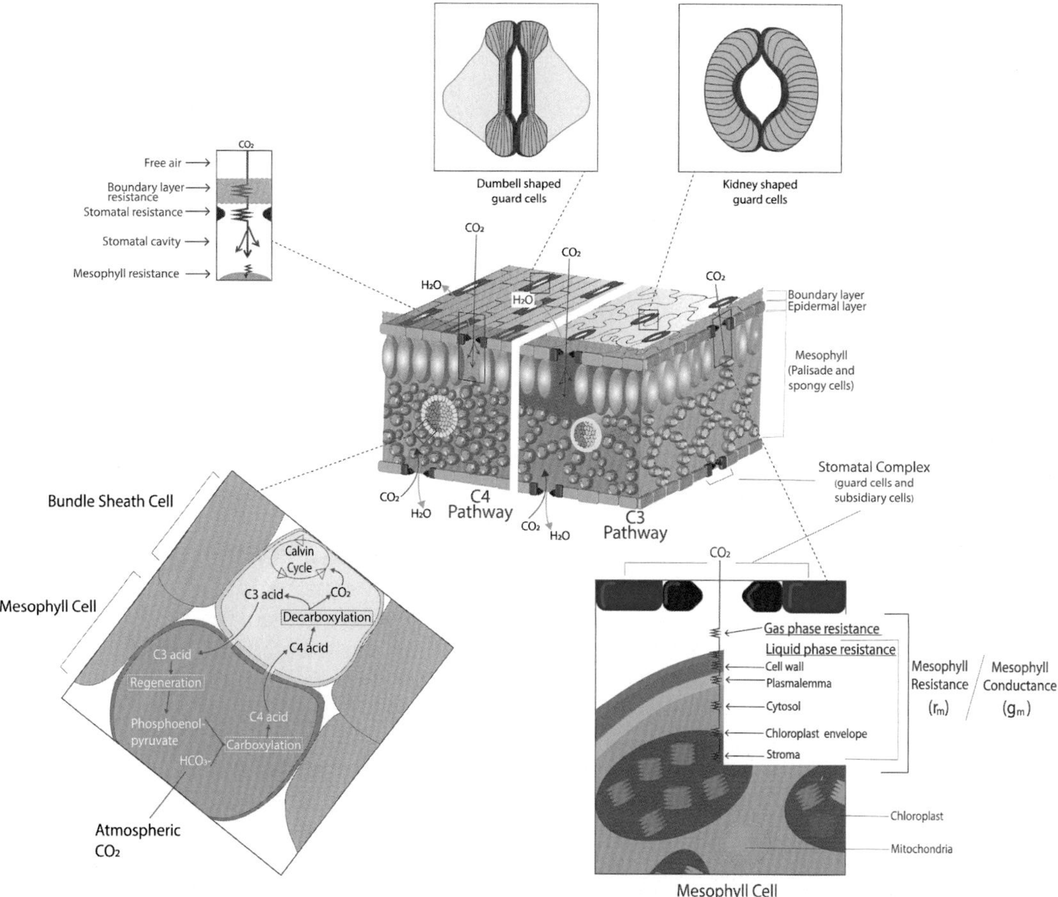

FIG. 1 Schematic diagram of a cross section of a C3 and C4 leaf illustrating the diffusion pathway from ambient air to the site of fixation. The drawings show differences in anatomy in C3 and C4 species, with variations in cell morphology including Kranz anatomy found in all C4 species. The two different guard cell types (dumbbell and kidney) with microfibrils can be found at the top of the figure and the differences in patterning across the leaf surface are illustrated, with dumbbell stomata found in files, whilst the patterning for kidney shaped stomata is less well defined. The enlarged upper left section explains the diffusion pathway in air, whilst the section on the right describes the components of the liquid phase or mesophyll conductance (g_m). Details of the C4 pathway are provided in the lower left enlarged section of the C4 leaf.

The next layer of resistance is the leaf cuticle (the protective layer covering the epidermis of the leaf) that is almost impermeable to gases, which is why 95% of all gaseous exchange takes place through the stomatal pores. Stomata are formed on the epidermis of the leaf between two specialised cells known as guard cells that may or may not be surrounded by adjacent subsidiary cells, which are morphologically distinct from general epidermal cells. Together these are known as the stomatal complex. Stomatal conductance is a measure of the ease with which gas diffuses through stomata, measured as a mole flux per unit area (mol m^{-2} s^{-1}). At the leaf level, the maximum possible stomatal conductance (g_{smax}) is determined by anatomical characteristics, including the size and density of stomata (SD) on the leaf surface and maximum pore area (Eq. 1), although this is rarely achieved physiologically (McElwain et al., 2016).

$$g_{\max} = \frac{\frac{dw}{v} \cdot SD \cdot pa_{\max}}{pd + \frac{\pi}{2}\sqrt{pa_{\max}/\pi}} \quad (1)$$

Eq. (1): dw = diffusivity of water vapour at 25°C ($0.0000249\,m^2\,s^{-1}$) and ν = molar volume of air ($0.0224\,m^3\,mol^{-1}$) are both constants, SD is stomatal density (m^2), pa_{max} is maximum stomatal pore area (m^2) calculated as an ellipse using stomatal pore length (m) as the long axis and ½ as the short axis, and pd is stomatal pore depth (m) considered to be equivalent to the width of an inflated, fully turgid guard cell (Franks and Beerling, 2009).

The size and density of stomata are species-specific (Tichá, 1982) and depend on surrounding environmental conditions, for example, high light generally increases stomatal density (Gay and Hurd, 1975), whilst higher than current ambient atmospheric CO_2 concentration has been shown to decrease the number of stomata in most species (Woodward, 1998). Stomatal aperture adjusts via changes in guard cell turgor driven by the uptake or loss of water and is regulated by both internal physiological and external environmental cues to ensure optimal CO_2 uptake for photosynthesis whilst also conserving water to avoid tissue dehydration and metabolic disruption (Lawson and Morison, 2004). In general, stomata open in response to high or increasing photosynthetic photon flux density (PPFD), low internal CO_2 concentration (C_i), and under conditions of low vapour pressure deficit (VPD) and close under opposite situations. Guard cell movements brought about by changes in turgor are the result of uptake or loss of ions (K^+) and organic solutes, such as malate and sucrose, which reduce the water potential of the cell, driving osmotic water uptake. Osmoregulation in guard cells has received considerable attention in the last century due to the importance of stomatal conductance for gas exchange and carbon fixation. There are several potential osmoregulatory pathways described for stomatal movements (see reviews, Lawson et al., 2014; Santelia and Lawson, 2016); however, complex interactions exist and the relative contribution of each pathway depends upon species, time of day, and environmental stimuli (Talbott and Zeiger, 1996). The first and oldest is the starch sugar hypothesis in which starch stored in the guard cell and present early in the morning is broken down to produce sugars and/or other metabolites used as osmotica or energy for stomatal opening. Now the most commonly reported osmoregulatory pathway is the potassium malate theory—which relies on the uptake of K^+ through ion channels along with the production of malate as a counterion. This pathway replaced the starch sugar hypothesis and was for a long time thought to be the only osmoregulatory pathway responsible for stomatal movements (see Lawson, 2009). A third and less well accepted process involves the production of sucrose from guard cell photosynthesis itself (Lawson, 2009); however, it has been argued that if any sucrose is produced in this way, the quantity is too low to contribute to stomatal function (Outlaw, 1989). Alternatively, guard cell electron transport could provide the ATP required for activation of the plasma membrane proton pumps (Lawson, 2009; Vialet-Chabrand et al., 2021), which are important in all possible osmoregulatory processes for the activation of ion channels.

There are generally two types of guard cells, defined by their shape, which are kidney- or elliptical-shaped and dumbbell or graminaceous guard cells (see Fig. 1). Although these two groups are morphologically distinct and there is considerable variation in size, density, and shape of these cells, there are also common features such as thickened cell walls in places and radial cellulose microfibrils, which are essential for function. Kidney-shaped guard cells generally have a thicker pore wall and thinner dorsal wall that is in contact with surrounding epidermal cells, and the microfibrils are arranged across the cells, which causes the cells to curve outward when volume is increased (Taiz and Zeiger, 1998). These two features are important for the way the pore operates. Dumbbell-shaped guard cells are typically found in grass species, including the majority of our major crops. These have bulbous ends and microfibrils that are arranged along their long axis, and when volume increases the bulbous ends swell which pulls the two cells apart, widening the slit opening between the two cells and increasing the pore aperture (Taiz and Zeiger, 1998, 2002). Another common but variable feature is the presence of subsidiary cells, the number of which is species specific, not always associated with kidney-shaped guard cells but always present with dumbbell-shaped guard cells (Lawson and Matthews, 2020; Nunes et al., 2020). The conventional definition of subsidiary cells are cells that directly surround guard cells and are distinct from other epidermal cells (in terms of dimension, structure, or form (Metcalfe and Chalk, 1950)). In order for stomata to open, the turgor pressure of guard cells must exceed that of the adjacent epidermal/subsidiary cells. Recently, the importance of subsidiary cells in stomatal responses has been acknowledged as a reservoir for solutes and water exchange between these cells and the guard cells, and additionally, this exchange process reduces the surrounding back pressure facilitating stomatal opening (Franks and Farquhar, 2007; Raissig et al., 2016).

2.2 Stomatal behaviour

There is a close relationship between photosynthetic carbon fixation (A) and stomatal conductance (g_s) (Wong et al., 1979); however, the majority of studies that explored this relationship have focussed on steady state conditions and overlooked the implications of dynamic responses, which are now becoming increasingly recognised (Murchie et al., 2018). Fig. 2 shows a typical response between g_s and A as stomata open, with the initial opening response and increase in g_s concurrent with a parallel increase in A up to a certain level when A is no longer diffusionally constrained. In some situations and depending on species, stomata can continue to open even though A has reached a maximum value (Lawson and Blatt, 2014). Under dynamic conditions such as those experienced in the field, stomatal responses to changing environmental conditions, e.g., changes in light intensity, are an order of magnitude slower than the rapid changes in photosynthesis (Lawson et al., 2010; Lawson and Blatt, 2014). Slow stomatal responses can thus cause a disconnect and nonsynchronised behaviour between A and g_s (Matthews et al., 2018). For example, when irradiance increases, slow stomatal opening will limit g_s, restricting CO_2 diffusion and reducing A (McAusland et al., 2016; Matthews et al., 2018). On the other hand, when surrounding conditions result in a decrease in A and stomatal closure lags behind this, this causes unnecessary water loss relative to carbon gain, which means intrinsic water use efficiency (A/g_s) is far from optimal (Matthews et al., 2017; Vialet-Chabrand et al., 2017) (see Fig. 3). Considerable variation in both the rapidity and magnitude of stomatal responses to changing environmental conditions has been reported between and within species (Elliott-Kingston et al., 2016; McAusland et al., 2016; Barradas and Jones, 1996; Qu et al., 2016; Faralli et al., 2019; Vialet-Chabrand et al., 2021). From these studies, it is clear that rapidity of g_s depends on guard cell shape and size (Hetherington and Woodward, 2003; Franks and Farquhar, 2007), photosynthetic type (McAusland et al., 2016), and environmental stimuli (Hepworth et al., 2018).

The rapidity of g_s responses is determined by both anatomical and functional characteristics of the stomatal complex (Lawson and Weyers, 1999; Lawson et al., 2010; Lawson and Matthews, 2020). In general, smaller stomata

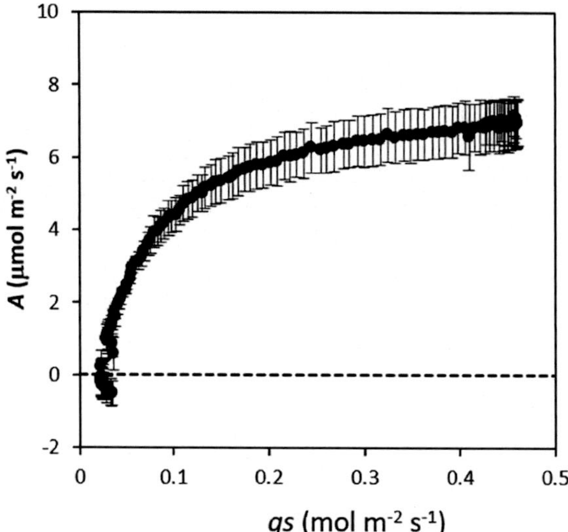

FIG. 2 Assimilation (A) as a function of stomatal conductance (g_s) (A/g_s curve). As g_s increases from a low value, a linear increase in A is observed as the diffusional constraints on photosynthesis are removed. Above a certain g_s value (which is species specific) further increases in g_s are not matched with similar increases in A as factors other than [CO_2] begin to limit A. *Unpublished data of Emmerson & Lawson.*

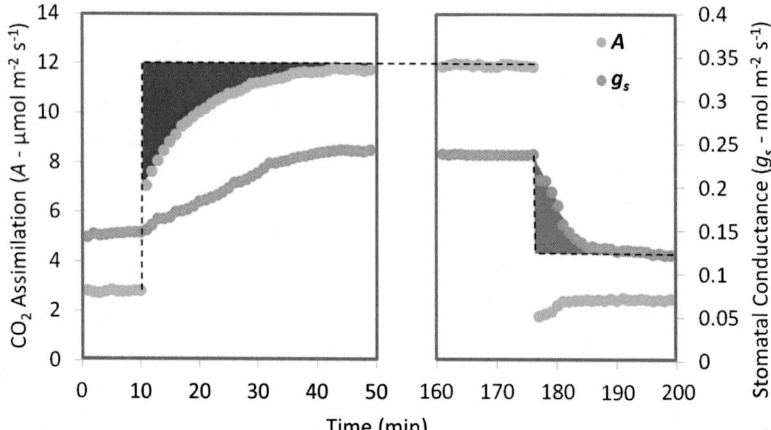

FIG. 3 Temporal response of stomatal conductance (g_s; *green*) and net CO_2 assimilation (A; *orange*) to a step change in PPFD from 100 to 1000 μmol m^{-2} s^{-1} (left figure) followed by a step decrease from 1000 back to 100 PPFD μmol m^{-2} s^{-1} (right figure). Area in *red* represents the amount of CO_2 that could have been gained if stomatal had opened more rapidly, whilst the area shaded in *blue* represents unnecessary water loss from slow stomatal closure. *Unpublished data of Emmerson & Lawson.*

exhibit faster responses due to a larger surface area to volume ratio and the smaller solute requirement to drive movement (Hetherington and Woodward, 2003; Franks et al., 2009; Drake et al., 2013). Furthermore, species with dumbbell-shaped guard cells (e.g., grasses including many key crops) have been reported to be much faster in their response compared to kidney-shaped guard cells (McAusland et al., 2016; Matthews et al., 2018; Hetherington and Woodward, 2003). Franks and Farquhar (2007) assigned this to the interaction between dumbbell-shaped guard cells and their surrounding subsidiary cells, which enables the exchange of osmolites and 'see-sawing' of turgor pressure.

2.3 Possible approaches to improve carbon fixation by manipulating g_s

Both stomatal anatomy and function determine g_s, and therefore, both these aspects represent potential targets for improving carbon fixation. Plants alter stomatal density in response to changing environmental conditions (Franks and Farquhar, 2007). The most well known and most studied is the change in stomatal density with atmospheric [CO_2] (Woodward, 1987). In general, increasing [CO_2] results in a decrease in SD in most species (Hetherington and Woodward, 2003; Woodward et al., 2002) although there are exceptions (McAusland et al., 2016). Much is known about the genes involved in the stomatal developmental pathway and the regulation of stomatal numbers providing potential novel targets for manipulation (Harrison et al., 2020). Several studies have shown that altering genes (such as members of the Epidermal Patterning Factor (EPFs) and the negative regulator of Stomatal Density and Distribution 1 (SDD-1) involved in the development and patterning of stomata) can increase (Schluter et al., 2003) or decrease density (Doheny-Adams et al., 2012). Decreasing SD in model and crop species has been shown to result in decreased g_s, which in turn has often led to an increase in water use efficiency (Bertolino et al., 2019); however, this is often at the expense of A, which was also reduced in these plants (Franks et al., 2015; Hughes et al., 2017; Caine et al., 2019; Dunn et al., 2019), although small decreases in SD in rice and wheat decreased g_s with no impact on A (Caine et al., 2019; Dunn et al., 2019). On the other hand, studies on plants in which SD had been increased through GM approaches demonstrated greater CO_2 diffusion that resulted in higher rates of photosynthesis in some (Tanaka et al., 2013) but not all cases (Franks et al., 2015). However, SD cannot be increased indefinitely as several studies have shown that stomatal patterning and spacing is important for optimal functioning. Stomatal patterning must obey the one cell spacing rule (i.e., that there should be at least one cell between two stomatal complexes), as in high density mutants (TMM), which had stomatal clustering, reductions in g_{smax} were observed along with lower assimilation rates (Dow et al., 2014). The lowered diffusion rate has been associated with a reduced supply or movement of solutes from surrounding cells for guard cell osmoregulation; physical restrictions from increased turgor pressure from surrounding cells; and/or interrupted signalling pathways that alter stomatal function (see Lawson and Matthews, 2020).

An alternative to manipulating morphology is altering signalling or biochemistry within the guard cells themselves or that of the surrounding mesophyll cells, resulting in altered function. For example, Arabidopsis mutants with impaired function of the vesicle trafficking protein SYP121 showed compromised stomatal re-opening after closing was initiated (Eisenach et al., 2012). Similarly, slow stomatal opening in response to increasing light intensity has been reported in the high leaf temperature 1 mutants (HT1 (Hashimoto et al., 2006)), whilst disruption to the closing anion channel SLAC1 in guard cells caused slow stomatal closure.

Mutation of the SLAC gene in rice has been shown to increase induction of both g_s and A (Yamori et al., 2020) demonstrating the potential of this approach to manipulate g_s to improve carbon fixation. On the other hand, increasing the expression of hexokinase (an enzyme important in starch breakdown) in guard cells decreased g_s and improved water use efficiency but at the expense of carbon assimilation (Kelly et al., 2013). Plants with manipulations in SUC2 [which increases invertase activity in guard cells (Antunes et al., 2012)] and with alterations to succinate dehydrogenase levels exhibited greater g_s and A (Araújo et al., 2011). The studies outlined above are just a few examples from hundreds of findings that have shown alteration to stomatal conductance via perturbations in metabolism in either the guard cells or underlying mesophyll cells. Although such studies have reported changes to steady-state g_s, with increases in g_s supporting greater A and decreases supporting increased WUE, g_s is dynamic and changes in response to surrounding environmental conditions, and mesophyll demands for the photosynthetic substrate CO_2. Therefore, manipulating the speed of stomatal responses may be a more appropriate target (Lawson et al., 2010, 2014; Lawson and Blatt, 2014; Lawson and Vialet-Chabrand, 2019; Lawson and Matthews, 2020). Such an approach would help to co-ordinate stomatal behaviour more closely with mesophyll demands for CO_2, resulting in reduced diffusional constraints as well as unnecessary water loss. As outlined above, stomatal movements are brought about through the flux of solutes into the guard cells from surrounding cells (with the subsidiary cells playing a key role in solute exchange (Franks and Farquhar, 2007; Raissig et al., 2017)), involving many transport channels on the plasma membrane and tonoplast (Lawson and Blatt, 2014). Transport capacity and guard cell turgor, responsible for pore opening and closing, are determined by density and activity of such transport proteins (Drake et al., 2013; Franks and Beerling, 2009; Hetherington and Woodward, 2003). However, the relationship between the rapidity of stomatal response and speed of solute fluxes is complex (Lawson and Matthews, 2020), and manipulating individual channels is unlikely to be sufficient to alter functional responses as all channel and movement of solutes interact with each other (Lawson and Blatt, 2014). The success of such an approach has recently been demonstrated by Petersen et al. (2019), who expressed a synthetic light activated K^+ channel to enhance solute fluxes, stomatal aperture and stomatal speed in response to blue light that resulted in greater biomass and WUE in plants grown under fluctuating light. Arabidopsis plants with greater SD have also been shown to have more rapid g_s responses to increasing light intensity that resulted in higher plant biomass (Sakoda et al., 2020). It has also been suggested that smaller stomata move more rapidly than larger stomata (Drake et al., 2013; Franks and Farquhar, 2007; Raven, 2014) and that surrounding subsidiary cells (Raissig et al., 2017) and guard cell type (Hetherington and Woodward, 2003; McAusland et al., 2016) may all play a key role in stomatal speed. Therefore, a combined anatomical and physiological approach may be the most promising for optimising stomatal kinetics to improve carbon assimilation.

3 Diffusion of CO_2 inside the leaf

3.1 The liquid phase

Mesophyll conductance (g_m) refers specifically to the movement of CO_2 through the liquid phase of the diffusion pathway from the intercellular airspaces to the rubisco site of C-fixation in the chloroplast stroma. Mesophyll conductance incorporates and can be subdivided into CO_2 conductance of the intercellular airspace (g_{ias}), cell wall and plasmalemma (g_w), and the liquid phase of the mesophyll cell and chloroplast (cellular cytoplasm and chloroplast stroma; g_{liq}) (von Caemmerer and Evans, 2015). CO_2 must also pass the chloroplast envelope and some models for estimating g_m use the conductance of the chloroplast envelope and stroma (g_c) in place of g_{liq}, assuming that the chloroplast is near enough to the cell plasmalemma and therefore the cellular cytoplasm plays little role in CO_2 diffusion (Cousins et al., 2020). As with g_s, it should be noted that this series of diffusion barriers, which make up mesophyll conductance, are sometimes presented as a series of resistances; however, the inverse conductance is more convenient for use in modelling (Pons et al., 2009).

With so many barriers, it is unsurprising that mesophyll conductance significantly impedes CO_2 movement within the leaf, limiting the rate of carbon fixation (Evans et al., 1986). It has been suggested that g_m provides a similar, or even greater, limitation to photosynthesis than stomatal conductance (g_s) (Evans and Loreto, 2000; Warren, 2006). Mesophyll conductance is determined and affected by leaf anatomy, biochemistry, and environmental factors (Heckwolf et al., 2011; Terashima et al., 2011; Flexas et al., 2012; Lundgren et al., 2019) and is highly variable between different plants, with significant variation observed between, but also within, species (Flexas et al., 2008). Whilst herbaceous annual plants have, on average, the highest g_m and CAM plants (see the section below) on average the lowest, there is broad variation within all groups as well as significant outliers (Flexas et al., 2008). Additionally, wider g_m variation is observed in cultivated species than in wild plants (Flexas et al., 2008). These factors suggest that g_m is a rapidly adapting trait that can be more tightly grouped by anatomical features, such as mesophyll structure, cell wall thickness and chloroplast distribution, than by evolutionary relationship (Syvertsen et al., 1995; Evans et al., 2009; Scafaro et al., 2011; Terashima et al., 2011; Tosens et al., 2012a,b; Tomás et al., 2013; Carriquí et al., 2015; Flexas and Diaz-Espejo, 2015).

Key anatomical characteristics that greatly influence g_m are mesophyll thickness (T_m), mesophyll cell wall thickness (T_w), and surface area of the mesophyll cells (S_{mes}), which depends on the number, size, and spacing of the palisade and spongy mesophyll cells (Evans et al., 1994, 2009; Terashima et al., 2011; Giuliani et al., 2013). The more cells the greater the surface area, which is important as the size of the cells influences the surface area:volume ratio with greater values facilitating greater diffusion, whilst spacing influences the amount of mesophyll surface area exposed (Lundgren et al., 2019). Therefore, the ratio of exposed mesophyll area (S_{mes}) to the total leaf area (S) (S_{mes}/S) has often been used to quantify the number of diffusion pathways (Flexas et al., 2012) as a proxy to g_m. However, this assumes that the area of the exposed mesophyll cell surface is covered by chloroplasts, which is not the case. For this reason, the ratio of the exposed surface area of chloroplasts (S_c) to the exposed surface area of mesophyll cells (S_m) (S_c/S_m) is a more reliable measure of anatomically determined g_m and could be used to explain genotypic differences within and across species (Flexas et al., 2012; Ouyang et al., 2017). Chloroplast movements and repositioning in response to light is another component that influences g_m (Gillon and Yakir, 2000; Flexas et al., 2012; Ogée et al., 2018; Earles et al., 2019) as light (and particular blue light) is known to stimulate chloroplasts to move towards the edge of the cells and therefore close to the membrane. Such movements and the consequential change in the distance of the diffusion pathway may account for distinctions in dynamic g_m responses (see below). Several reports have suggested that mesophyll cell wall thickness accounts for about 25% of mesophyll resistance (Evans et al., 2009; Terashima et al., 2011; Tholen and Zhu, 2011), and strong negative correlations have been reported between g_m and cell wall thickness

(Flexas et al., 2012), but not always and not in all species (Ouyang et al., 2017).

It is not only physical structures within the leaf and mesophyll cells that determine species specific g_m, it has also been suggested that leaf biochemistry and in particular the localisation, abundance, and activity of the enzyme carbonic anhydrase (CA) also plays a key role (Momayyezi et al., 2020). CA has been reported to be the second most abundant enzyme in the leaf after rubisco, representing about 2% of the total leaf protein (Coleman, 2000). CA catalyses the interconversion of CO_2 into bicarbonate (Badger and Price, 1994; Moroney et al., 2001). There are three families of CA isoforms in plants, α, β, and γ, and these are found in different amounts and cellular compartments where they are believed to perform slightly different functions (Momayyezi et al., 2020). The dominant isoform in plants is β-CA (Moroney et al., 2001), which is found free in the cytosol and stroma as well as bound to organelle membranes (Fabre et al., 2007; Wang et al., 2016a) including the mitochondria. The majority of CA activity in C3 plants has been reported in the chloroplasts, where it facilitates CO_2 diffusion though the mesophyll and chloroplast stroma (the liquid phase), which results in a steady CO_2 supply to rubisco (Coleman, 2000; Ogée et al., 2018), increasing g_m. Cytosolic CA activity (which can be between 10 and 15% of total activity) and mitochondrial CA activity are thought to play a role in preventing leakage of CO_2 (as HCO_3) through membranes and in the recapture of CO_2 lost from respiration and photorespiration (Hodges et al., 2013; Sherlock and Raven, 2001). Although there are many reports that speculate on the influence of CA on g_m, there are only a few studies that have examined this (see Momayyezi et al., 2020) and no consistent and clear relationship has been shown between CA activity, g_m or carbon fixation in higher plants (Momayyezi et al., 2020). The role of CA in C_4 photosynthesis (see below) is much clearer, where it plays an essential part in primary carbon fixation by converting CO_2 to bicarbonate, which is required for PEPc activity and initial fixation (Badger and Price, 1994).

The permeability of biological membranes to CO_2 has been, and still is, debated (Flexas et al., 2012) and diffusion of CO_2 across membranes is also facilitated by aquaporins (Terashima and Ono, 2002), which are intrinsic proteins that form pores in biological membranes and aid in the transport of water across these barriers (King et al., 2004). Altered expression of aquaporins has been shown to alter CO_2 diffusion across membranes, and therefore, these water channels are thought to play a crucial role in g_m (Flexas et al., 2006) and could represent key genetic targets for exploitation to increase g_m and photosynthetic capacity.

3.2 Dynamics of mesophyll conductance

Although g_m is dependent on the physical structure of the leaf, it is also highly dynamic, with rapid, short-term alterations observed in response to a wide range of stressors, including virus infection, light, temperature, CO_2 concentration, salinity, nitrogen availability, drought, and water logging (Flexas et al., 2008). Increasing CO_2 concentration around the leaf has been shown to cause a rapid decrease in g_m, even quicker than stomatal responses (Tazoe et al., 2011), with some tobacco species presenting an up to ninefold decrease in g_m in the substomatal cavity (C_i) within a matter of minutes (Flexas et al., 2007). These rapid responses have been suggested to be linked to gating of CO_2 diffusion through aquaporins (Terashima et al., 2006; Uehlein et al., 2008). However, it has also been shown that whilst g_m responses to increasing [CO_2] are rapid, subsequent responses to lowering [CO_2] are significantly slower, suggesting that a separate mechanism may be involved the reversal of the g_m response (Tazoe et al., 2011). g_m has also been observed to rapidly respond to increases in intensity of blue light, decreasing as light intensity increases (Loreto et al., 2009). This response is believed to be partially caused by chloroplast movements (to avoid photodamage) to the edge of the cell, decreasing the length of the CO_2 diffusion path through the cytoplasm to reach the chloroplast, although the speed of the g_m response is greater than that of chloroplast movement, suggesting that another factor is likely to be involved (Loreto et al., 2009).

Some anatomical features are linked to variation in g_m with changing growth environments, for example, shade leaves have been shown to have lower g_m than sun leaves in several species (Hanba et al., 2002; Piel et al., 2002; Laisk et al., 2005; Warren et al., 2007). It has also been demonstrated that short-term modification of g_m can be induced via exogenous application of the plant hormone abscisic acid (ABA) (Flexas et al., 2006), which is involved in many plant responses to various stresses, linking g_m variation to the wider network of plant stress responses. Longer term adjustments of g_m have also been observed in response to extended periods of stress, typically through modification of leaf anatomy, such as increased cell wall thickness (Tw) and reduction of mesophyll:cell surface area (S_{mes}/S) in response to extended drought stress (Han et al., 2016). Additionally, changes in the expression of genes linked to the synthesis of aquaporins in response to drought have been linked to improvements in g_m (Mahdieh et al., 2008; Sade et al., 2014). Changes in the expression or activity of carbonic anhydrase (CA) enzymes have also been associated with dynamic changes in g_m (Price et al., 1994; Gillon and Yakir, 2000; Flexas et al., 2008).

As g_m is determined by both anatomical features and biochemical processes there is scope for manipulation in order to reduce diffusional constraints imposed by the liquid phase to facilitate greater CO_2 uptake for carbon fixation. However, the fact that g_m cannot be directly measured and as such must be estimated using a combination of measurements fitted to one or more models (Pons et al., 2009; Flexas et al., 2013; Cousins et al., 2020) makes it more difficult to determine targets for improving this key component to CO_2 diffusion. Table 1 provides an overview of the benefits and limitations of several established methods for estimating g_m. (For a more detailed review of these methods and the equations supporting them, see Pons et al., 2009.)

3.3 Possible approaches to improve carbon fixation by manipulating g_m

Of the barriers to CO_2 movement incorporated into g_m, the liquid phase (g_{liq}) has been shown to be most limiting (Meyer and Genty, 1998; Piel et al., 2002). As such, when seeking to improve g_m, the route often taken involves improving the rate at which CO_2 can dissolve into the cytoplasm. Increasing g_m increases carbon fixation (A) but does not increase water loss through transpiration because the diffusion pathway involving g_m is not the same (Ouyang et al., 2017), and therefore, improvements in g_m are a key target for improving plants water use efficiency (WUE) (Barbour et al., 2011; Flexas et al., 2013).

Three major approaches for improving g_m have been identified; increasing the exposed mesophyll cell surface area (S_{mes}) by altering mesophyll cell size, shape, division, and separation to maximise CO_2 flux, improving cell wall conductance, such as by reducing cell wall thickness (Tw) or altering cell wall biochemistry, in order to improve CO_2 diffusion into the liquid phase, and modifying the features which control dynamic acclimation of g_m, such as CO_2 channels and chloroplast movement, in order to increase conductivity and reduce diffusion path length (Pons et al., 2009; Lundgren and Fleming, 2020). It should be noted however that the mechanics behind these approaches are still poorly understood, making the improvement of g_m a difficult prospect (Kromdijk et al., 2020; Lundgren and Fleming, 2020). Alternative approaches, such as

TABLE 1 Overview of methods used to estimate mesophyll conductance.

Method	Required measurements	Advantages	Disadvantages
Curve-fitting method (Ethier and Livingston, 2004; Sharkey et al., 2007)	• Gas exchange	• Relatively simple protocol • Equipment is relatively inexpensive • Performable with only portable equipment • Simultaneously estimates g_m, V_{cmax}, J_{max}, and R_L	• Requires known values for rubisco kinetics • Requires known C_i transition point from V_{cmax} to J_{max} • Assumes partial pressure of CO_2 does not influence g_m • Can only estimate g_m in C3 plants
Variable *J* method (Di Marco et al., 1990; Harley et al., 1992)	• Gas exchange • Chlorophyll fluorescence	• Performable with only portable equipment • Allows measurement responses of g_m to CO_2 concentration	• Requires known CO_2 compensation point of rubisco • Sensitive to photorespiration • Highly sensitive to changing environmental conditions • Assumes that measurements at zero or low O_2 and ambient O_2 concentrations are the same • Can only estimate g_m in C3 plants
Constant *J* method (Bongi and Loreto, 1989; Harley et al., 1992; Loreto et al., 1992)	• Gas exchange • Chlorophyll fluorescence	• Performable with only portable equipment • Makes no assumptions about environmental conditions other than CO_2 concentration	• Requires known CO_2 compensation point of rubisco • Sensitive to photorespiration • Assumes that g_m and J are constant regardless of CO_2 concentration • Sensitive to changes in rate of mitochondrial respiration, especially at high g_m • Requires truly constant J, which must be confirmed via chlorophyll fluorescence • Can only estimate g_m in C3 plants
Online carbon (and oxygen) isotope discrimination method (Barbour et al., 2016; Evans et al., 1986; Gillon and Yakir, 2000; Lloyd et al., 1992; von Caemmerer and Evans, 1991)	• Gas exchange • Carbon isotope discrimination	• Can simultaneously measure gas exchange and carbon isotope discrimination (Cousins et al., 2006) • Can separate g_m in the cell wall and plasma membrane from g_m in the chloroplast membrane • Can measure g_m in C3, C4, and CAM plants	• Requires expensive, high-maintenance equipment • Sensitive to differences between isotopic composition of gases used during plant growth and testing • Requires precise isotopic fractionation factor measurements at multiple steps • Field work requires gas samples to be trapped and cryogenically stored for isotopic discrimination analysis in the laboratory

modifying the patterning of cells within the leaf to replace much of the intercellular air space with small mesophyll cells, have also indicated an improvement in g_m, although again the mechanisms behind this remain poorly understood (Lehmeier et al., 2017).

Of these approaches, modifying gene expression is the most straight forward and well-studied (Flexas et al., 2006; Perez-Martin et al., 2014; Kromdijk et al., 2020). By altering expression of genes related to CO_2 channels, such as the *Nicotiana tabacum AQUAPORIN1* (*NtAQP1*) gene in tobacco, cell wall, and membrane conductance to CO_2, and by extension g_m, can be improved (Flexas et al., 2006). Due to the significance of g_{liq} to total g_m, overexpression of aquaporins has also resulted in increased g_m (Flexas et al., 2006, 2008; Uehlein et al., 2008; Perez-Martin et al., 2014; Sade et al., 2014; Xu et al., 2019). However, the exact mechanism by which aquaporins improve g_m is not well understood and some aquaporins have also been shown to have little or no effect on g_m (Flexas et al., 2008; Kromdijk et al., 2020). Along with aquaporins, the expression, localisation, and activity of the different carbonic anhydrases found in higher plants may represent another potentially unexploited target for manipulating g_m and improving carbon fixation and or water use efficiency. Whilst such an approach is possible, greater knowledge on the influence of the various CA isoforms on g_m and their interactions with other biochemical processes and the influence of environmental conditions on performance is required (for more information, see review by Momayyezi et al., 2020).

4 Carbon fixation pathways

4.1 C3 carbon fixation cycle

The primary photosynthetic pathway is the Calvin–Benson–Bassham Cycle (Fig. 4), often referred to as the C3 cycle, as the first stable component of the pathway is a three carbon molecule. As mentioned above, the pathway is autocatalytic meaning that the substrates required to keep the pathway running are generated as part of the process (Raines, 2003). The C3 cycle utilises the end products of the electron transport chain (ATP and NADPH) to fix CO_2 that has diffused from the atmosphere as detailed above to produce all the carbon skeletons required for metabolism and plant growth. The cycle can be divided into three phases: (1) carboxylation, (2) reduction, and finally (3) regeneration defining the key activities that take place during these reactions. This cycle requires 11 different enzymes to carry out 13 reactions (Raines, 2003). The cycle begins with the carboxylation phase in which the enzyme ribulose-1,5-bisphosphate carboxylase oxygenase (rubisco) catalyses the fixation or carboxylation of CO_2 by the acceptor molecule rubulose-1,5-bisphosphate (RuBP). RuBP is a 5 carbon molecule, and with the addition of 1 carbon from CO_2 forms an unstable 6 carbon molecule that immediately breaks down into 2 molecules of the 3C molecule 3-phosphoglycerate (3-PGA), the first stable product in the pathway (see Taiz and Zeiger, 1998, 2002). The second phase of the cycle involves the reduction of 3-PGA to triose phosphate with the use of ATP and NADPH produced during electron transport in the light dependent reaction (see Chapter 2). 3-PGA is first phosphorylated to 1,3-bisphosphoglycerate (using the enzyme phosphoglycerate kinase and ATP), after which glyceraldehyde 3 phosphate dehydrogenase catalyses the reduction of 1,3-bisphosphoglycerate using NADPH to produce the triose phosphate glyceraldehyde 3-phosphate (G3P or GAP). Most of the G3P produced by these steps remains in the cycle and is used in the regeneration of the original acceptor molecular RuBP (in the regeneration phase*)*. For every 6 molecules of G3P produced, only 1 exits the cycle to produce sucrose and starch, essential for plant growth and development, whilst the remaining 5 are used to generate RuBP and keep the cycle functioning. The majority of

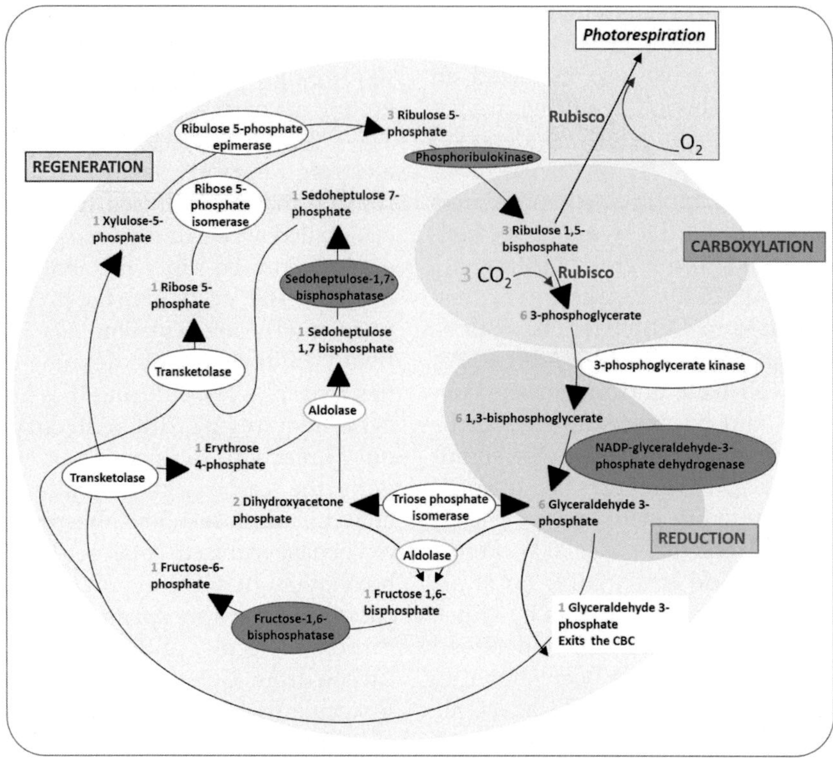

Calvin-Benson Cycle

Regeneration. Glyceraldehyde-3-phosphate is converted to dihydroxyacetone-3-phosphate by triose phosphate isomerase.
Dihydroxyacetone-3-phosphate along with a second molecule of G3P undergoes aldol condensation to produce fructose-1,6-bisphosphate, catalysed by the enzyme aldolase.

Fructose-1,6-bisphosphate is hydrolysed to fructose-6-phosphate which then reacts with the enzyme transketolase, along with another molecule of G3P in a reaction that splits the pathway producing erythrose-4-phosphate and xylulose 5-phosphate.

Erythrose-4-phosphate combines with a another dihydroxyacetone-3-phosphate molecule catalysed by aldolase to produce the seven carbon sugar sedoheptulose-1,7-bisphosphate.

Sedoheptulose,1,7-bisphosphate is hydrolysed to sedoheptulose 7-phosphate and with an additional molecular of G3P and the assistance of transketolase produced a molecule of ribose 5-phosphate and a molecule of xylulose 5-phosphate.

Each of the xylulose 5-phosphate molecules produced from the above reaction and the earlier reactions are converted to a molecule of ribulose 5-phosphate with the aid of the enzyme ribulose 5-phosphate 3-epimerase.

A third molecule of ribulose 5-phosphate is formed from ribose 5-phosphate with the enzyme ribulose 5-phosphate isomerase.

In the final step the 3 molecules of ribulose 5-phosphate are phosphorylated to 3 molecules of RuBP by ribulose-5-phosphate kinase (also known as phosphoribulokinase) and ATP.

Carboxylation. Rubisco catalyzes the addition of a single CO_2 molecule to the five-carbon acceptor RuBP. The resulting 6-carbon intermediate is highly unstable, and rapidly degrades into two 3-carbon molecules of 3-PGA.

Reduction. 3-PGA is phosphorylated (using ATP) via phosphoglycerate kinase after which Glyceraldehyde 3 phosphate dehydrogenase reduces the 1,3-bisphosphoglycerate using NADPH produced by electron transport to form G3P.

FIG. 4 Schematic representation of the photosynthetic Calvin–Benson–Bassham cycle. The three phases of the cycle: carboxylation *(blue)*, reduction *(green)*, and regeneration *(yellow)* are illustrated along with the different enzymes (ovals) and intermediates. Photorespiration is highlighted in the *red box*. Enzymes activated by the ferredoxin–thioredoxin system are shaded *blue. Redrawn from Chapter 5 with descriptive text in lower box provided from Taiz, L., Zeiger, E., 2002. Plant Physiology, third ed. Sinauer Associates, Sunderland, MA, USA.*

enzymes within the C3 cycle are involved in the regeneration of RuBP and the resulting products that exit the pathway for secondary metabolism (see CARs chapter for details of the secondary metabolism exits). This part of the cycle involves reshuffling of carbons from the 5 triose phosphate molecules that remain within the regeneration phase of the cycle to prevent depletion of the intermediates. An overview of the regeneration phase is provided below; for details of the individual steps, see Fig. 4.

Triose phosphate isomerase drives the reversible conversion of glyceraldehyde 3-phosphate into dihydroxyacetone-3-phosphate (DHAP). One molecule of each substrate come together in an aldol condensation, with the loss of a phosphate ion into solution. This forms fructose 1,6-bisphosphate, which is subsequently hydrolysed to fructose-6-phosphate by fructose-1,6-bisphosphatase. This, and a third molecule of G3P, then serve as donor substrates for the enzyme transketolase, which rearranges to form erythrose-4-phosphate and xylulose-5-phosphate. These molecules now diverge in the pathway, but their ultimate fate remains the same with each once again forming ribulose 5-phosphate to regenerate RuBP. Erythrose-4-phosphate and DHAP can also undergo aldol condensation, which yields sedoheptulose-1,7-bisphosphate. Sedoheptulose-1,7-bisphosphatase releases a phosphate anion into solution. Sedoheptulose 7-phosphate reacts with a fourth G3P molecule in another transketolase-mediated two-carbon transfer producing ribose 5-phosphate and an additional molecule of xylulose 5-phosphate. Each molecule of xylulose 5-phosphate is reversibly epimerised to ribulose 5-phosphate. Ribose 5-phosphate isomerase converts each molecule of ribose 5-phosphate to ribulose 5-phosphate. These molecules are each phosphorylated by ATP via phosphoribulokinase (PRK) to once again form RuBP.

4.2 Light regulation of Calvin cycle enzymes

Several enzymes within the C3 cycle are light regulated via the ferredoxin/thioredoxin (Fd/TRX) system which results in the redox activation of these enzymes. Four key enzymes in the cycle, glyceraldehyde-3-phosphate dehydrogenase; fructose-1,6-bisphosphatase; sedoheptulose-1,7-bisphosphatase; and ribulose-5-phosphate kinase or PRK contain one or more disulphide (—S—S—) groups that are light regulated by the Fd/TRX system via an oxidation–reduction mechanism (Buchanan, 1980; Schürmann and Jacquot, 2000; Wolosiuk et al., 1980). In the dark, the enzymes are inactive and in the oxidised (—S—S—) state. In the light, these residues are reduced to the sulphhydryl or dithiol (—SH HS—) state by reduced thioredoxin leading to the activation of the enzymes. In the dark, oxygen converts thioredoxin and the enzymes back to the inactive oxidised state (see Fig. 5; Taiz and Zeiger, 1998; Buchanan et al., 2000). The initial reduction of thioredoxin by ferredoxin (the electron acceptor at the end of the electron transport chain) is catalysed by the enzyme ferredoxin:thioredoxin reductase (Michelet et al., 2013).

Rubisco is also light activated but does not rely on the above ferredoxin/thioredoxin system and instead involves a process called carbamylation in which a CO_2 molecule reacts with a specific lysine residue on the rubisco active site (Lorimer and Hartman, 1988). The carbamate derivative then rapidly binds Mg^{2+} to produce the active complex, which then binds another molecule of CO_2 in the carboxylation reactions. It should be noted that the CO_2 molecule used in the carbamylation process is a different molecule to that fixed by the active rubisco molecule in the carboxylation reaction. During the

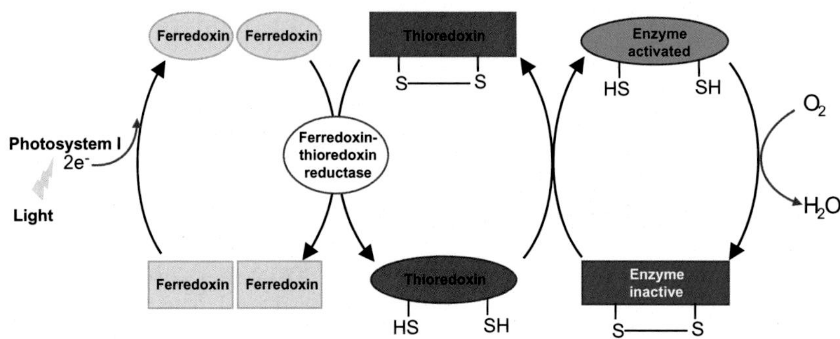

FIG. 5 The ferredoxin–thioredoxin system. Illumination and reduced ferredoxin drives the reduction of thioredoxin via ferredoxin–thioredoxin reductase that eventually reduces the disulphide bond (—S—S—) to the reduced form (—SH HS—) triggering the activation of the target enzymes of the CBC (highlighted in *blue* in Fig. 4). *Squares* represent oxidised states and *circles* reduced. *Figure redrawn with permission from Jones, R., Ougham, H., Thomas, H., Waaland, S. 2012. The Molecular Life of Plants. John Wiley and Sons.*

formation of the active carbamylated rubisco, two protons are released; therefore, this process is supported by higher Mg^{2+} concentrations and high pH, both of which are facilitated by the electron transport process (see Chapter 2). Inactive rubisco is also inhibited by the binding of sugar phosphates (including RuBP), which prevents carbamylation from taking place (see Degen et al., 2021). These inhibitors are detached by the enzyme rubisco activase (Rca), which removes these sugar phosphates from the active site in an energy (ATP) dependent manner (Spreitzer and Salvucci, 2002). Rubisco activase activity is thermally sensitive (Perdomo et al., 2017), influenced by changes in redox status and ADP/ATP ratio in the chloroplast, and is regulated by the light dependent Fd/TRX enzyme activation system described above, and therefore, this enzyme is believed to play a key role in the regulation of rubisco activity and photosynthesis itself (Salvucci et al., 1985; Mott and Woodrow, 2000; Carmo-Silva and Salvucci, 2013; Scales et al., 2014). The light activation of the key enzymes in the C3 pathway provides a mechanism of regulatory control of the C3 cycle that links carbon fixation directly with light activation and the light driven electron transport process, preventing unnecessary activation of enzymes and/or futile cycling of cycle intermediates.

4.3 Photorespiration

Rubisco is the main enzyme of the Calvin cycle; however, in addition to carboxylation, rubisco also catalyses the oxygenation of RuBP in which O_2 rather that CO_2 is fixed, in a process known as photorespiration (Fig. 4). As these are competing reactions, photorespiration results in a loss of CO_2 and a general lowering of photosynthetic efficiency by about 25% (Amthor et al., 2019). The amount of photorespiration that takes place in C3 plants depends on several factors including the concentration of substrates (CO_2 or O_2) at the site of fixation (which also depends on temperature and stomatal behaviour) and the specificity and catalytic properties of rubisco itself.

4.4 Alternative carbon fixation pathways that reduce photorespiration

In addition to CBBC, two further photosynthetic pathways exist; C4 [or the Hatch-Slack pathway, after the scientists who discovered it

>50 years ago (Hatch and Slack, 1966)] and Crassulacean acid metabolism (CAM). C4 and CAM plants are able to reduce photorespiration by including a CO_2 concentrating mechanism that separates the initial fixation of atmospheric CO_2 from rubisco activity and the C3 cycle, either spatially or temporally (Edwards and Ogburn, 2012). However, such adaptation only represents a competitive advantage over C3 photosynthesis when environmental conditions demand it. For example, higher temperatures promote significant photorespiration (Ehleringer and Björkman, 1977), and therefore, C4 plants tend to be found in warmer conditions; however, these plants also require more irradiance because of the higher energy (ATP) requirement for the C4 process. CAM plants are found in arid environments where the driving force for water loss is great and via the utilisation of temporal nocturnal gas exchange (see below) CAM plants have vastly higher water use efficiency than C3 or C4 plants. CAM plants are reported to lose ca. 10 x less water than C3 plants which lose about 400–500 molecules of H_2O for every CO_2 molecule gained, whilst C4 plants fall in between the CAM and C3 with 250–300 molecules of water loss for CO_2 gain (Taiz and Zeiger, 1998). Evolution of C4 and CAM has occurred independently many times, with C4 having evolved from at least 62 independent origins (Sage, 2016). It has been estimated that the origins of C4 photosynthesis occurred between 25 and 35 Mya driven by the decline in atmospheric $[CO_2]$ to present day levels from 800 $\mu mol\,mol^{-1}$ (Zhang et al., 2013). The ability to tolerate drier hotter conditions whilst maintaining yield is an attractive quality of plants with C4 metabolism, which is prevalent amongst crop species including maize, sugarcane, sorghum and millet (Sage and Monson, 1998), as well as representing ~3% of flowering plant species (Sage, 2004). As a result, the mechanisms of concentrating CO_2 at the site of rubisco is a target for genetic manipulation of C3 crop plants, with the aim of increasing photosynthesis and productivity, particularly relevant given the climatic changes predicted for the future.

4.5 C4 photosynthesis

C4 photosynthesis is thought to have evolved in response to environments where the oxygenation reaction of rubisco was high (Sage et al., 2012), as a mechanism to reduce photorespiration by increasing CO_2 concentration at the site of carboxylation. C4 plants achieve this by separating the Calvin cycle (and rubisco) in the bundle sheath from the initial fixation of atmospheric CO_2 in the mesophyll cells (Welkie and Caldwell, 1970). Leaf anatomy in C4 plants differs (see Fig. 1) from that of C3, facilitating the spatial separation of the concentrating mechanism. C4 plants possess a specialised grouping of cells around the vasculature, termed the bundle sheath (BS) cells, which contain chloroplasts and all the C3 enzymes including rubisco, but lack the key electron transport component photosystem II (PSII), the action of which results in the production of O_2 from photolysis (see Chapter 2). Bundle sheath cells are surrounded by an outer layer of mesophyll (M) cells (Welkie and Caldwell, 1970) an arrangement known as Kranz ('wreath') anatomy. Unlike C3 plants, the mesophyll cells have no rubisco activity (Way, 2012), and instead use the enzyme phosphoenolpyruvate (PEP) carboxylase (PEPc), which has no affinity for O_2 to carry out the initial fixation of CO_2 from the atmosphere. The close contact between BS and M cells allows for the fast, effective exchange of metabolites (Lundgren et al., 2014) between these two cell types. To ensure there is a short distance between M and BS cells, many C4 plants limit the number of cells separating these two compartments, achieving this by organising mesophyll cells into layers around the bundle sheath cells (Lundgren et al., 2014). As in C3 plants, stomata of C4 plants open during the day, although often with a lower conductance (Taylor et al., 2010), possibly driven by a lower density and smaller pore size (Taylor et al., 2012). The lower g_s does not limit CO_2 diffusion to the same extent as C3 plants because initial fixation is not reduced with a competing oxygenase reaction (Ludwig, 2016); however, water loss is reduced (Taylor et al., 2010) facilitating greater water use efficiency. There are four basic stages to the C4 cycle: First, CO_2 molecules react with carbonic anhydrase to generate a molecule of bicarbonate, the substrate for PEPc. PEP carboxylase catalyses the reaction between the bicarbonate and phosphoenolpyruvate to make oxaloacetate, which is a highly reactive molecule and generally does not accumulate in the mesophyll cell, but is instead converted into a C_4 carbon acid, malate, or aspartate (Ludwig, 2016) which can act as an intermediate store for carbon dioxide (Hatch, 1971). This C4 acid is subsequently transported from the mesophyll cells to the bundle sheath cells, via the plasmodesmata that connect the two cells. Once inside the bundle sheath, the C4 acids are decarboxylated to release CO_2. This mechanism acts to concentrate CO_2 to a much greater level than could be achieved by simple diffusion of CO_2 from ambient air (Ludwig, 2016). The high concentration of CO_2 at the rubisco site of carboxylation and the lack of O_2 production with the absence of PSII greatly reduces O_2 fixation and photorespiration. Once decarboxylated, the remaining C3 acid (pyruvate or alanine) is transported out of the bundle sheath back to the mesophyll cells where in a series of reactions PEP is regenerated (see Fig. 1). The additional steps in the C4 pathway come at an increased energetic costs utilising ATP for PEP regeneration; however, the benefits of the concentrating mechanism outweigh the costs due to the reduction in photorespiration (Sage, 2004).

C4 photosynthesis can be categorised into different subtypes (see Fig. 6), depending on the primary decarboxylation enzyme utilised in the bundle sheath: the NADP-malic enzyme (NADP-ME), the NAD-malic enzyme (NAD-ME), and the PEP carboxykinase (PEPCK) subtypes (Sage et al., 2012). These subtypes also differ in the C4 acid transported into the BS cell, with NADP-ME C4 plants utilising malate, and both NAD-ME and PEPCK C4 plants utilising aspartate (Garner et al., 2016). Furthermore, the location of the primary decarboxylation enzyme differs between the 3 C4 types. In NADP-ME plants, including sorghum, maize, and sugarcane, NADP-ME is located in the BS cell chloroplasts whilst NAD-ME of the NAD-ME subtype, such as millet, is localised in the

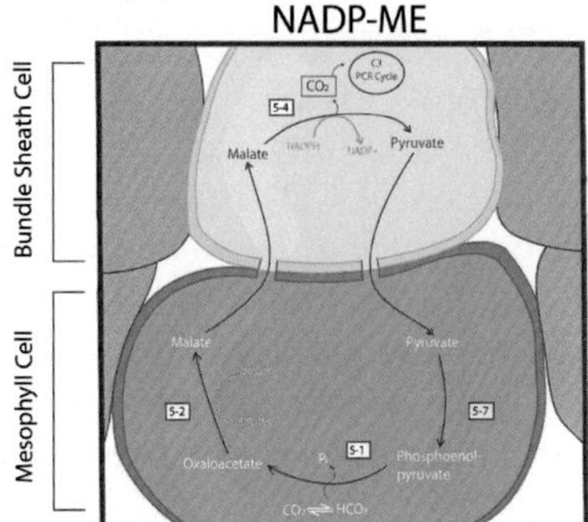

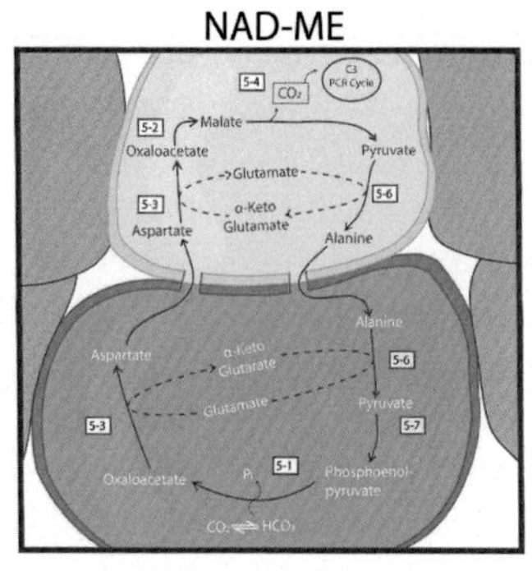

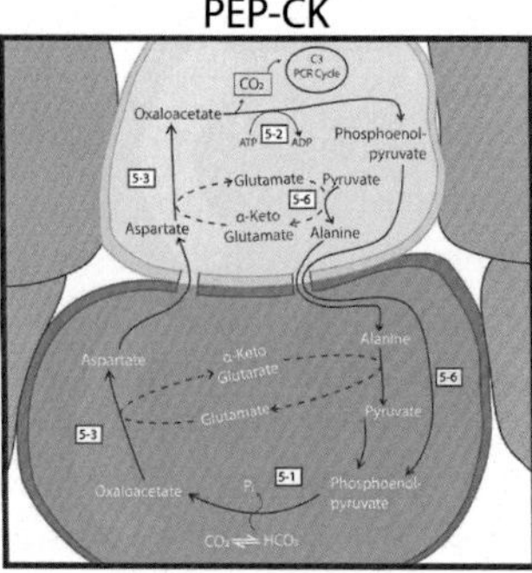

FIG. 6 Schematic representation of the different C4 subtypes: NADP-ME, NAD-ME, and PEP-CK. Bundle sheath cells are shaded *yellow* and mesophyll cells *green*. Decarboxylation colours represent the different enzymes and location; *green* for NADP-ME located in the chloroplast; *red* for the NAD-ME located in the mitochondria; and *blue* for PEP-CK located in the cytoplasm. *Figure redrawn with permission from Jones, R., Ougham, H., Thomas, H., Waaland, S. 2012. The Molecular Life of Plants. John Wiley and Sons.*

BS cell mitochondria (Garner et al., 2016). Finally, in PEPCK-type plants, including many grass species, the enzyme is localised in the cytoplasm (Garner et al., 2016). These differences in enzyme localisation may account for the slight differences in the C4 cycle between plant types. The NADP-ME enzyme utilises malate produced in the M cells, which is transported to the BS chloroplasts and decarboxylated to CO_2 and pyruvate. This pyruvate is transported back to the M cells and converted to PEP by pyruvate orthophosphate dikinase (PPdK). In comparison, NADP-ME plants transport aspartate to BS cells, first for transamination to oxaloacetate (OAA) followed by reduction to malate. This is then transported into the BS mitochondria for decarboxylation. Similarly, PEPCK-type plants transport aspartate into BS cells for transamination to OAA, but can then be directly decarboxylated by PEPCK to CO_2 and PEP. For completion of the C4 cycle, both NAD-ME and PEPCK-type plants produce alanine from PEP, creating the C3 acid transported back into M cells, which is subsequently converted back to PEP by PPdK for use in carbon fixation (Garner et al., 2016).

4.6 Crassulacean acid metabolism

Crassulacean acid metabolism (CAM) plants also exhibit a CO_2 concentrating mechanism, separating initial fixation of atmospheric CO_2 from rubisco and the Calvin cycle by time rather than space, with stomatal opening and CO_2 fixation taking place at night (Fig. 7). CAM is most common in dry climates and is particularly prevalent in cacti as well as commercially significant crops such as pineapple (*Ananas comosus* (L.) Merr.) (Davis et al., 2019), representing ~6% of plant species (Winter and Smith, 1996).

CAM photosynthesis can be divided into four classical phases, phases I–IV, that are interdependent (Osmond, 1978). Phase I consists of initial CO_2 fixation in the cytosol into oxaloacetate (Edwards and Ogburn, 2012) at night time using PEP and PEPc (as in C4 metabolism) with open stomata (Osmond, 1978). Nocturnal stomatal opening in CAM plants greatly reduces water loss by up to 80% as evaporative demands are significantly less at night (DePaoli et al., 2014). Oxaloacetate is then converted to malate by NAD(P)-malate dehydrogenase and stored in the vacuole as malic acid until daylight (Borland et al., 2014). Phase II consists of a short transition period where stomata can remain open for a short time (up to about 2h (Osmond, 1978) depending on the species and the environmental conditions) and atmospheric CO_2 diffusing into the leaf through the stomata can be fixed through typical C3 photosynthesis. In phase III, in the light stomata close so preventing water loss and it is under these conditions that the stored malate in the vacuole is decarboxylated and CO_2 released behind closed stomata (Borland et al., 2014). This release of CO_2 in a 'sealed' compartment results in an increased CO_2 concentration, which represents the CAM CO_2 concentrating mechanism for fixation by rubisco as part of the Calvin cycle, reducing photorespiration. Finally, in phase IV, late in the light period stomata can re-open once again allowing direct fixation of CO_2 by rubisco (Osmond, 1978; Luttge, 2004). Many genes associated with CAM demonstrate a circadian element, such as NAD-ME and PEPC (Wai et al., 2017), indicating circadian regulation may play a vital role in CAM photosynthesis.

Stomatal kinetics are particularly important in CAM photosynthesis, with timings of opening and closure allowing for plants to separate carbon assimilation and initial CO_2 fixation whilst also improving water use efficiency. Stomatal conductance is usually highest during the nocturnal phase I, where the uptake of CO_2 is essential for the later CO_2 concentrating mechanism around rubisco, and lowest during phase III, the light period (Males and Griffiths, 2017), reducing water loss. The timing of stomatal closure in CAM plants has been linked to genes

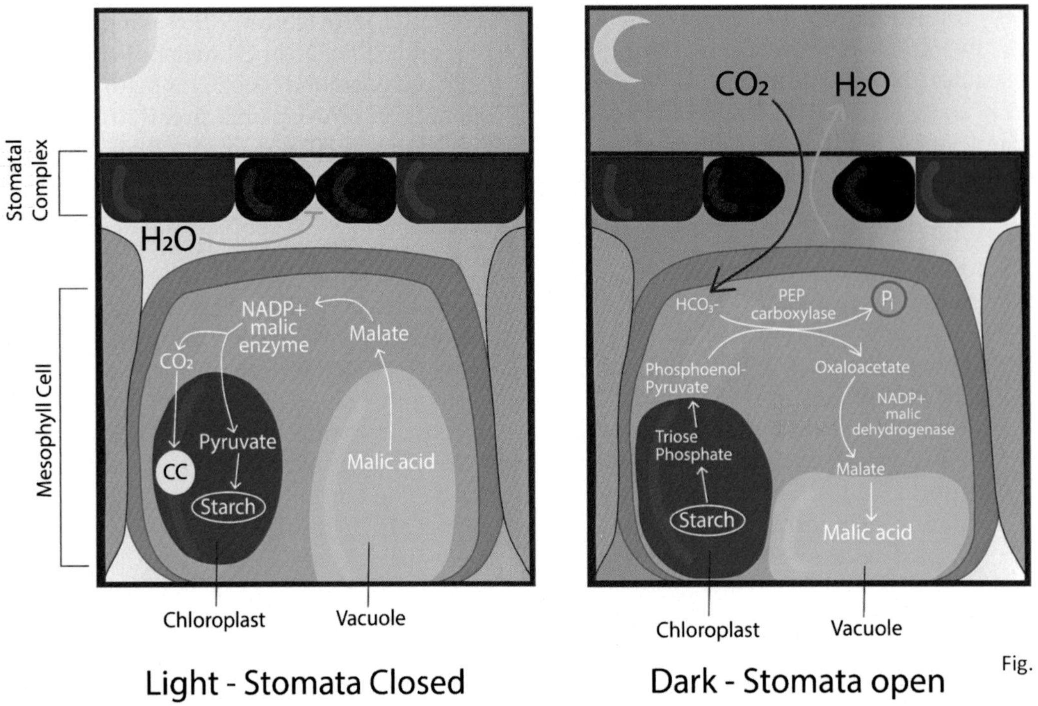

FIG. 7 Crassulacean acid metabolism (CAM), illustrating the temporal separation of initial uptake of CO_2 from the atmosphere from rubisco activity. CO_2 is taken up at night-time and stored in the vacuole as malic acid, which is decarboxylated in the light behind closed stomata and CO_2 is released for fixation by rubisco in the chloroplast. The closed stomata through the day help CAM plants to maintain water. *Redrawn from Taiz, L., Zeiger, E., 2002. Plant Physiology, third ed. Sinauer Associates, Sunderland, MA, USA, with permission.*

associated with CO_2 and ABA-signalling, whilst the specific stomatal blue light response, which is important for early morning and rapid stomatal opening (Matthews et al., 2020) is absent in CAM (Abraham et al., 2016), suggesting that environmental cues aside from light are central to stomatal regulation in CAM.

CAM plants can exhibit varying degrees of CAM photosynthesis. In most CAM plants, expression of the CAM pathway is obligate or constitutive, meaning the CAM photosynthetic pathway is always expressed in mature photosynthetic tissue (Winter, 2019). However, plants in which CAM photosynthesis can be switched on and off have also been described, termed facultative CAM. Facultative CAM, in which expression of CAM photosynthesis is inducible, is thought to be present in over 5% of vascular plant species, and most commonly occurs in plants also expressing C3 photosynthesis (Winter, 2019). Facultative CAM was first described in the C3 plant *Mesembryanthemum crystallinum* under high salinity, in which water deficit stress induced conversion from C3 to CAM photosynthesis whilst removal of this stress resulted in reversion to C3 photosynthesis (Winter and von Willert, 1972). Since then, facultative CAM has been identified in at least 54 species, mainly from the order Caryophyllales, but is suspected to be present in over 1000 species

(Winter, 2019). Nearly all cases of facultative CAM have been noted to occur in response to stress, where a reversible increase in nocturnal CO_2 uptake and acidification occurs (Winter, 2019). The transition from C3 to CAM photosynthesis in these facultative CAM plants has been associated with specific changes in gene expression and transcriptional networks. A global gene expression study in *Sedum album* noted a 73-fold higher expression in the core CAM genes PEPc and PPCK (phosphoenolpyruvate carboxylase kinase) in droughted plants compared to well-watered C3 plants, whilst the networks regulating C3 and CAM co-expression had little overlap (Wai et al., 2019). This demonstrates that there is a large degree of reprogramming associated with the C3-CAM transition.

4.7 Genetic manipulation

These photosynthetic adaptations allow for C4 and CAM plants to survive in hot, dry climates. At higher temperatures with decreased soil water content, C4 plants have higher net assimilation rates than C3 plants as well as performing better under high light intensities (Zhou et al., 2018) due to their extra energy requirement. The ability to be productive under higher temperature conditions and lower water availability is a desirable trait for crop plants, with current interest in engineering C4 photosynthesis into rice (*Oryza sativa* L.) to increase yield and improve resource use efficiency. Rice is normally a tropical C3 plant with high levels of photorespiration above 30°C, reducing its photosynthetic efficiency (Karki et al., 2013), so current efforts are aimed at generating rice plants with Kranz anatomy and C4 photosynthesis (Wang et al., 2016b). By introducing genes specifically associated with C4 anatomy and biochemistry, there is the potential to generate more efficient rice lines. For example, introduction of maize *GOLDEN2-LIKE* genes into rice resulted in the development of mitochondria and chloroplasts in bundle sheath cells, with no fitness cost compared to wild type plants (Wang et al., 2017), representing an important step towards C4 rice lines.

5 Conclusion

This chapter has provided an overview of primary carbon assimilation, including the pathway of CO_2 diffusion from the atmosphere to the site of fixation and the relative contribution of the various resistances reducing this flow and limiting photosynthesis. An outline of the biochemical limitation to C3 metabolism due to rubisco photorespiratory reaction is provided along with plants' evolutionary strategies to overcome this and increase photosynthetic efficiency. A selection of studies that explore approaches to overcome diffusional constraints in an attempt to improve plant productivity have also been highlighted. The subsequent chapter will give an in-depth account of studies that exploit genetic targets to increases C3 photosynthesis and crop yield.

References

Abraham, P.E., Yin, H., Borland, A.M., Weighill, D., Lim, S.-D., De Paoli, H.C., Engle, N., Jones, P.C., Agh, R., Weston, D.J., Wullschleger, S.D., Tschaplinski, T., Jacobson, D., Cushman, J.C., Hettich, R.L., Tuskan, G.A., Yang, X., 2016. Transcript, protein and metabolite temporal dynamics in the CAM plant Agave. Nat. Plants 2, 16178.

Amada, G., Onoda, Y., Ichie, T., Kitayama, K., 2017. Influence of leaf trichomes on boundary layer conductance and gas-exchange characteristics in *Metrosideros polymorpha* (Myrtaceae). Biotropica 49, 482–492.

Amthor, J.S., Bar-Even, A., Hanson, A.D., Millar, A.H., Stitt, M., Sweetlove, L.J., Tyerman, S.D., 2019. Engineering strategies to boost crop productivity by cutting respiratory carbon loss. Plant Cell 31, 297–314.

Antunes, W.C., Provart, N.J., Williams, T.C.R., Loureiro, M.-E., 2012. Changes in stomatal function and water use efficiency in potato plants with altered sucrolytic activity. Plant Cell Environ. 35, 747–759.

Araújo, W.L., Nunes-Nesi, A., Osorio, S., Usadel, B., Fuentes, D., Nagy, R., Balbo, I., Lehmann, M., Studart-

Witkowski, C., Tohge, T., Martinoia, E., Jordana, X., DaMatta, F.M., Fernie, A.R., 2011. Antisense inhibition of the iron-sulphur subunit of succinate dehydrogenase enhances photosynthesis and growth in tomato via an organic acid–mediated effect on stomatal aperture. Plant Cell 23, 600–627.

Badger, M.R., Price, G.D., 1994. The role of carbonic anhydrase in photosynthesis. Annu. Rev. Plant Physiol. Plant Mol. Biol. 45, 369–392.

Barbour, M.M., Evans, J.R., Simonin, K.A., von Caemmerer, S., 2016. Online CO_2 and H_2O oxygen isotope fractionation allows estimation of mesophyll conductance in C_4 plants, and reveals that mesophyll conductance decreases as leaves age in both C_4 and C_3 plants. New Phytol. 210, 875–889.

Barbour, M.M., Tcherkez, G., Bickford, C.P., Mauve, C., Lamothe, M., Sinton, S., Brown, H., 2011. δ13C of leaf-respired CO2 reflects intrinsic water-use efficiency in barley. Plant Cell Environ. 34, 792–799.

Barradas, V.L., Jones, H.G., 1996. Responses of CO2 assimilation to changes in irradiance: laboratory and field data and a model for beans (*Phaseolus vulgaris* L.). J. Exp. Bot. 47, 639–645.

Bertolino, L.T., Caine, R.S., Gray, J.E., 2019. Impact of stomatal density and morphology on water-use efficiency in a changing world. Front. Plant Sci. 10, 225.

Bongi, G., Loreto, F., 1989. Gas-exchange properties of salt-stressed olive (*Olea europea* L.) leaves. Plant Physiol. 90, 1408–1416.

Borland, A.M., Hartwell, J., Weston, D.J., Schlauch, K.A., Tschaplinski, T.J., Tuskan, G.A., Yang, X., Cushman, J.C., 2014. Engineering crassulacean acid metabolism to improve water-use efficiency. Trends Plant Sci. 19, 327–338.

Buchanan, B.B., 1980. Role of light in the regulation of chloroplast enzymes. Annu. Rev. Plant Physiol. 31, 341–374.

Buchanan, B.B., Gruissem, W., Jones, R.L., 2000. Biochemistry and Molecular Biology of Plants. American Society of Plant Biologists, Rockville, MD.

Caine, R.S., Yin, X., Sloan, J., Harrison, E.L., Mohammed, U., Fulton, T., Biswal, A.K., Dionora, J., Chater, C.C., Coe, R.-A., Bandyopadhyay, A., Murchie, E.H., Swarup, R., Quick, W.P., Gray, J.E., 2019. Rice with reduced stomatal density conserves water and has improved drought tolerance under future climate conditions. New Phytol. 221, 371–384.

Carmo-Silva, A.E., Salvucci, M.E., 2013. The regulatory properties of rubisco activase differ among species and affect photosynthetic induction during light transitions. Plant Physiol. 161, 1645–1655.

Carriquí, M., Cabrera, H.M., Conesa, M., Coopman, R.E., Douthe, C., Gago, J., Gallé, A., Galmés, J., Ribas-Carbo, M., Tomás, M., Flexas, J., 2015. Diffusional limitations explain the lower photosynthetic capacity of ferns as compared with angiosperms in a common garden study. Plant Cell Environ. 38, 448–460.

Coleman, J.R., 2000. Carbonic anhydrase and its role in photosynthesis. In: Photosynthesis. vol. 9. Springer, Dordrecht, pp. 353–367.

Cousins, A.B., Badger, M.R., von Caemmerer, S., 2006. Carbonic anhydrase and its influence on carbon isotope discrimination during C4 photosynthesis. Insights from antisense RNA in *Flaveria bidentis*. Plant Physiol. 141, 232–242.

Cousins, A.B., Mullendore, D.L., Sonawane, B.V., 2020. Recent developments in mesophyll conductance in C3, C4, and crassulacean acid metabolism plants. Plant J. 101, 816–830.

Davis, S.C., Simpson, J., Gil-Vega, K.D.C., Niechayev, N.A., van Tongerlo, E., Castano, N.H., Dever, L.V., Búrquez, A., 2019. Undervalued potential of crassulacean acid metabolism for current and future agricultural production. J. Exp. Bot. 70, 6521–6537.

Degen, G.E., Orr, D.J., Carmo-Silva, E., 2021. Heat-induced changes in the abundance of wheat Rubisco activase isoforms. New Phytol. 229 (3), 1298–1311.

DePaoli, H.C., Borland, A.M., Tuskan, G.A., Cushman, J.C., Yang, X., 2014. Synthetic biology as it relates to CAM photosynthesis: challenges and opportunities. J. Exp. Bot. 65, 3381–3393.

Di Marco, G., Manes, F., Tricoli, D., Vitale, E., 1990. Fluorescence parameters measured concurrently with net photosynthesis to investigate chloroplastic CO_2 concentration in leaves of *Quercus ilex* L. J. Plant Physiol. 136, 538–543.

Doheny-Adams, T., Hunt, L., Franks, P.J., Beerling, D.J., Gray, J.E., 2012. Genetic manipulation of stomatal density influences stomatal size, plant growth and tolerance to restricted water supply across a growth carbon dioxide gradient. Philos. Trans. R. Soc. B Biol. Sci. 367, 547–555.

Dow, G.J., Berry, J.A., Bergmann, D.C., 2014. The physiological importance of developmental mechanisms that enforce proper stomatal spacing in *Arabidopsis thaliana*. New Phytol. 201, 1205–1217.

Drake, P.L., Froend, R.H., Franks, P.J., 2013. Smaller, faster stomata: scaling of stomatal size, rate of response, and stomatal conductance. J. Exp. Bot. 64, 495–505.

Dunn, J., Hunt, L., Afsharinafar, M., Al Meselmani, M., Mitchell, A., Howells, R., Wallington, E., Fleming, A.J., Gray, J.E., 2019. Reduced stomatal density in bread wheat leads to increased water-use efficiency. J. Exp. Bot. 70, 4737–4748.

Earles, J.M., Buckley, T.N., Brodersen, C.R., Busch, F.A., Cano, F.J., Choat, B., Evans, J.R., Farquhar, G.D., Harwood, R., Huynh, M., John, G.P., Miller, M.L., Rockwell, F.E., Sack, L., Scoffoni, C., Struik, P.C., Wu, A., Yin, X., Barbour, M.M., 2019. Embracing 3D complexity in leaf carbon–water exchange. Trends Plant Sci. 24, 15–24.

Edwards, E.J., Ogburn, R.M., 2012. Angiosperm responses to a low-CO2 world: CAM and C4 photosynthesis as parallel evolutionary trajectories. Int. J. Plant Sci. 173, 724–733.

Ehleringer, J., Björkman, O., 1977. Quantum yields for CO2 uptake in C3 and C4 plants. Plant Physiol. 59, 86–90.

Ehleringer, J.R., Mooney, H.A., 1978. Leaf hairs: effects on physiological activity and adaptive value to a desert shrub. Oecologia 37, 183–200.

Eisenach, C., Chen, Z., Grefen, C., Blatt, M.R., 2012. The trafficking protein SYP121 of Arabidopsis connects programmed stomatal closure and K+ channel activity with vegetative growth. Plant J. 69, 241–251.

Elliott-Kingston, C., Haworth, M., Yearsley, J.M., Batke, S.P., Lawson, T., McElwain, J.C., 2016. Does size matter? Atmospheric CO2 may be a stronger driver of stomatal closing rate than stomatal size in taxa that diversified under low CO2. Front. Plant Sci. 7, 1253.

Evans, J., Loreto, F., 2000. Acquisition and diffusion of CO2 in higher plant leaves. In: Leegood, R.C., Sharkey, T.T., von Cammerer, S. (Eds.), Photosynthesis: Physiology and Metabolism. Kluwer Academic Publishers, Dordrecht, pp. 321–351.

Evans, J., Sharkey, T., Berry, J., Farquhar, G., 1986. Carbon isotope discrimination measured concurrently with gas exchange to investigate CO2 diffusion in leaves of higher plants. Funct. Plant Biol. 13, 281.

Evans, J.R., von Caemmerer, S., Setchell, B.A., Hudson, G.S., 1994. The relationship between CO2 transfer conductance and leaf anatomy in transgenic tobacco with a reduced content of Rubisco. Aust. J. Plant Physiol. 21, 475–495.

Ethier, G.J., Livingston, N.J., 2004. On the need to incorporate sensitivity to CO_2 transfer conductance into the Farquhar–von Caemmerer–Berry leaf photosynthesis model. Plant Cell Environ. 27, 137–153.

Evans, J.R., Kaldenhoff, R., Genty, B., Terashima, I., 2009. Resistances along the CO2 diffusion pathway inside leaves. J. Exp. Bot. 60, 2235–2248.

Fabre, N., Reiter, I.M., Becuwe-Linka, N., Genty, B., Rumeau, D., 2007. Characterization and expression analysis of genes encoding? And? Carbonic anhydrases in Arabidopsis. Plant Cell Environ. 30, 617–629.

Faralli, M., Matthews, J., Lawson, T., 2019. Exploiting natural variation and genetic manipulation of stomatal conductance for crop improvement. Curr. Opin. Plant Biol. 49, 1–7.

Flexas, J., Diaz-Espejo, A., 2015. Interspecific differences in temperature response of mesophyll conductance: food for thought on its origin and regulation. Plant Cell Environ. 38, 625–628.

Flexas, J., Ribas-Carbó, M., Hanson, D.T., Bota, J., Otto, B., Cifre, J., McDowell, N., Medrano, H., Kaldenhoff, R., 2006. Tobacco aquaporin NtAQP1 is involved in mesophyll conductance to CO2 in vivo. Plant J. 48, 427–439.

Flexas, J., Diaz-Espejo, A., Galmes, J., Kaldenhoff, R., Medrano, H., Ribas-Carbo, M., 2007. Rapid variations of mesophyll conductance in response to changes in CO_2 concentration around leaves. Plant Cell Environ. 30, 1284–1298.

Flexas, J., Ribas-Carbó, M., Diaz-Espejo, A., Galmés, J., Medrano, H., 2008. Mesophyll conductance to CO2: current knowledge and future prospects. Plant Cell Environ. 31, 602–621.

Flexas, J., Barbour, M.M., Brendel, O., Cabrera, H.M., Carriquí, M., Díaz-Espejo, A., Douthe, C., Dreyer, E., Ferrio, J.-P., Gago, J., Gallé, A., Galmés, J., Kodama, N., Medrano, H., Niinemets, Ü., Peguero-Pina, J.J., Pou, A., Ribas-Carbó, M., Tomás, M., Tosens, T., Warren, C.R., 2012. Mesophyll diffusion conductance to CO2: an unappreciated central player in photosynthesis. Plant Sci. 193–194, 70–84.

Flexas, J., Scoffoni, C., Gago, J., Sack, L., 2013. Leaf mesophyll conductance and leaf hydraulic conductance: an introduction to their measurement and coordination. J. Exp. Bot. 64, 3965–3981.

Franks, P.J., Beerling, D.J., 2009. Maximum leaf conductance driven by CO2 effects on stomatal size and density over geologic time. Proc. Natl. Acad. Sci. 106, 10343–10347.

Franks, P.J., Farquhar, G.D., 2007. The mechanical diversity of stomata and its significance in gas-exchange control. Plant Physiol. 143, 78–87.

Franks, P.J., Drake, P.L., Beerling, D.J., 2009. Plasticity in maximum stomatal conductance constrained by negative correlation between stomatal size and density: an analysis using *Eucalyptus globulus*. Plant Cell Environ. 32, 1737–1748.

Franks, P.J., Doheny-Adams, T.W., Britton-Harper, Z.J., Gray, J.E., 2015. Increasing water-use efficiency directly through genetic manipulation of stomatal density. New Phytol. 207, 188–195.

Garner, D.M., Mure, C.M., Yerramsetty, P., Berry, J.O., 2016. Kranz Anatomy and the C4 Pathway. ELS, John Wiley & Sons, Chichester, UK, pp. 1–10.

Gay, A.P., Hurd, R.G., 1975. The influence of light on stomatal density in the tomato. New Phytol. 75, 37–46.

Gillon, J.S., Yakir, D., 2000. Internal conductance to CO2 diffusion and C18OO discrimination in C3 leaves. Plant Physiol. 123, 201–213.

Giuliani, R., Koteyeva, N., Voznesenskaya, E., Evans, M.A., Cousins, A.B., Edwards, G.E., 2013. Coordination of leaf photosynthesis, transpiration, and structural traits in rice and wild relatives (genus oryza). Plant Physiol. 162, 1632–1651.

Han, J.-M., Meng, H.-F., Wang, S.-Y., Jiang, C.-D., Liu, F., Zhang, W.-F., Zhang, Y.-L., 2016. Variability of mesophyll conductance and its relationship with water use efficiency in cotton leaves under drought pretreatment. J. Plant Physiol. 194, 61–71.

Hanba, Y.T., Kogami, H., Terashima, I., 2002. The effect of growth irradiance on leaf anatomy and photosynthesis in Acer species differing in light demand. Plant Cell Environ. 25, 1021–1030.

Harley, P.C., Loreto, F., Di Marco, G., Sharkey, T.D., 1992. Theoretical considerations when estimating the mesophyll conductance to CO_2 flux by analysis of the response of photosynthesis to CO_2. Plant Physiol. 98, 1429–1436.

Harrison, E.L., Arce Cubas, L., Gray, J.E., Hepworth, C., 2020. The influence of stomatal morphology and distribution on photosynthetic gas exchange. Plant J. 101, 768–779.

Hashimoto, M., Negi, J., Young, J., Israelsson, M., Schroeder, J.I., Iba, K., 2006. Arabidopsis HT1 kinase controls stomatal movements in response to CO2. Nat. Cell Biol. 8, 391–397.

Hatch, M.D., 1971. The C4-pathway of photosynthesis. Evidence for an intermediate pool of carbon dioxide and the identity of the donor C4-dicarboxylic acid. Biochem. J. 125, 425–432.

Hatch, M., Slack, C., 1966. Photosynthesis by sugar-cane leaves. A new carboxylation reaction and the pathway of sugar formation. Biochem. J. 101, 103–111.

Haworth, M., McElwain, J., 2008. Hot, dry, wet, cold or toxic? Revisiting the ecological significance of leaf and cuticular micromorphology. Palaeogeogr. Palaeoclimatol. Palaeoecol. 262, 79–90.

Heckwolf, M., Pater, D., Hanson, D.T., Kaldenhoff, R., 2011. The *Arabidopsis thaliana* aquaporin AtPIP1;2 is a physiologically relevant CO2 transport facilitator. Plant J. 67, 795–804.

Hepworth, C., Caine, R.S., Harrison, E.L., Sloan, J., Gray, J.E., 2018. Stomatal development: focusing on the grasses. Curr. Opin. Plant Biol. 41, 1–7.

Hetherington, A.M., Woodward, F.I., 2003. The role of stomata in sensing and driving environmental change. Nature 424, 901–908.

Hodges, M., Jossier, M., Boex-Fontvieille, E., Tcherkez, G., 2013. Protein phosphorylation and photorespiration. Plant Biol. 15, 694–706.

Hughes, J., Hepworth, C., Dutton, C., Dunn, J.A., Hunt, L., Stephens, J., Waugh, R., Cameron, D.D., Gray, J.E., 2017. Reducing stomatal density in barley improves drought tolerance without impacting on yield. Plant Physiol. 174, 776–787.

Joel, G., Aplet, G., Vitousek, P.M., 1994. Leaf morphology along environmental gradients in Hawaiian *Metrosideros polymorpha*. Biotropica 26, 17.

Johnson, H.B., 1975. Plant pubescence: an ecological perspective. Bot. Rev. 41, 233–258.

Jones, H.G., 1992. Plants and Microclimate: A Quantitative Approach to Environmental Plant Physiology, second ed. Cambridge University Press, Cambridge, UK.

Karki, S., Rizal, G., Quick, W.P., 2013. Improvement of photosynthesis in rice (*Oryza sativa* L.) by inserting the C4 pathway. Rice 6, 28.

Kelly, G., Moshelion, M., David-Schwartz, R., Halperin, O., Wallach, R., Attia, Z., Belausov, E., Granot, D., 2013. Hexokinase mediates stomatal closure. Plant J. 75, 977–988.

King, L.S., Kozono, D., Agre, P., 2004. From structure to disease: the evolving tale of aquaporin biology. Nat. Rev. Mol. Cell Biol. 5, 687–698.

Kromdijk, J., Głowacka, K., Long, S.P., 2020. Photosynthetic efficiency and mesophyll conductance are unaffected in *Arabidopsis thaliana* aquaporin knock-out lines. J. Exp. Bot. 71, 318–329.

Laisk, A., Eichelmann, H., Oja, V., Rasulov, B., Padu, E., Bichele, I., Pettai, H., Kull, O., 2005. Adjustment of leaf photosynthesis to shade in a natural canopy: rate parameters. Plant Cell Environ. 28, 375–388.

Lawson, T., 2009. Guard cell photosynthesis and stomatal function. New Phytol. 181, 13–34.

Lawson, T., Blatt, M.R., 2014. Stomatal size, speed, and responsiveness impact on photosynthesis and water use efficiency. Plant Physiol. 164, 1556–1570.

Lawson, T., Matthews, J., 2020. Guard cell metabolism and stomatal function. Annu. Rev. Plant Biol. 71, 273–302.

Lawson, T., Morison, J.I.L., 2004. Stomatal function and physiology. In: The Evolution of Plant Physiology. Elsevier, Amsterdam, pp. 217–242.

Lawson, T., Vialet-Chabrand, S., 2019. Speedy stomata, photosynthesis and plant water use efficiency. New Phytol. 221, 93–98.

Lawson, T., Weyers, J., 1999. Spatial and temporal variation in gas exchange over the lower surface of *Phaseolus vulgaris* L. primary leaves. J. Exp. Bot. 50, 1381–1391.

Lawson, T., von Caemmerer, S., Baroli, I., 2010. Photosynthesis and stomatal behaviour. In: Progress in Botany. Springer, Berlin, Heidelberg, pp. 265–304.

Lawson, T., Simkin, A.J., Kelly, G., Granot, D., 2014. Mesophyll photosynthesis and guard cell metabolism impacts on stomatal behaviour. New Phytol. 203, 1064–1081.

Lehmeier, C., Pajor, R., Lundgren, M.R., Mathers, A., Sloan, J., Bauch, M., Mitchell, A., Bellasio, C., Green, A., Bouyer, D., Schnittger, A., Sturrock, C., Osborne, C.P., Rolfe, S., Mooney, S., Fleming, A.J., 2017. Cell density and airspace patterning in the leaf can be manipulated to increase leaf photosynthetic capacity. Plant J. 92, 981–994.

Lloyd, J., Syvertsen, J.P., Kriedemann, P.E., Farquhar, G.D., 1992. Low conductances for CO_2 diffusion from stomata to the sites of carboxylation in leaves of woody species. Plant Cell Environ. 15, 873–899.

Loreto, F., Harley, P.C., Di Marco, G., Sharkey, T.D., 1992. Estimation of mesophyll conductance to CO_2 flux by three different methods. Plant Physiol. 98, 1437–1443.

Loreto, F., Tsonev, T., Centritto, M., 2009. The impact of blue light on leaf mesophyll conductance. J. Exp. Bot. 60, 2283–2290.

Lorimer, G.H., Hartman, F.C., 1988. Evidence supporting lysine 166 of *Rhodospirillum rubrum* ribulosebisphosphate carboxylase as the essential base which initiates catalysis. J. Biol. Chem. 263, 6468–6471.

Ludwig, M., 2016. The roles of organic acids in C4 photosynthesis. Front. Plant Sci. 7, 1–11.

Lundgren, M.R., Fleming, A.J., 2020. Cellular perspectives for improving mesophyll conductance. Plant J. 101, 845–857.

Lundgren, M.R., Osborne, C.P., Christin, P.-A., 2014. Deconstructing Kranz anatomy to understand C4 evolution. J. Exp. Bot. 65, 3357–3369.

Lundgren, M.R., Mathers, A., Baillie, A.L., Dunn, J., Wilson, M.J., Hunt, L., Pajor, R., Fradera-Soler, M., Rolfe, S., Osborne, C.P., Sturrock, C.J., Gray, J.E., Mooney, S.J., Fleming, A.J., 2019. Mesophyll porosity is modulated by the presence of functional stomata. Nat. Commun. 10, 2825.

Luttge, U., 2004. Ecophysiology of crassulacean acid metabolism (CAM). Ann. Bot. 93, 629–652.

Mahdieh, M., Mostajeran, A., Horie, T., Katsuhara, M., 2008. Drought stress alters water relations and expression of PIP-type aquaporin genes in *Nicotiana tabacum* plants. Plant Cell Physiol. 49, 801–813.

Males, J., Griffiths, H., 2017. Stomatal biology of CAM plants. Plant Physiol. 174, 550–560.

Matthews, J.S.A., Vialet-Chabrand, S.R.M., Lawson, T., 2017. Diurnal variation in gas exchange: the balance between carbon fixation and water loss. Plant Physiol. 174, 614–623.

Matthews, J.S.A., Vialet-Chabrand, S., Lawson, T., 2018. Acclimation to fluctuating light impacts the rapidity of response and diurnal rhythm of stomatal conductance. Plant Physiol. 176, 1939–1951.

Matthews, J.S.A., Vialet-Chabrand, S., Lawson, T., 2020. Role of blue and red light in stomatal dynamic behaviour. J. Exp. Bot. 71, 2253–2269.

McAusland, L., Vialet-Chabrand, S., Davey, P., Baker, N.R., Brendel, O., Lawson, T., 2016. Effects of kinetics of light-induced stomatal responses on photosynthesis and water-use efficiency. New Phytol. 211, 1209–1220.

McElwain, J.C., Yiotis, C., Lawson, T., 2016. Using modern plant trait relationships between observed and theoretical maximum stomatal conductance and vein density to examine patterns of plant macroevolution. New Phytol. 209, 94–103.

Metcalfe, C.R., Chalk, L., 1950. Anatomy of the Dicotyledons, first ed. Clarendon Press, Oxford, UK.

Meyer, S., Genty, B., 1998. Mapping intercellular CO2 mole fraction (Ci) in *Rosa rubiginosa* leaves fed with abscisic acid by using chlorophyll fluorescence imaging: significance of Ci estimated from leaf gas exchange. Plant Physiol. 116, 947–957.

Michelet, L., Zaffagnini, M., Morisse, S., Sparla, F., Pérez-Pérez, M.E., Francia, F., Danon, A., Marchand, C.H., Fermani, S., Trost, P., Lemaire, S.D., 2013. Redox regulation of the Calvin-Benson cycle: something old, something new. Front. Plant Sci. 4, 1–22.

Momayyezi, M., McKown, A.D., Bell, S.C.S., Guy, R.D., 2020. Emerging roles for carbonic anhydrase in mesophyll conductance and photosynthesis. Plant J. 101, 831–844.

Moroney, J.V., Bartlett, S.G., Samuelsson, G., 2001. Carbonic anhydrases in plants and algae: invited review. Plant Cell Environ. 24, 141–153.

Mott, K.A., Woodrow, I.E., 2000. Modelling the role of Rubisco activase in limiting non-steady-state photosynthesis. J. Exp. Bot. 51, 399–406.

Murchie, E.H., Kefauver, S., Araus, J.L., Muller, O., Rascher, U., Flood, P.J., Lawson, T., 2018. Measuring the dynamic photosynthome. Ann. Bot. 122, 207–220.

Nunes, T.D.G., Zhang, D., Raissig, M.T., 2020. Form, development and function of grass stomata. Plant J. 101, 780–799.

Ogée, J., Wingate, L., Genty, B., 2018. Estimating mesophyll conductance from measurements of C18OO photosynthetic discrimination and carbonic anhydrase activity. Plant Physiol. 178, 728–752.

Osmond, C.B., 1978. Crassulacean acid metabolism: a curiosity in context. Annu. Rev. Plant Physiol. 29, 379–414.

Outlaw, W.H., 1989. Critical examination of the quantitative evidence for and against photosynthetic CO2 fixation by guard cells. Physiol. Plant. 77, 275–281.

Ouyang, W., Struik, P.C., Yin, X., Yang, J., 2017. Stomatal conductance, mesophyll conductance, and transpiration efficiency in relation to leaf anatomy in rice and wheat genotypes under drought. J. Exp. Bot. 68, 5191–5205.

Papanatsiou, M., Petersen, J., Henderson, L., Wang, Y., Christie, J.M., Blatt, M.R., 2019. Optogenetic manipulation of stomatal kinetics improves carbon assimilation, water use, and growth. Science 363, 1456–1459.

Perdomo, J.A., Capó-Bauçà, S., Carmo-Silva, E., Galmés, J., 2017. Rubisco and rubisco activase play an important role in the biochemical limitations of photosynthesis in rice, wheat, and maize under high temperature and water deficit. Front. Plant Sci. 8, 1–15.

Perez-Martin, A., Michelazzo, C., Torres-Ruiz, J.M., Flexas, J., Fernández, J.E., Sebastiani, L., Diaz-Espejo, A., 2014. Regulation of photosynthesis and stomatal and mesophyll conductance under water stress and recovery in olive trees: correlation with gene expression of carbonic anhydrase and aquaporins. J. Exp. Bot. 65, 3143–3156.

Piel, C., Frak, E., Le Roux, X., Genty, B., 2002. Effect of local irradiance on CO2 transfer conductance of mesophyll in walnut. J. Exp. Bot. 53, 2423–2430.

Pons, T.L., Flexas, J., von Caemmerer, S., Evans, J.R., Genty, B., Ribas-Carbo, M., Brugnoli, E., 2009. Estimating mesophyll conductance to CO2: methodology, potential errors, and recommendations. J. Exp. Bot. 60, 2217–2234.

Price, G.D., von Caemmerer, S., Evans, J.R., Yu, J.-W., Lloyd, J., Oja, V., Kell, P., Harrison, K., Gallagher, A., Badger, M.R., 1994. Specific reduction of chloroplast carbonic anhydrase activity by antisense RNA in transgenic tobacco plants has a minor effect on photosynthetic CO2 assimilation. Planta 193, 331–340.

Qu, M., Hamdani, S., Li, W., Wang, S., Tang, J., Chen, Z., Song, Q., Li, M., Zhao, H., Chang, T., Chu, C., Zhu, X., 2016. Rapid stomatal response to fluctuating light: an under-explored mechanism to improve drought tolerance in rice. Funct. Plant Biol. 43, 727–738.

Raines, C.A., 2003. The Calvin cycle revisited. Photosynth. Res. 75, 1–10.

Raissig, M.T., Abrash, E., Bettadapur, A., Vogel, J.P., Bergmann, D.C., 2016. Grasses use an alternatively wired bHLH transcription factor network to establish stomatal identity. Proc. Natl. Acad. Sci. 113, 8326–8331.

Raissig, M.T., Matos, J.L., Anleu Gil, M.X., Kornfeld, A., Bettadapur, A., Abrash, E., Allison, H.R., Badgley, G., Vogel, J.P., Berry, J.A., Bergmann, D.C., 2017. Mobile MUTE specifies subsidiary cells to build physiologically improved grass stomata. Science 355, 1215–1218.

Raven, J.A., 2014. Speedy small stomata? J. Exp. Bot. 65, 1415–1424.

Rumble, J., 2020. CRC Handbook of Chemistry and Physics, 101st ed. CRC Press, Boca Raton, FL.

Sade, N., Shatil-Cohen, A., Attia, Z., Maurel, C., Boursiac, Y., Kelly, G., Granot, D., Yaaran, A., Lerner, S., Moshelion, M., 2014. The role of plasma membrane aquaporins in regulating the bundle sheath-mesophyll continuum and leaf hydraulics. Plant Physiol. 166, 1609–1620.

Sage, R.F., 2004. The evolution of C4 photosynthesis. New Phytol. 161, 341–370.

Sage, R.F., 2016. A portrait of the C4 photosynthetic family on the 50th anniversary of its discovery: species number, evolutionary lineages, and Hall of Fame. J. Exp. Bot. 67, 4039–4056.

Sage, R., Monson, R., 1998. C4 Plant Biology, first ed. Academic Press, Cambridge, MA.

Sage, R.F., Sage, T.L., Kocacinar, F., 2012. Photorespiration and the evolution of C4 photosynthesis. Annu. Rev. Plant Biol. 63, 19–47.

Sakoda, K., Yamori, W., Shimada, T., Sugano, S.S., Hara-Nishimura, I., Tanaka, Y., 2020. Higher stomatal density improves photosynthetic induction and biomass production in Arabidopsis under fluctuating light. Front. Plant Sci. 11, 1609. https://doi.org/10.1101/2020.02.20.958603.

Salvucci, M.E., Portis, A.R., Ogren, W.L., 1985. A soluble chloroplast protein catalyzes ribulosebisphosphate carboxylase/oxygenase activation in vivo. Photosynth. Res. 7, 193–201.

Santelia, D., Lawson, T., 2016. Rethinking guard cell metabolism. Plant Physiol. 172, 1371–1392.

Scafaro, A.P., Von Caemmerer, S., Evans, J.R., Atwell, B.J., 2011. Temperature response of mesophyll conductance in cultivated and wild Oryza species with contrasting mesophyll cell wall thickness. Plant Cell Environ. 34, 1999–2008.

Scales, J.C., Parry, M.A.J., Salvucci, M.E., 2014. A non-radioactive method for measuring Rubisco activase activity in the presence of variable ATP: ADP ratios, including modifications for measuring the activity and activation state of Rubisco. Photosynth. Res. 119, 355–365.

Schluter, U., Muschak, M., Berger, D., Altmann, T., 2003. Photosynthetic performance of an Arabidopsis mutant with elevated stomatal density (sdd1-1) under different light regimes. J. Exp. Bot. 54, 867–874.

Schuepp, P.H., 1993. Leaf boundary layers. New Phytol. 125, 477–507.

Schürmann, P., Jacquot, J.-P., 2000. Plant thioredoxin systems revisited. Annu. Rev. Plant Physiol. Plant Mol. Biol. 51, 371–400.

Sharkey, T.D., Bernacchi, C.J., Farquhar, G.D., Singsaas, E.L., 2007. Fitting photosynthetic carbon dioxide response curves for C_3 leaves. Plant Cell Environ. 30, 1035–1040.

Sherlock, D.J., Raven, J.A., 2001. Interactions between carbon dioxide and oxygen in the photosynthesis of three species of marine red macroalgae. Bot. J. Scotl. 53, 33–43.

Simkin, A.J., Faralli, M., Ramamoorthy, S., Lawson, T., 2020. Photosynthesis in non-foliar tissues: implications for yield. Plant J. 101, 1001–1015.

Spreitzer, R.J., Salvucci, M.E., 2002. Rubisco: structure, regulatory interactions, and possibilities for a better enzyme. Annu. Rev. Plant Biol. 53, 449–475.

Stokes, V.J., Morecroft, M.D., Morison, J.I., 2006. Boundary layer conductance for contrasting leaf shapes in a deciduous broadleaved forest canopy. Agric. For. Meteorol. 139, 40–54.

Syvertsen, J.P., Lloyd, J., McConchie, C., Kriedmann, P.E., Farquha, G.D., 1995. On the relationship between leaf anatomy and CO2 diffusion through the mesophyll of hypostomatous leaves. Plant Cell Environ. 18, 149–157.

Taiz, L., Zeiger, E., 1998. Plant Physiology, second ed. Sinauer Associates, Sunderland, MA, USA.

Taiz, L., Zeiger, E., 2002. Plant Physiology, third ed. Sinauer Associates, Sunderland, MA, USA.

Talbott, L.D., Zeiger, E., 1996. Central roles for potassium and sucrose in guard-cell osmoregulation. Plant Physiol. 111 (4), 1051–1057.

Tanaka, Y., Sugano, S.S., Shimada, T., Hara-Nishimura, I., 2013. Enhancement of leaf photosynthetic capacity through increased stomatal density in Arabidopsis. New Phytol. 198, 757–764.

Taylor, S.H., Hulme, S.P., Rees, M., Ripley, B.S., Ian Woodward, F., Osborne, C.P., 2010. Ecophysiological traits in C3 and C4 grasses: a phylogenetically controlled screening experiment. New Phytol. 185, 780–791.

Taylor, S.H., Franks, P.J., Hulme, S.P., Spriggs, E., Christin, P.A., Edwards, E.J., Woodward, F.I., Osborne, C.P., 2012. Photosynthetic pathway and ecological adaptation explain stomatal trait diversity amongst grasses. New Phytol. 193, 387–396.

Tazoe, Y., Von Caemmerer, S., Estavillo, G.M., Evans, J.R., 2011. Using tunable diode laser spectroscopy to measure carbon isotope discrimination and mesophyll conductance to CO2 diffusion dynamically at different CO2 concentrations. Plant Cell Environ. 34, 580–591.

Terashima, I., Ono, K., 2002. Effects of HgCl2 on CO2 dependence of leaf photosynthesis: evidence indicating involvement of aquaporins in CO2 diffusion across the plasma membrane. Plant Cell Physiol. 43, 70–78.

Terashima, I., Hanba, Y.T., Tanzoe, Y., Vyas, P., Yano, S., 2006. Irradiance and phenotype: comparative eco-development of sun and shade leaves in relation to photosynthetic CO_2 diffusion. J. Exp. Bot. 57, 343–354.

Terashima, I., Hanba, Y.T., Tholen, D., Niinemets, Ü., 2011. Leaf functional anatomy in relation to photosynthesis. Plant Physiol. 155, 108–116.

Tholen, D., Zhu, X.-G., 2011. The mechanistic basis of internal conductance: a theoretical analysis of mesophyll cell photosynthesis and CO2 diffusion. Plant Physiol. 156, 90–105.

Tholen, D., Ethier, G., Genty, B., Pepin, S., Zhu, X.G., 2012. Variable mesophyll conductance revisited: theoretical background and experimental implications. Plant Cell Environ. 35, 2087–2103.

Tichá, I., 1982. Photosynthetic characteristics during ontogenesis of leaves. 7. Stomata density and sizes. Photosynthetica 16, 375–471.

Tomás, M., Flexas, J., Copolovici, L., Galmés, J., Hallik, L., Medrano, H., Ribas-Carbó, M., Tosens, T., Vislap, V., Niinemets, Ü., 2013. Importance of leaf anatomy in determining mesophyll diffusion conductance to CO2 across species: quantitative limitations and scaling up by models. J. Exp. Bot. 64, 2269–2281.

Tosens, T., Niinemets, Ü., Vislap, V., Eichelmann, H., Castro Díez, P., 2012a. Developmental changes in mesophyll diffusion conductance and photosynthetic capacity under different light and water availabilities in *Populus tremula*: how structure constrains function. Plant Cell Environ. 35, 839–856.

Tosens, T., Niinemets, Ü., Westoby, M., Wright, I.J., 2012b. Anatomical basis of variation in mesophyll resistance in eastern Australian sclerophylls: news of a long and winding path. J. Exp. Bot. 63, 5105–5119.

Uehlein, N., Otto, B., Hanson, D.T., Fischer, M., McDowell, N., Kaldenhoff, R., 2008. Function of *Nicotiana tabacum* aquaporins as chloroplast gas pores challenges the concept of membrane CO2 permeability. Plant Cell 20, 648–657.

Vialet-Chabrand, S., Lawson, T., 2020. Thermography methods to assess stomatal behaviour in a dynamic environment. J. Exp. Bot. 71, 2329–2338.

Vialet-Chabrand, S., Matthews, J.S.A., Simkin, A.J., Raines, C.A., Lawson, T., 2017. Importance of fluctuations in light on plant photosynthetic acclimation. Plant Physiol. 173, 2163–2179.

Vialet-Chabrand, S., Matthews, J.S., Lawson, T., 2021. Light, power, action! Interaction of respiratory energy and blue light induced stomatal movements. New Phytol. 231, 2231–2246.

von Caemmerer, S., Evans, J.R., 1991. Determination of the average partial pressure of CO_2 in chloroplasts from leaves of several C3 plants. Funct. Plant Biol. 18, 287–305.

von Caemmerer, S., Evans, J.R., 2015. Temperature responses of mesophyll conductance differ greatly between species. Plant Cell Environ. 38, 629–637.

Wai, C.M., VanBuren, R., Zhang, J., Huang, L., Miao, W., Edger, P.P., Yim, W.C., Priest, H.D., Meyers, B.C., Mockler, T., Smith, J.A.C., Cushman, J.C., Ming, R., 2017. Temporal and spatial transcriptomic and microRNA dynamics of CAM photosynthesis in pineapple. Plant J. 92, 19–30.

Wai, C.M., Weise, S.E., Ozersky, P., Mockler, T.C., Michael, T.P., Vanburen, R., 2019. Time of day and network reprogramming during drought induced CAM photosynthesis in *Sedum album*. PLoS Genet. 15, e1008209.

Wang, L., Jin, X., Li, Q., Wang, X., Li, Z., Wu, X., 2016a. Comparative proteomics reveals that phosphorylation of β carbonic anhydrase 1 might be important for adaptation to drought stress in *Brassica napus*. Sci. Rep. 6, 1–16.

Wang, P., Vlad, D., Langdale, J.A., 2016b. Finding the genes to build C4 rice. Curr. Opin. Plant Biol. 31, 44–50.

Wang, P., Khoshravesh, R., Karki, S., Tapia, R., Balahadia, C.-P., Bandyopadhyay, A., Quick, W.P., Furbank, R., Sage, T.L., Langdale, J.A., 2017. Re-creation of a key step in the evolutionary switch from C3 to C4 leaf anatomy. Curr. Biol. 27, 3278–3287.e6.

Warren, C., 2006. Estimating the internal conductance to CO2 movement. Funct. Plant Biol. 33, 431.

Warren, C.R., Löw, M., Matyssek, R., Tausz, M., 2007. Internal conductance to CO2 transfer of adult *Fagus sylvatica*: variation between sun and shade leaves and due to free-air ozone fumigation. Environ. Exp. Bot. 59, 130–138.

Way, D.A., 2012. What lies between: the evolution of stomatal traits on the road to C4 photosynthesis. New Phytol. 193, 291–293.

Welkie, G.W., Caldwell, M., 1970. Leaf anatomy of species in some dicotyledon families as related to the C3 and C4 pathways of carbon fixation. Can. J. Bot. 48, 2135–2146.

Weyers, J.D.B., Meidner, H., 1990. Methods in Stomatal Research. Longman Scientific & Techical, Harlow.

Winter, K., 2019. Ecophysiology of constitutive and facultative CAM photosynthesis. J. Exp. Bot. 70, 6495–6508.

Winter, K., Smith, J.A.C., 1996. Crassulacean acid metabolism: current status and perspectives. In: Crassulacean Acid Metabolism. Springer, Berlin, Heidelberg, pp. 389–426.

Winter, K., von Willert, D.J., 1972. NaCl induced crassulacean acid metabolism in mesembryanthemum crysallinum. Z. Pflanzenphysiol. 67, 166–170.

Wolosiuk, R.A., Schürmann, P., Buchanan, B.B., 1980. Thioredoxin and ferredoxin-thioredoxin reductase of spinach chloroplasts. In: San Pietro, A. (Ed.), Methods in Enzymology. vol. 69. Academic Press, Cambridge, MA, pp. 382–391.

Wong, S.C., Cowan, I.R., Farquhar, G.D., 1979. Stomatal conductance correlates with photosynthetic capacity. Nature 282, 424–426.

Woodward, F.I., 1987. Stomatal numbers are sensitive to increases in CO2 from pre-industrial levels. Nature 327, 617–618.

Woodward, F.I., 1998. Do plants really need stomata? J. Exp. Bot. 49, 471–480.

Woodward, F.I., Lake, J.A., Quick, W.P., 2002. Stomatal development and CO2: ecological consequences. New Phytol. 153, 477–484.

Xu, F., Wang, K., Yuan, W., Xu, W., Liu, S., Kronzucker, H.J., Chen, G., Miao, R., Zhang, M., Ding, M., Xiao, L., Kai, L., Zhang, J., Zhu, Y., 2019. Overexpression of rice aquaporin OsPIP1;2 improves yield by enhancing mesophyll CO2 conductance and phloem sucrose transport. J. Exp. Bot. 70, 671–681.

Yamori, W., Kusumi, K., Iba, K., Terashima, I., 2020. Increased stomatal conductance induces rapid changes to photosynthetic rate in response to naturally fluctuating light conditions in rice. Plant Cell Environ. 43, 1230–1240.

Zhang, Y.G., Pagani, M., Liu, Z., Bohaty, S.M., DeConto, R., 2013. A 40-million-year history of atmospheric CO_2. Philos. Trans. R. Soc. A Math. Phys. Eng. Sci. 371, 20130096.

Zhou, H., Helliker, B.R., Huber, M., Dicks, A., Akçay, E., 2018. C4 photosynthesis and climate through the lens of optimality. Proc. Natl. Acad. Sci. 115, 12057–12062.

Zhou, H., Whalley, W.R., Hawkesford, M.J., Ashton, R.W., Atkinson, B., Atkinson, J.A., Sturrock, C.J., Bennett, M.J., Mooney, S.J., 2020. The interaction between wheat roots and soil pores in structured field soil. J. Exp. Bot. 72, 1–27.

PART II

Adaptations

CHAPTER

4

Abiotic stress and adaptation in light harvesting

Jun Minagawa

National Institute for Basic Biology, Division of Environmental Photobiology, Okazaki, Japan

Abbreviations

Chl	chlorophyll
Cyt b_6f	cytochrome b_6f complex.
DCMU	3(3,4-dichlorophenyl)-1,1-dimethyl urea
DCCD	dicyclohexylcarbodiimide
Ddx	diadinoxanthin
Dtx	diatoxanthin
EM	electron microscopy
FCP	fucoxanthin-chlorophyll protein
FFEM	freeze-fracture electron microscopy
Fo	minimal fluorescence in the dark
Fo′	minimal fluorescence at the quenched state
Fm	maximal fluorescence in the dark
Fm′	maximal fluorescence at the quenched state
FRP	fluorescence recovery protein
Ft	fluorescence in the actinic light
Fx	fucoxanthin
HL	high light
LHC	light-harvesting complex
LL	low light
Lut	lutein
NPQ	nonphotochemical quenching
φ_{II}	quantum yield of photochemistry at photosystem II
OCP	orange carotenoid protein
PBS	phycobilisome
PQ	plastoquinone
PSI	photosystem I
PSII	photosystem II
RC	reaction centre
RCII	reaction centre of photosystem II
Vx	violaxanthin
VDE	violaxanthin de-epoxidase
Zx	zeaxanthin

1 Introduction: Why photosynthetic organisms adapt to environmental light?

Photosynthesis is initiated when sunlight is captured by light-harvesting antennae, and this leads to electron transport amongst the protein complexes and the concomitant generation of proton motive forces, in and across the thylakoid membranes, respectively. NADPH is generated from the electron flow and ATP is produced by the proton motive force, and these are utilised for the assimilation of carbon dioxide in the Calvin–Benson cycle. Photosystem I (PSI) and photosystem II (PSII) are charge separation devices on the thylakoid membranes that energise the flow of the electrons using energy harvested from the light. Both photosystems originated from a common prototype during evolution, but they have each since become specialised and have differences in terms of organisation of their light-harvesting antennae, pigment compositions, electron acceptors and donors, and several other features. We begin here with an overview of the components of the two

Photosynthesis in Action
https://doi.org/10.1016/B978-0-12-823781-6.00004-6

photosystems and then describe how their light-harvesting machinery adapts to their ever-changing light environments.

PSII and its light-harvesting complex proteins (LHCIIs) are made up of a large chlorophyll (Chl)–protein supercomplex that consists of more than 30 subunits. Light energy captured by LHCIIs is transferred to the central dimeric core complex where it is trapped and used to drive the electron flow from water to plastoquinone (PQ). In green plants, the light-harvesting antennae are formed by two layers, the major trimeric and the minor monomeric LHCII proteins (Fig. 1). In vascular plants, there are three major trimeric LHCII polypeptides, Lhcb1–3, and three minor monomeric LHCII polypeptides, CP29 (Lhcb4), CP26 (Lhcb5), and CP24 (Lhcb6). The Lhcb proteins have several isoforms, as, for example, in *Arabidopsis thaliana*, Lhcb1–6 have five, three, one, three, one, and one isoforms, respectively. Lhcb3 diversified in parallel with CP24 after the emergence of land plants, and both were found to have eventually disappeared from some clades (Kouřil et al., 2016), suggesting that they have structural interactions (Fig. 1). In the green alga *Chlamydomonas reinhardtii*, the major trimeric LHCIIs are differentiated into four types: type I (LhcbM3/4/6/8/9), II (LhcbM5), III (LhcbM2/7), and IV (LhcbM1), and there are two minor monomeric LHCII polypeptides: CP29 and CP26. All the major LHCII types in algae, except for type II, have been found to be associated with PSII; type II, on the other hand, has only been detected with PSI in state 2 conditions (Takahashi et al., 2006) (a process termed state transitions is a short-term adaptation mechanism that balance the power between the two photosystems and will be discussed in detail in Section 3).

KEY TERM 1. Plants and green plants

When asked to think of *plants*, many typically think of land plants, namely, embryophytes. However, the Plantae kingdom is comprised of all photosynthetic eukaryotes, meaning not only land plants but also all eukaryotic algae. Viridiplantae, literally means *green plants*, are defined as green photosynthetic eukaryotes; it is a clade formed by Streptophyta including land plants and Chlorophyta including green algae, both of which have Chl *a* and Chl *b*, and hence, present a green colour.

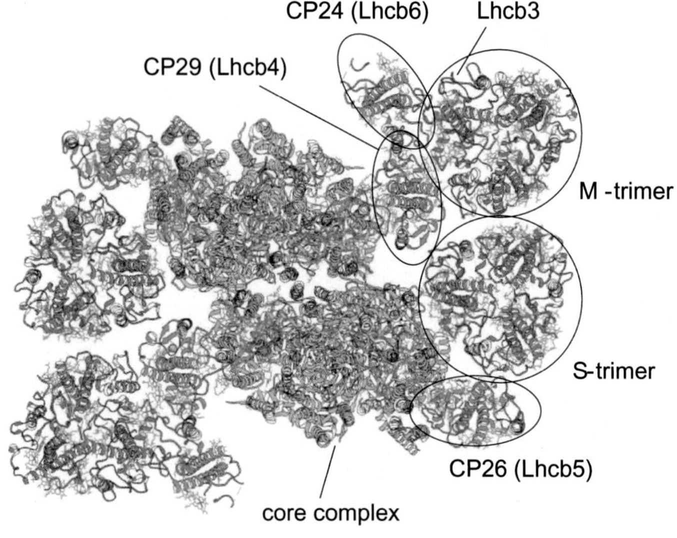

FIG. 1 **Pea PSII–LHCII supercomplex in the $C_2S_2M_2$ configuration.** The largest stable PSII–LHCII supercomplex reported in land plants. The central dimeric core complex *(brown)* is surrounded by two layers of LHCII proteins: an inner layer with monomer LHCIIs *(blue)* and an outer layer with trimer LHCIIs *(pink)*. The view is from the top stromal side. Chls are in *green*. PDB, 5XNL.

KEY TERM 2. Reaction centre, core complex, and supercomplex

In relation to the photosynthetic machinery, a minimum protein subcomplex that is required for charge separation is traditionally referred to as the *reaction centre*. It is now known to be composed of five subunits including D1 (PsbA), D2 (PsbD), Cyt b_{559} α (PsbE), Cyt b_{559} β (PsbF), and PsbI. For O_2 evolving reactions, many other small and large subunits are required in addition to the reaction centre. Such a holo-complex for O_2 evolution is referred to as the *core complex*. In thylakoid membranes, the core complex is surrounded by many light-harvesting antennae subunits. Their association is relatively weak, but researchers are working to develop methods for the biochemical isolation of the entirety of these complexes. The entire complex, consisting of the core complex and the light-harvesting antennae, is referred to as a *supercomplex*.

The LHCII proteins were found to be bound to both sides of the central dimeric core complex, where the reaction centre (RC) and several auxiliary subunits are located; the LHCII trimers composed of the major trimeric LHCII proteins were bordered by 2–3 minor LHCII monomers. In spinach, one LHCII trimer was found to be strongly bound to each side of the core (S-trimer), and this type of organisation is called C_2S_2 (Cao et al., 2020). In *A. thaliana*, the most observed organisation type was $C_2S_2M_2$ (Cao et al., 2020), where a moderately bound trimer (M-trimer) was also bound to each side (Fig. 1). Additionally, a loosely bound trimer (L-trimer) was found to be in contact with the core complex near CP26; however, in spinach, it was found to be between the two adjacent PSII and in *C. reinhardtii*, in the place of CP24, where this minor antenna is absent (Fig. 2). Cryo-electron microscopy (EM) has revealed several PSII–LHCII supercomplex structures at near atomic resolutions. These include the $C_2S_2M_2$ structure in pea, the $C_2S_2M_2$ structure in *Chaetoceros gracilis*, and the $C_2S_2M_2L_2$ structure in *C. reinhardtii* (Fig. 2; Cao et al., 2020). In cyanobacteria, the antenna organisation is unique, when compared with that in plants, as they harvest light via a large extra-membrane complex called a phycobilisome (PBS) (Fig. 2). The thylakoid membranes are packed with these large supercomplexes.

The supercomplex formed by PSI and its light-harvesting complex proteins (LHCI) is also a large Chl–protein complex that is made up of approximately 20 subunits. The PSI supercomplex collects light energy and drives the electron flow from the plastocyanin to ferredoxin. The crystal structure of the pea PSI–LHCI supercomplex shows the side of the PsaF subunit core that is occupied by the crescent-shaped 'LHCI belt' (Fig. 3; Suga and Shen, 2020). The other side of the core is unoccupied under normal conditions, exposing the PsaH/L subunits to membrane lipids. However, it can dock a 'mobile LHCII' in special circumstances, called 'State 2', as described in Section 3. In vascular plants, the LHCI belt is formed by the four LHCI proteins (Fig. 2). In other organisms, however, there are a variable number of LHCI proteins surrounding the core: 5 in red algae, 10 in green algae, 16 in diatoms, 12 in mosses, and 18 in dinoflagellates (Iwai et al., 2018; Kato et al., 2020; Suga and Shen, 2020). The composition of the LHCI proteins is also known to be variable in response to environmental changes, such as downregulation, degradation, or the processing of LHCI polypeptides (Moseley et al., 2002).

The core complex conducts charge separations (Fig. 4), and its components are almost completely conserved amongst oxygenic photosynthetic organisms, including cyanobacteria, eukaryotic algae, and land plants. This is because the photosynthetic process was sufficiently optimised when the cyanobacteria originally acquired the ability to oxidise water 2.7 billion years ago (Brocks et al., 1999; Summons et al., 1999). Furthermore, this core complex is

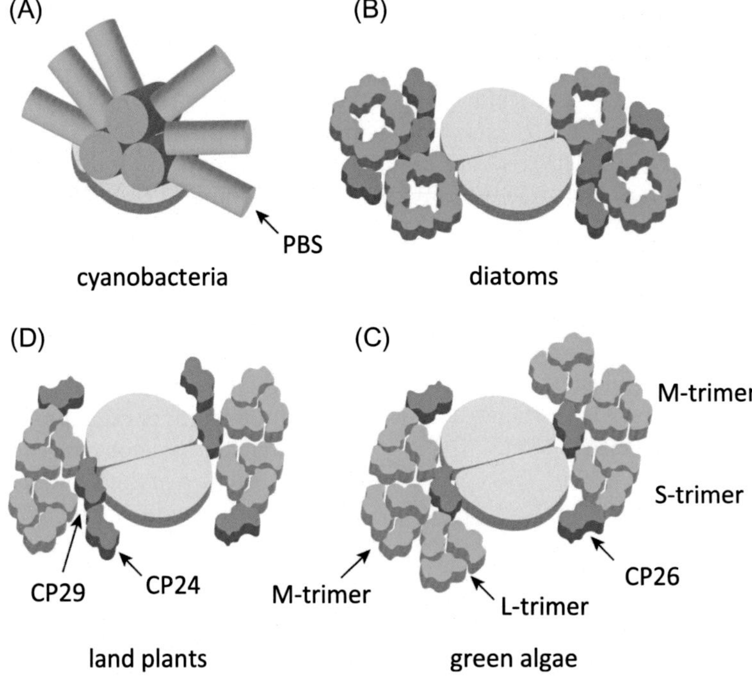

FIG. 2 **Diversity of the supramolecular organisation of the PSII supercomplex.** Photosynthetic organisms in different clades show different supramolecular organisation of their PSII supercomplexes. Here are examples of the cyanobacteria carrying extra-membrane PBS (A), a diatom with the $C_2S_2M_2$ organisation (B), green algae with the $C_2S_2M_2L_2$ organisation (C), and land plants with the $C_2S_2M_2$ organisation (D). The view is from the top stromal side. LHCII monomers, trimers, and tetramers are in *blue*, *green*, and *dark green*, respectively.

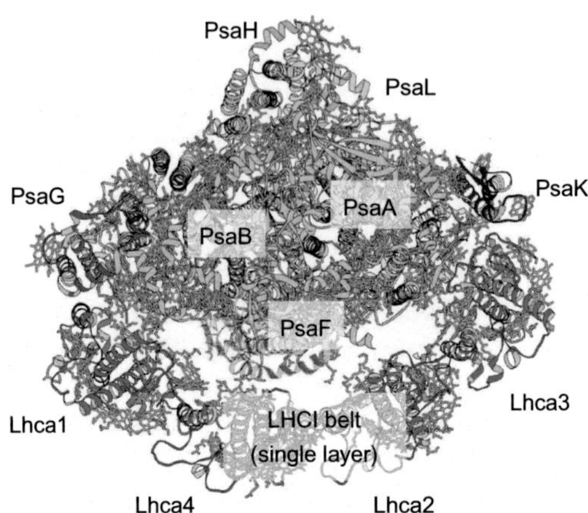

FIG. 3 **Pea PSI–LHCI supercomplex.** In the plant PSI–LHCI supercomplex, the 'LHCI belt' at the PsaF *(orange)* side of the core complex *(grey)* is composed of four Lhca proteins *(blue)*. The side between PsaG *(pink)* and PsaH *(sky blue)*/PsaL *(yellow)* is open in the plant supercomplex whilst it is occupied by two Lhca proteins in the red algal and the green algal supercomplexes. The side between PsaH/PsaL and PsaK *(red)* is to accept a 'mobile' LHCII when the state transition occurs. See also Fig. 15. The view is from the top stromal side. Chls are in *green*. PDB, 5L8R.

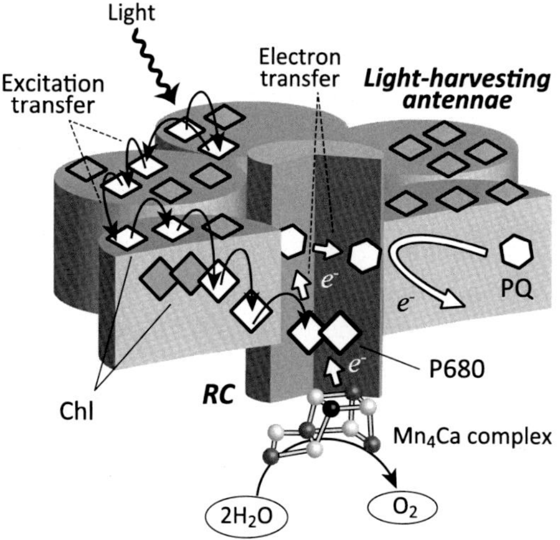

FIG. 4 **Photosynthetic supercomplex: reaction centre and light-harvesting antennae.** The reaction centre (RC) is the place where the photochemistry takes place. Its structure is conserved across species. Surrounding the RC are the light-harvesting antennae, where a number of specialised Chls and carotenoids for light-harvesting and energy dissipation are bound. Subunit organisation and pigmentation vary by species.

invariable in changing environments. The only exception is during photoinhibition, which is a multistep process of protein degradation upon exposure to high light (HL) illumination and the replacement of damaged polypeptides with newly synthesised polypeptides (Nixon et al., 2010). Whilst on one hand, photoinhibition is a deleterious event for PSII as it is demolished, on the other hand, it can also save the downstream components from more devastating events, such as the production of dangerous reactive oxygen species. Photoinhibition could therefore be considered a safety valve for photosynthetic electron transport.

In contrast to the core complex, the surrounding light-harvesting antennae are diverse and dynamic (Fig. 4). Different photosynthetic organisms have distinct antennae to take advantage of their specific niches, and are consequently, visually different. Moreover, the light-harvesting efficiency of the antennae in a single species can be dynamically adjusted to different light regimes. The capabilities of the light-harvesting antennae to adapt to the environment are especially important for photosynthetic organisms in the natural environment, where the quality and quantity of light fluctuates over time.

We now review the two representative short-term adaptation events of the light-harvesting antennae of the photosystems: nonphotochemical quenching (NPQ) and state transitions, in Sections 2 and 3, respectively.

2 Nonphotochemical quenching

2.1 Introduction: NPQ is a process to dump excess energy

In nature, photosynthetic organisms experience fluctuations in the intensity of incident light, with quantitative changes over several orders of magnitude. The absorption of light more than their photosynthetic capacity increases their potential to produce reactive oxygen species, which leads to pigment bleaching and death in extreme cases, a phenomenon experienced when indoor plants are moved outdoors and exposed to direct sunlight. Photosynthetic organisms employ mechanisms, such as NPQ, to cope with these abiotic stresses. The NPQ mechanism, for the rapid downregulation of light harvesting efficiency, involves the thermal de-excitation of the excited states of Chl in the antennae. When the antennae are in the NPQ state, a fraction of the absorbed energy is dissipated before arriving at the RC, to ease the overexcitation of the photosystems. NPQ is induced by the acidification of the thylakoid lumen, in which protons are injected from the photosynthetic electron transfer chain, so that it is only active when organisms are exposed to excess light. NPQ is therefore a natural feedback mechanism for the de-excitation of the overexcited light-harvesting antennae (Fig. 5). The extent and kinetics of the NPQ are regulated by several effectors including the xanthophyll cycle carotenoid, zeaxanthin (Zx), and photoprotective proteins such as PsbS. Knowledge of these effectors has considerably increased during the last few decades and will be discussed in the following sections.

KEY TERM 3. NPQ

Care should be taken when using the term *NPQ*, which literally means 'quenching in the absence of photochemistry'. In this book, we refer to the 'thermal quenching of the excited state of Chl without photochemistry' as NPQ. However, NPQ has also been widely used to indicate the phenomenon of the 'quenching of Chl fluorescence without photochemistry'. The 'latter NPQ' can be experimentally measured as fluorescence lowering. In fact, the 'latter NPQ' is sometimes almost exclusively composed of the 'former NPQ'. To distinguish the thermal quenching component of the fluorescence quenching, the term 'q_E quenching' can be used. It is also confusing that there is a fluorescence parameter called NPQ that can be obtained by a PAM-type fluorometer; it is a numerical value representing the overall level of the quenching of the Chl fluorescence without photochemistry, which is computed based on the formula, $(Fm - Fm')/Fm'$. If we assume that only q_E quenching is occurring in the system, the computed NPQ parameter should be proportional to the number of quenchers.

2.2 Xanthophylls

There are many xanthophylls in plants and algae, including lutein (Lut), violaxanthin (Vx), Zx, and fucoxanthin (Fx). They primarily act as light-harvesting pigments, but some are involved in NPQ: Zx in the case of green plants and diatoxanthin (Dtx) in the case of the secondary endosymbionts such as diatoms and dinoflagellates. Here, we review Zx as an example pigment involved in NPQ.

KEY TERM 4. Xanthophyll

Xanthophylls (*xantho*, 'yellow' + *phyllon*, 'leaf' from Greek) are yellow and polar carotenoids that are found widely in nature. They are collectively

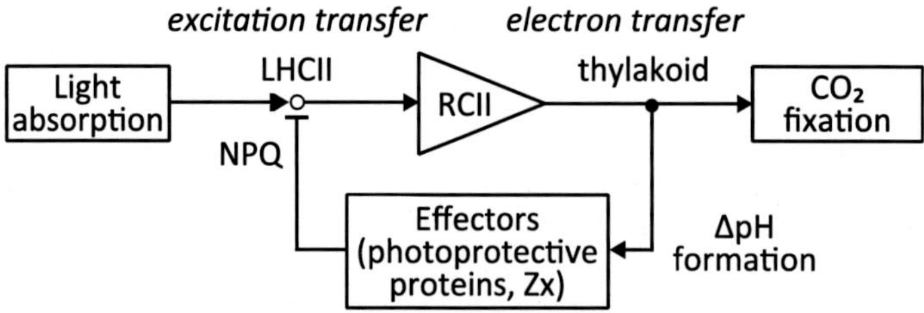

FIG. 5 **NPQ is a feedback mechanism for de-excitation of Chls.** When plants are under HL, the thylakoid lumen is acidified by protons that are released from the electron transfer chain. NPQ is activated under such conditions and de-excites Chls in the light-harvesting antennae to prevent photochemistry at the RC and alleviate the oxidative stress.

defined as hydrocarbons containing oxygen that are typically made of isoprene chains. In many cases, xanthophylls have their absorption maxima between 350 and 500 nm, due to their electronic transitions from a ground state (S_0) to an excited state (S_2). The longer the central double bonds, the longer the wavelength of the absorption maxima.

In green plants, the molecular role of Zx is clear, as its content is correlated with the extent of the NPQ (Niyogi, 1999). The majority of the Zxs are located in the thylakoid membranes and are bound to LHC proteins, which typically have three transmembrane helices and coordinate, 3–4 xanthophylls as well as 14 Chls at the most per monomer. There are four carotenoid binding sites on the LHC, including L1, L2, V1, and N1 (Fig. 6). L1 is always occupied by Lut. L2 is occupied by Lut in LHCII, CP26, LHCA1, and LHCA3, and by Vx in the others. N1 is occupied by neoxanthin in LHCII, CP29, and CP26 and by β-carotene in the others. Only LHCII and LHCA1 have an additional loose binding site, V1, at the periphery for binding Vx.

Upon exposure to HL, the Zx content is increased by de-epoxidation of the Vx, in a process called the xanthophyll cycle (Fig. 7; Yamamoto et al., 1962). The enzyme driving the xanthophyll cycle, Vx deepoxidase, is activated by the acidification of the thylakoid lumen; absence of this enzyme showed significant impairment of the NPQ in *npq1*, the mutant lacking this enzyme in *A. thaliana*. Zx has thus been associated with the formation of a quenching centre for NPQ. Energy dissipation has been reported to occur via the charge-transfer between a Chl dimer and Zx at L2 in CP29 (Ahn et al., 2008). Alternatively, Zx is proposed to promote LHCII aggregation (Goral et al., 2012), which could also induce thermally dissipation of the excitation energy, via excitation energy transfer from a nearby Chl cluster to the Lut at L1 (Ruban et al., 2007). The details of this mechanism will be discussed in Section 2.4.

In contrast to the *npq1* mutant in vascular plants, the same mutant, deficient in the

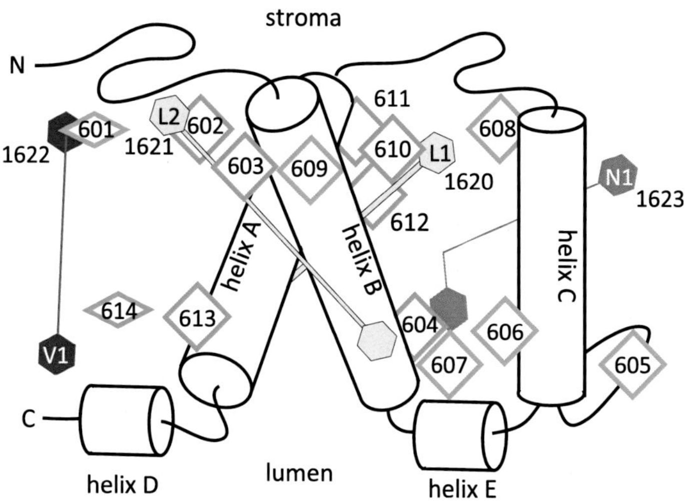

FIG. 6 **Pigment binding in LHCII.** Proteins in the LHC superfamily typically have three transmembrane helices (A–C) and 3–4 carotenoid binding sites, including L1, L2, V1, and N1. Lut at L1 in LHCII with a few Chls in the vicinity and Zx at L2 in CP29 with a few Chls in the vicinity are proposed as the quenching sites for NPQ. Chl *a* and Chl *b* are in *green* and *yellow green*, respectively.

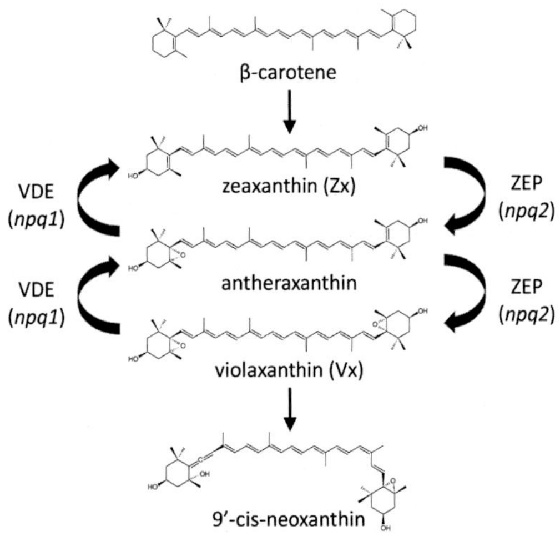

FIG. 7 The xanthophyll cycle.

conversion of Vx to Zx in *C. reinhardtii*, had little effect on the capacity of the NPQ, indicating that most photoprotection in green algae can be achieved without Zx (Niyogi, 1999). Although little photoinhibition was observed in the *npq1* mutant in *C. reinhardtii* under HL conditions, an additional mutation that inhibits Lut synthesis (*lor1*) caused severe photoinhibition, suggesting that Lut largely contributes to the NPQ in green algae (Niyogi, 1999).

Secondary symbiotic algae with red algal origins, such as diatoms, are able to generate a very high NPQ (Ruban et al., 2004), but they do not possess Vx. Instead, diatoms have several Fx and a few diadinoxanthin (Ddx). The latter is also subjected to a de-epoxidation cycle to form Dtx upon HL exposure (Olaizola et al., 1994). As the Dtx content is correlated with the level of NPQ, it has also been suggested that this xanthophyll is involved in the NPQ, like the Vx is in plants (Goss et al., 2006). However, the light-harvesting antennae for the diatoms PSII, Fx-Chl protein complex (FCP), binds seven Fx per monomer, which is greater than the typical two Luts found in LHCII, suggesting that Fx contributes to the NPQ (Wang et al., 2019).

2.3 Photoprotective proteins

Genetic screenings have identified various protein effectors that are required for NPQ. The *npq4* mutant was identified as it lacked the rapid induction of NPQ, despite the normal xanthophyll cycle and normal lumenal acidification, in both *A. thaliana* and *C. reinhardtii*, but the alleles were mapped to different genes: the four-helix LHC family protein PsbS in the former and the three-helix LHC family protein light-harvesting complex stress related protein 3 (LHCSR3) in the latter (Niyogi and Truong, 2013). The existence of the same phenotype from different genotypes suggests that there are fundamental differences in the NPQ strategies of terrestrial and aquatic photosynthesis.

Although one Chl molecule was identified in the PsbS crystal structure (Fan et al., 2015), none of the canonical pigment binding sites identified in the LHCII were conserved. This means that it is unlikely that PlabS forms a quenching site. In fact, PsbS does not change the fluorescence lifetime itself but changes the level and the initial induction rate for NPQ (Sylak-Glassman et al., 2014). It is thus more likely that it serves as a pH sensor. After sensing the low pH of the thylakoid lumen, the PsbS is not stably associated with the PSII–LHCII supercomplex (Caffarri et al., 2009), but shows stronger associations with the LHCII trimer upon its own monomerisation under HL conditions (Correa-Galvis et al., 2016). It controls the macro-organisation of the thylakoid membranes. Such membrane 'phase transitions' may facilitate the dissociation of LHCII proteins from PSII; moreover, the dissociated LHC proteins may aggregate themselves (Kiss et al., 2008), in turn inducing the conformational change(s) necessary to form energy-quenching centre(s) in major LHCII (Ruban et al., 2007) and/or minor LHCII (Ahn et al., 2008). Although both Zx and PsbS are thought to be crucial for NPQ in vascular plants, the green alga *C. reinhardtii*, which also harbours a homologue of the *PsbS* gene, only transiently expresses it, and a mutant deficient in *psbS* does not show a phenotype. Instead, the LHCSR proteins are essential photoprotective proteins in green algae.

The *C. reinhardtii npq4* mutant was mapped to *Lhcsr3.1* (Niyogi and Truong, 2013). LHCSR is an ancient member of the LHC superfamily, whose orthologues are found throughout eukaryotic algae, including diatoms (Niyogi and Truong, 2013). The genes for LHCSR3 (*Lhcsr3.1* and *Lhcsr3.2*) encode a 25–26 kDa integral membrane protein, having a low similarity to LHC. Although PsbS binds few pigments, a recombinant LHCSR3 polypeptide can bind Chls and xanthophylls and dissipate excitation energy at low pH, suggesting that this photoprotective protein could serve as a quenching site for itself (Niyogi and Truong, 2013). In *C. reinhardtii*, there are additional photoprotective proteins. LHCSR1, a paralogue of LHCSR3, induces excitation energy quenching in LHCII (Dinc et al., 2016) as well as its transfer to PSI upon lumenal acidification (Kosuge et al., 2018). One of the major trimeric LHCII proteins, LhcbM1, is also crucial for NPQ, as its mutant *npq5* loses the capacity for NPQ (Niyogi and Truong, 2013). The LHCII trimer containing LhcbM1 does have the capacity for NPQ when aggregated at a low pH (Kim et al., 2020).

2.4 Nonphotochemical quenching in terrestrial photosynthesis

NPQ in higher plants has been most intensely studied initially by Chl fluorescence. It was found that NPQ correlated with ΔpH and later proposed that it only acts as a trigger of the process. Indeed, the character of relationship between NPQ and the gradient was variable and kinetically ΔpH was faster to form and recover than NPQ (Wraight and Crofts, 1970). Horton and co-workers discovered that the shape of this relationship was xanthophyll cycle- and PsbS-dependent (for review, see Ruban, 2016). Fig. 8 (top) shows strong alterations in NPQ vs ΔpH (lumen pH) in plant chloroplasts with and without PsbS protein and with Vx or Zx. It appears that the protein and the xanthophyll cycle pigments exert regulatory role upon NPQ, making NPQ most responsive to lumen pH changes in the presence of Zx. The shape of this relationship also changes from strongly sigmoidal (Zx or Vx only) to almost hyperbolic (Zx + PsbS). This bares an allosteric character the current model of NPQ is based upon (see the following paragraphs) (Ruban et al., 2012). Interestingly, NPQ can be attained even without Zx and PsbS protein, provided lumen pH is low enough (Fig. 8). However, since in nature maximum attained lumen pH was proposed to be not much lower than 5.5 (Kramer et al., 1999), a certain amount of NPQ (but not qE) is always present even in the absence of Vx but not PsbS protein (Fig. 8). Hence, the data for Vx and Zx on the figure have only been obtained on isolated chloroplasts in the presence of artificial ΔpH enhancer, diaminodurene, in order to show that neither PsbS nor Zx are actually directly responsible to carry or be NPQ excess energy quenchers (Ruban et al., 2012). Xanthophyll cycle plays a role of modulator of not only NPQ amplitude but kinetics too. Fig. 8 (bottom) shows that whilst Zx accelerates the quenching formation rate slows down its rate of dark recovery, working as illumination dosage memory—a remarkable physiological function. Indeed, the longer and stronger light exposure, the more Zx will be formed as a result of violaxanthin de-epoxidase (VDE) activation and NPQ will become bigger, last longer, and be quick to rise after another onset of HL exposure. In fact, NPQ will be 'sensitised' by light via the xanthophyll cycle.

The component(s) of PSII that actually host the excited state/fluorescence quencher were studied first from various biochemistry and spectroscopy methods. The mutagenesis proved to be the ultimate way of addressing the NPQ site question directly and in situ. Whilst removal of the major trimeric LHCII polypeptides led to

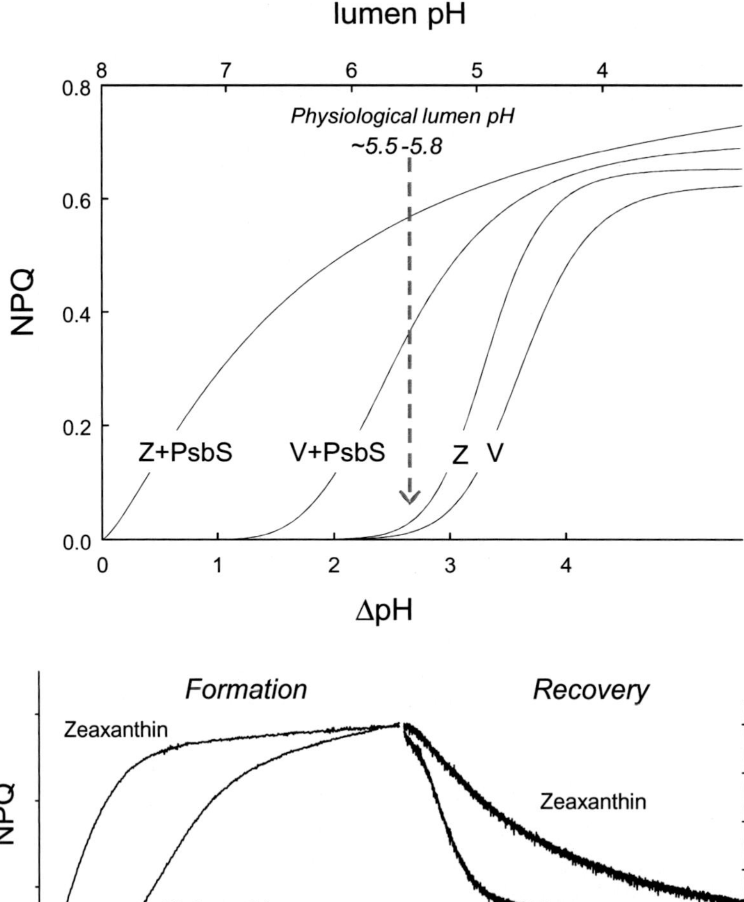

FIG. 8 **Allosteric control of NPQ.** *Top*, kinetics of NPQ formation and dark recovery in intact spinach chloroplasts containing either Vx or Zx. *Bottom*, △pH titrations of qE in intact *Arabidopsis* chloroplasts containing Zx and PsbS (Z+Psbs), Vx and PsbS (V+PsbS), lacking PsbS but containing Zx (Z) and lacking PsbS and containing Vx (V). The estimated average lumen pH in vivo is marked by a *dashed arrow*.

strong reduction in NPQ, neither depletion from the reaction centre of photosystem II (RCII) core antenna nor removal of all minor LHCII complexes caused any significant decrease in NPQ (Townsend et al., 2018). Indeed, Fig. 9 displays the schematic composition of the dimeric PSII in the *Arabidopsis* WT (top) and mutant lacking all three minor antenna complexes (CP24, 26, and 29) and possessing very much reduced amount of core antenna and RCII proteins (bottom). Not

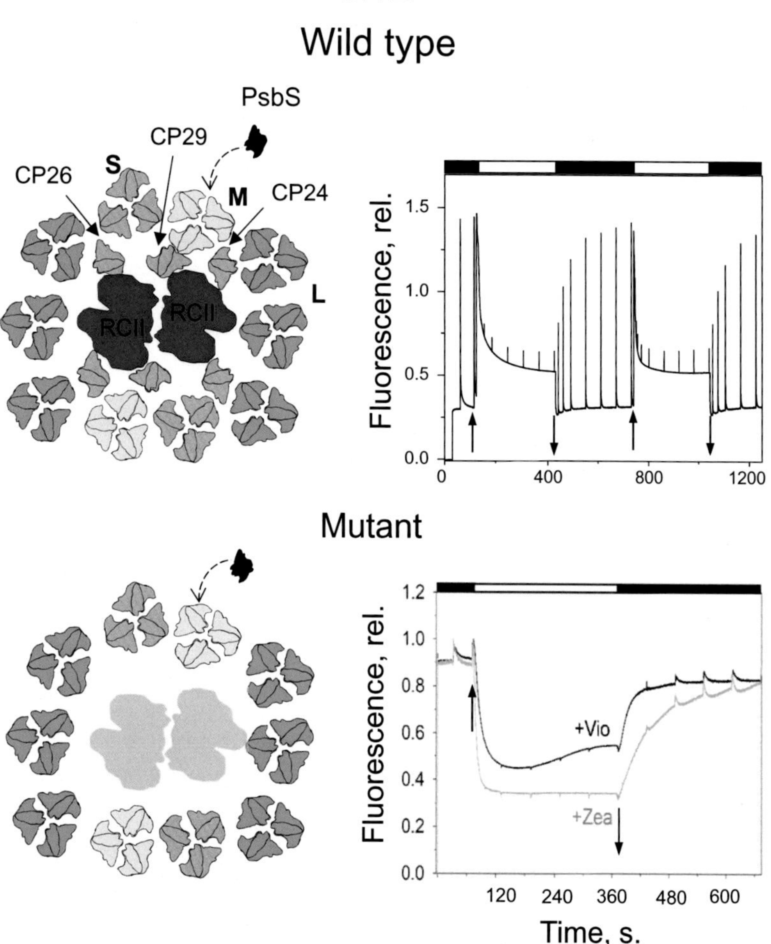

FIG. 9 **The localisation of NPQ in PSII.** PSII unit structure *(left)* and PAM fluorescence quenching traces *(right)* of the WT *(top)* and mutant lacking all the minor antenna complexes with depleted RC and core antenna proteins *(bottom)*.

only NPQ and qE were of a similar amplitude in the mutant but also, most importantly, the kinetic character behaviour of the quenching related to the xanthophyll cycle activity was the same as in the WT plants (see the previous paragraph). This remarkable feature of allosteric regulation of NPQ is indeed present in the major LHCII complex that actually carries almost all Vx available for conversion into Zx (see for review Ruban, 2016). In addition, PsbS protein was shown to possess higher affinity of binding to LHCII in NPQ state, meaning that upon protonation it sends a signal to LHCII that initiates NPQ in it under somewhat moderate level of lumen pH (see Fig. 9, top; Sacharz et al., 2017). However, as was discussed above, PsbS was shown not to be a direct carrier of the NPQ quencher, hence cannot be part of the quenching site.

COLUMN 1. Pulsed-amplitude modulation (PAM) fluorometer

The PAM fluorometer is a widely used device for evaluating photosynthetic performance, both in the laboratory and field. It uses a weak measuring light, which can be rapidly switched on and off. Since its amplitude is modulated in a pulsed manner, the fluorescence signals can be isolated by a detector with a lock-in amplifier [hence named *pulse amplitude modulation* (Schreiber et al., 1986)], allowing for an accurate separation of the signals in the background of strong continuous light, and most importantly in sunlight. Today's commercial PAM fluorometers employ the so-called 'saturation pulse'. This technique is used to close all RCIIs by applying a high-intensity, short-duration flash of light to transiently eliminate the contributions of the photochemical quenching. Of particular importance is the fact that Fm′ (quenched maximal yield of fluorescence) and Fo′ (quenched minimal yield of fluorescence) can be measured in addition to Fm (maximal yield of fluorescence after dark adaptation) and Fo (minimal yield of fluorescence after dark adaptation), which are required to estimate the two widely used parameters: the effective quantum yield (real time quantum yield), $\varphi_{II} = (Fm - Ft)/Fm'$, and an NPQ parameter, $NPQ = (Fm - Fm')/Fm'$, where Ft represents fluorescence yield in actinic light. Typical fluorescence transients measured by PAM fluorometer are shown in Fig. 7 (right).

Thus, what happens to the trimeric LHCII in the NPQ state? Horton and co-workers proposed that ΔpH triggers lateral aggregation of this protein (Horton et al., 1991). Indeed, they observed that isolated aggregates of LHCII are highly quenched (up to 10 times compared to unaggregated complex). Further, work on in situ using freeze-fracture electron microscopy (FFEM) enabled to find a proof of such aggregation (Johnson et al., 2011). Fig. 10 shows the fragment of FFEM view of the fractured PSII membrane showing fracture surfaces (EFs) carrying particles of RCII and core complexes as well as surfaces (PFs) that carry LHCII. The latter displayed clustering patterns that was enhanced in the presence of Zx, and ΔpH and this process was driven by PsbS (for review, see Ruban, 2016). An independent approach that used fluorescence recovery after photo-bleaching method discovered that in the presence of PsbS the population of LHCII available for the clustering was significantly higher than in the absence (Johnson et al., 2011), suggesting involvement of the protein as a membrane dynamics factor, a qE catalyst, prompting rapid NPQ recovery as well as formation. The molecular details of the action of PsbS are not known; however, recent theoretical research indicates its effect on LHCII structure and involvement in interactions with lipids that surround LHCII (Daskalakis et al., 2019).

The NPQ quencher and its mechanism of action in LHCII has been studied only in vitro and in silico. It was shown that the aggregation per se is not obligatory required for the quenching to occur (Ilioaia et al., 2008). Hence, the conformational change within LHCII unit—monomer—has become a clear focus of the further search for the event that generates the NPQ quencher. First, spectroscopic and thermodynamic studies discovered subtle specific changes in some LHCII Chls and xanthophylls (for review, see Ruban, 2012). The ultrafast transient absorption spectroscopy revealed the involvement of Lx (L1) in position close to the emitting Chl *a* (Chl610 and Chl612) (Ruban et al., 2007). L1 via S1 excited state was proposed to accept energy from these Chls and quickly (within 10ps) dissipate it into heat. Recent molecular dynamics and infrared spectroscopy study proposed the model of transition of the LHCII protein into the quenching state (Li et al., 2020). Fig. 11 displays that protonation of the lumen exposed glutamate (D94) triggers the chain of events on the LHCII luminal site that involve alteration in the hydrogen bonding, formation of the helix and movement of amphipathic helixes D and E towards the core of the

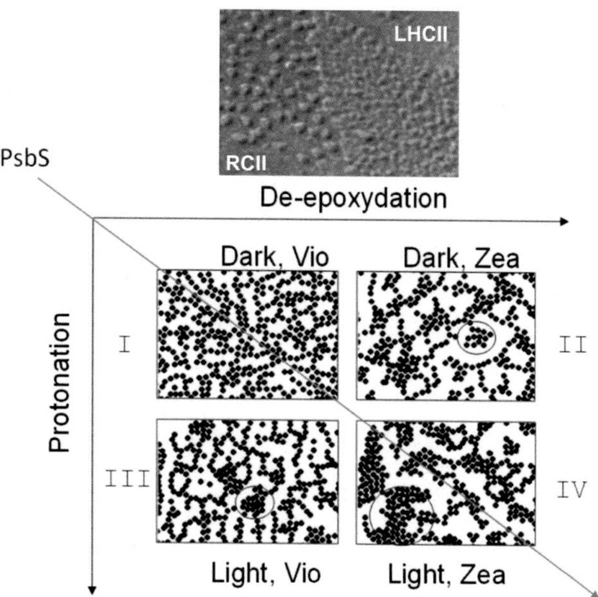

FIG. 10 **Lateral reorganisation of LHCII trimers in the photosynthetic membranes of chloroplasts.** 2D co-ordinates of LHCII trimers in the photosynthetic membrane determined by FFEM (each particle was fitted with a 50 nm^2 circle using image recognition software, and their positions are presented in the figure). The four states of organisation were observed in dark-adapted chloroplasts [Dark Vio, analogous to state I in the LHCII aggregation model by Horton et al. (1991)], chloroplasts frozen immediately after 5 min illumination at 350 μmol photons m^{-2} s^{-1} in the absence of Zx (Light Vio, analogous to state III), chloroplasts frozen immediately after 5 min illumination at 350 μmol photons m^{-2} s^{-1} in the presence of Zx (Light Zea, analogous to state IV) and chloroplasts frozen following 5 min illumination at 350 μmol photons m^{-2} s^{-1} in the presence of Zx and a further 5 min dark adaptation (Dark Zea, analogous to state II). *Blue circles* indicate the sizes of clustered particles. *From Ruban, A.V., Johnson, M.P., Duffy, C.D., 2012. The photoprotective molecular switch in the photosystem II antenna. Biochim. Biophys. Acta 1817, 167–181, © 2012, Elsevier.*

complex. These changes cause tilting of helixes A and B that leads to the overall flattening of the LHCII structure. Although this change is not big it is sufficient to bring Chl610 and Chl612 close enough to L1 to increase excitonic coupling and enable energy quenching of Chls by this xanthophyll.

COLUMN 2. Towards improving crop yield

As NPQ is a mechanism that thermally dissipates excitation energy and decreases the number of photons arriving at the RCs, it depresses the overall quantum yield of photosynthesis. This could be problematic when the incident light is lower (under the weak light conditions) than that required for photosynthetic capacity. Ideally, NPQ should only be activated when the thylakoid lumen is acidified, but in reality, the rate of NPQ activation/relaxation is not rapid enough to catch up with the fluctuating light; hence, a transient depression of photosynthetic quantum yield tends to occur. This point, in principle, became a target of bioengineering. Researchers have succeeded in providing a proof of concept that plant productivity could be improved by accelerating the response of the NPQ switch (Kromdijk et al., 2016).

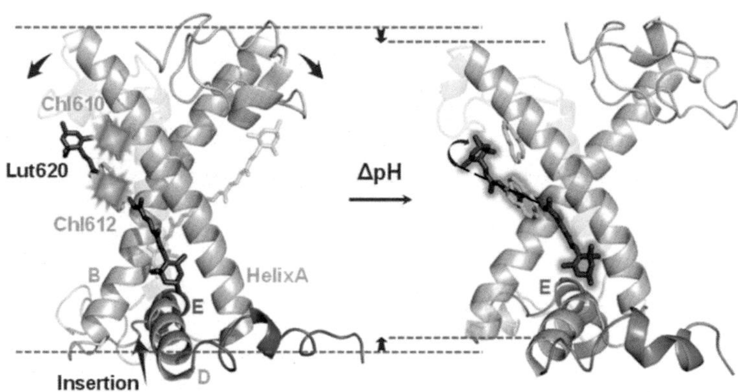

FIG. 11 **Formation of the NPQ quencher in the major LHCII complex.** The hydrophobic contacts connecting the helix D and E are the lever. The trigger, △pH, induces a local structural transition to convert the 310-helix E and C-terminal loop into two α-helices, which prompts a switch of the hydrogen bond of Glu94 from Lys99 in the light-harvesting state *(left)* to Gln103 in the photoprotection state *(right)*. In turn, α-helix D and E are pulled closer against transmembrane helices A and B, shifting its conformational equilibrium to a wider intercrossing angle. The net effect of the overall conformational change of transmembrane helices is to reduce the contact distance between Lut at L1 and Chl612, enhancing their electronic coupling and energy transfer from excited Chls to the S1 dark state of Lut at L1. *Courtesy of Dr. Yuxiang Weng, Laboratory of Soft Matter and Biophysics, Institute of Physics, Chinese Academy of Sciences, Beijing 100190, China.*

2.5 Nonphotochemical quenching in aquatic photosynthesis

The study of NPQ at the molecular level has occurred in aquatic phototrophs, namely, algae. *Chlamydomonas reinhardtii*, the model eukaryotic algae, has been used for forward and reverse genetic analyses. In *C. reinhardtii*, NPQ is induced after a prolonged exposure to HL (Niyogi, 1999). This is because the green algae express the LHCSRs first and then activate them by acidifying the lumen of the thylakoid membranes. This is in contrast with the observations made in land plants, where PsbS is constitutively expressed, so that the lumenal acidification can instantly activate NPQ (Demmig-Adams et al., 2006). These differences may be because land plants are sessile and constantly exposed to HL, whilst algae are mostly in LL environments and are only sporadically exposed to HL. Many algae, including *C. reinhardtii*, are also motile and only need to express photoprotective proteins when they are required. Stress-induced regulations of the photoprotective proteins need to be supported by an intimate signal transduction network. The expression of the *LHCSR* genes is regulated at the level of transcription by the CONSTANS complex, whose transcription is strictly controlled by the circadian clock (Tokutsu et al., 2019). The CONSTANS complex is in turn negatively regulated by COP1/SPA1-type ubiquitin E3 ligase, together with DET1-type ubiquitin E3 ligase (Aihara et al., 2019). These ubiquitin ligases are further regulated by photoreceptors, like UVR8 for LHCSR1 (Allorent et al., 2016), or phototropin for LHCSR3 (Petroutsos et al., 2016). Intriguingly, these regulatory components became useless for regulating photoprotection in land plants because their photoprotective protein, PsbS, is constitutively expressed. The kernel of the regulatory system seems to have been repurposed for flowering regulation in flowering plants (Tokutsu et al., 2019; Fig. 12).

When the PSII-LHCII supercomplex was purified from HL-grown cells, a fraction of the LHCSR3 was associated with the supercomplex.

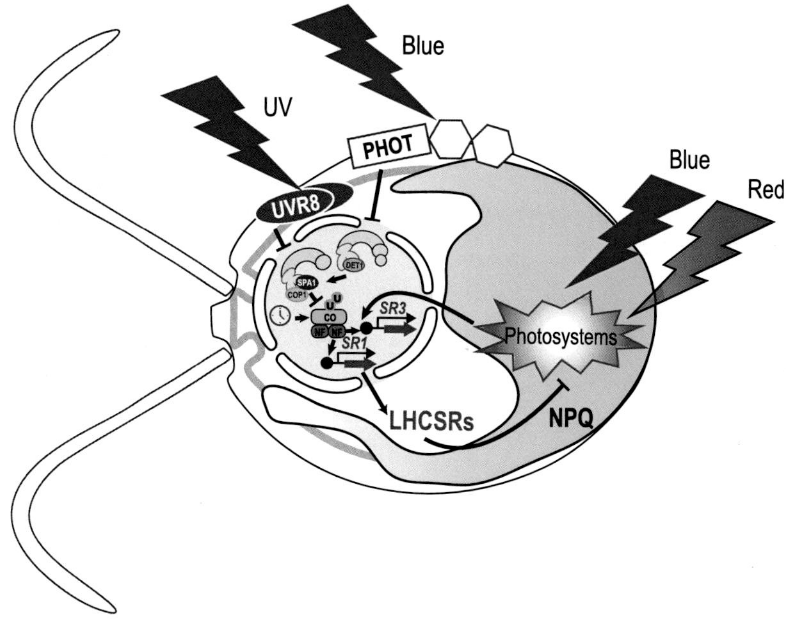

FIG. 12 **CONSTANS-dependent control of the photoprotective genes in *C. reinhardtii*.** Blue and UV lights perceived by the photoreceptors PHOT and UVR8, respectively, deactivate the E3 ubiquitin ligases. The circadian clock-regulated CONSTANS transcriptional module accumulates when its ubiquitin-dependent degradation is inactivated. The CONSTANS module then transcribes the photoprotective genes to promote photoprotection.

This PSII–LHCII–LHCSR3 supercomplex was in a light-harvesting state at a neutral pH but in an energy-dissipative state at an acidic pH (Tokutsu and Minagawa, 2013). As NPQ switching is sensitive to dicyclohexylcarbodiimide (DCCD), a protein-modifying agent specific to protonatable amino acid residues, LHCSR3 is likely to act as a sensor for low pH; the amino acid residues for its pH sensing ability are localised to its C-terminus (Liguori et al., 2013). LHCSR3 is proposed to be a low pH sensor, as well as a quenching site, which is in contrast with PsbS in land plants, which serves as a low pH sensor and energy-dissipation-inducer. Whether LHCSR3 also has this energy-dissipation-inducing function is currently unknown.

Diatoms are secondary endosymbionts that originated from red algae and are one of the most important phytoplankton groups as a primary producer. Their abundance is based on their capability to thrive in turbulent waters, where they can exploit the high levels of available nutrients. This is because they have variable NPQ capacities, depending on the light characteristics of their respective habitats. Similar to green plants, diatoms possess LHC-type antennae FCP for both PSI and PSII, which contain almost exclusively Fx (Cao et al., 2020; Suga and Shen, 2020). Whilst *C. reinhardtii* contains only two LHCSR proteins, many diatoms possess several LHCSR homologues (Lhcx), e.g., the model diatom *Phaeodactylum tricornutum* possesses 4 Lhcx

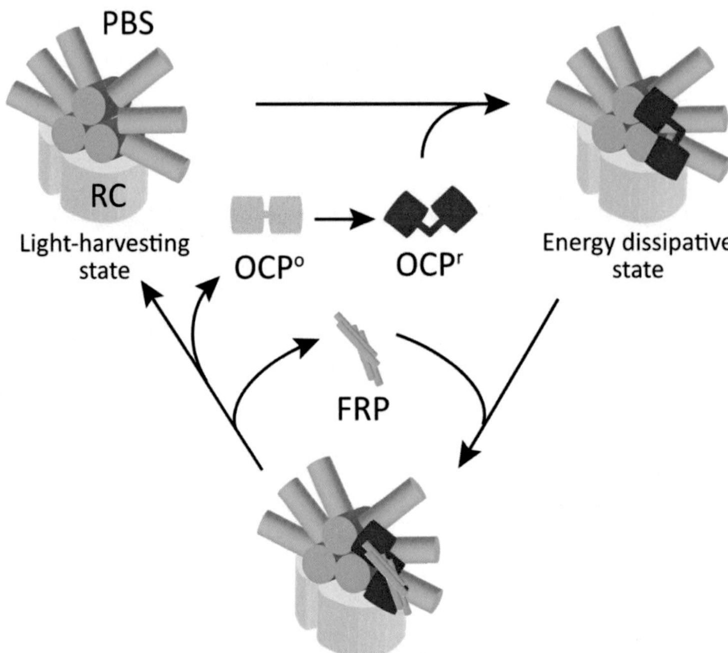

FIG. 13 **Schematic of the OCP-induced energy-dissipation mechanisms in cyanobacteria.** Under LL conditions, the absorbed energy by PBS is delivered to the RC. HL induces conformational changes in the OCP, converting its orange inactive form (OCP^o) to the red active form (OCP^r). Only OCP^r interacts with PBS or FRP. OCP^r interacts with PBS and induces the energy transfer to itself, where it is quenched. OCP is both the inducer of quenching and the quencher itself. FRP, under light and dark conditions, interacts with OCP^r bound to PBS, to induce its conversion back to OCP^o and the detachment from PBS.

proteins. The NPQ capacity in diatoms correlates with the expression of different Lhcx proteins, indicating the involvement of various Lhcx proteins in triggering NPQ under different environmental conditions, such as LL (Lhcx1), HL (Lhcx2, Lhcx3), blue light (Lhcx1–3), iron limitation (Lhcx2), and nitrogen starvation (Lhcx3, Lhcx4) (Taddei et al., 2016).

Photoprotection in cyanobacteria is unique when compared to that in plants, as they use PBS to harvest light (Fig. 2) and a specific photoprotective protein called the orange carotenoid protein (OCP). A representative photoprotective mechanism in cyanobacteria is the thermal dissipation induced by OCP and the decoupling of the PBS itself. The thermal dissipation is triggered by blue–green light, which is required to induce the expression of OCP (Fig. 13). Strong blue–green light causes conformational changes in the OCP, leading to the formation of its 'red active form', OCP^r. The OCP^r interacts with the PBS core to induce thermal dissipation, which decreases the energy arriving at the RCs. To recover the light-harvesting state of the PBS, a second protein, called the fluorescence recovery protein (FRP), is required, which plays a key role in undocking the OCP^r from the PBS and converting it back to its 'orange inactive form', OCP^o (Boulay et al., 2010). Although this mechanism is widespread in PBS-containing cyanobacteria, around 20% of the cyanobacteria species, including *Synechococcus elongatus* and *Thermosynechococcus elongatus*, lack it (Kirilovsky and Kerfeld, 2012).

3 State transitions

3.1 Introduction: State transitions balance the power between the two photosystems

Each of the two photosystems in the thylakoid membranes, PSI and PSII, has a distinct pigment system with unique absorption characteristics. PSII and its antennae have more Chl *b* and PSI and its antennae have more Chl *a*. Consequently, there tends to be an imbalance in the energy distribution between the two photosystems in natural environments. For instance, blue light preferentially excites PSII and far-red light preferentially excites PSI. As the two photosystems are functionally connected in series, oxygenic photosynthetic organisms must constantly balance their excitation levels to ensure the optimal efficiency of their electron flow. State transitions are induced by such situations to balance the light-harvesting capacities of the two photosystems. State 1 occurs when PSI is preferentially excited and the light-harvesting capacities of PSII and PSI are increased and decreased, respectively. State 2 occurs when PSII is preferentially excited and the light-harvesting capacities of PSII and PSI are decreased and increased, respectively (Minagawa, 2011). The actual balancing can be seen in the low temperature fluorescence spectra (Fig. 14). The core concept of state transitions was documented in 1969 as an instant redistribution of light-harvesting systems between the two photosystems (Bonaventura and Myers, 1969; Murata, 1969) and the molecular mechanisms that regulate the redistribution of the antennae have been intensively studied and debated for decades. Many of the recent findings regarding the molecular components and structural basis for state transitions have been advanced through molecular genetics and biochemical studies in two model organisms: a green alga *C. reinhardtii* and a higher plant *A. thaliana*. In particular, many findings were reported for *C. reinhardtii*. This is partly because up to 35%–38% of the LHCIIs are mobile during the state transitions in this green algae, whilst only 20%–25% are mobile in higher plants (Nawrocki et al., 2016). State transitions also occur in cyanobacteria, at a variance with the eukaryotic photosynthetic organisms that move LHC proteins; the physical movement of the PBS or PS I monomers leads to the redistribution of the energy absorbed by the PBS between the PS II and PS I.

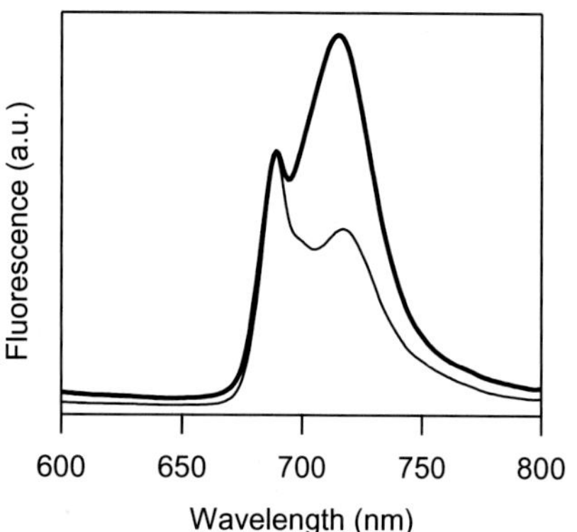

FIG. 14 **Changes in the fluorescence spectra landscape upon state transitions.** Relative fluorescence emission spectra of the thylakoid membranes at liquid N_2 temperatures are shown. The *thin and bold lines* indicate membranes in state 1 and state 2, respectively. The increased yield of the PSI fluorescence in the far-red region (~714 nm) as compared with the PSII fluorescence in the red region (~688 nm) indicates that state 2 increases the proportion of absorbed energy transferred to PSI.

3.2 Redox control

LHCII phosphorylation was first observed in illuminated pea chloroplasts supplied with [^{32}P] orthophosphate; the Thr residues in these proteins were reversibly phosphorylated upon illumination (Bennett, 1977). This light-dependent phosphorylation was dependent upon electron flow rather than a process mediated by direct activation by light. LHCII phosphorylation resulted in decreased yields for the PSII fluorescence, indicating that LHCII phosphorylation increased the proportion of absorbed energy transferred to PSI. In *A. thaliana*, amongst the three LHCII types, only Lhcb1 and Lhcb2 can be phosphorylated in the N-terminal region, and only the phosphorylation of Lhcb2 was found to be central in state transitions (Pietrzykowska et al., 2014; Longoni et al., 2015). This was confirmed by the Cryo-EM structure of the maize PSI–LHCI–LHCII supercomplex under state 2 conditions, where a trimer of phosphor-Lhcb2/(Lhcb1)$_2$ was associated with the PSI core (Pan et al., 2018; Fig. 15).

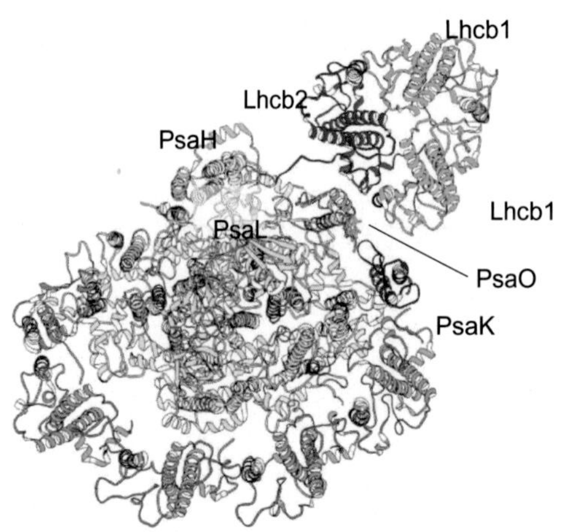

FIG. 15 **Plant PSI–LHCI supercomplex under state 2 conditions.** LHCII trimer-Lhcb1×2 *(yellow green)*/Lhcb2 *(magenta)* is associated with the PsaO *(orange)* on the edge between PsaH *(light blue)*/PsaL *(yellow)* and PsaK *(red)*. The phosphorylated Thr2 in the N-terminal region of Lhcb2 has several interactions between the amino acid residues in PsaH and PsaL. View from the stromal side.

Once the activity of the LHCII kinase was identified, biochemical attempts were made to isolate specific LHCII kinases, but these have, to date, been unsuccessful. Using the characteristic fluorescence quenching, due to a state 1-to-2 transition, several state transition mutants were isolated. One such mutant, *stt7*, was found to be deficient in a Ser/Thr kinase in the chloroplast thylakoid membranes (Rochaix, 2007). Subsequently, a mutant in *A. thaliana* lacking an orthologue of Stt7, STN7, showed an impairment in state transitions (Rochaix, 2007). The Stt7/STN7 kinase contains a single membrane-spanning domain and is localised in the thylakoid membranes. Immunoprecipitation assays indicated that Stt7 could interact with the LHCII, cytochrome b_6f complex (Cyt b_6f), and PSI subunits.

The reversible phosphorylation of LHCII observed during state transitions implies that protein phosphatases also participated in this process. The LHCII phosphatase activity was shown to depend on the presence of divalent cations, and it was not inhibited by either microcystin or okadaic acid, suggesting the involvement of a PP2C-type phosphatase in the dephosphorylation of phosphor-LHCII. TAP38/PPH1 is required for dephosphorylation of LHCII but not for that of the PSII core proteins (Pribil et al., 2010; Shapiguzov et al., 2010). This phosphatase, which belongs to a family of monomeric PP2C type phosphatases, is mainly associated with the stromal lamellae of the thylakoid membranes. Loss of the TAP38/PPH1 phosphatase leads to an increase in the antenna size of the PSI and impairs the state 2-to-1 transition, whilst overexpression of the gene decreases LHCII phosphorylation and inhibits a state 1-to-2 transition (Pribil et al., 2010; Shapiguzov et al., 2010).

In vascular plants, the protein kinase STN7 has a paralogue called STN8, which is involved in the phosphorylation of numerous thylakoid proteins, including the core subunits of PSII

(Rochaix, 2007). The protein phosphatase, PHOTOSYSTEM II CORE PHOSPHATASE (PBCP), is required for the dephosphorylation of proteins phosphorylated by STN8 (Samol et al., 2012; Puthiyaveetil et al., 2014). Whilst there is some overlap in their substrate specificities, the antagonistic pairs of kinases and phosphatases appear to have distinct roles. STN7 and PPH1/TAP38 are mainly involved in state transitions, whilst STN8 and PBCP influence the architecture of the thylakoid membranes and the repair cycle of the photoinhibited PSII. In *C. reinhardtii*, however, homologues of PPH1/TAP38 and PBCP were shown to play a role in the regulation of state transitions, with partially redundant functions, namely, a double mutation, [*pph1*, *pbcp*], to inhibit all the dephosphorylation activities, was necessary to lock it in state 2 (Cariti et al., 2020).

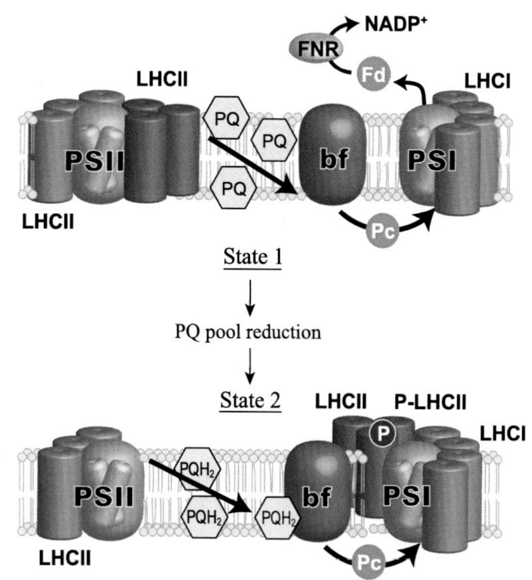

FIG. 16 **Schematic representation of state transitions: a transition from state 1 to state 2.** *Upper*: When PSI is preferentially excited, the stroma of the chloroplast and the PQ pool are oxidised. Under these conditions, LHCIIs are bound to PSII (State 1). *Lower*: When the PQ pool is reduced, mobile LHCIIs migrate from PSII to PSI to establish state 2. *bf*, Cyt b_6f; *Fd*, ferredoxin; *Pc*, plastocyanin.

3.3 LHCII phosphorylation/dephosphorylation

Imbalances in the excitations of the two photosystems are sensed by the redox state of the intersystem electron transfer carrier in the thylakoid membranes, which gives rise to LHCII phosphorylation (Fig. 16). Reduced ferredoxin, dithionite, and duroquinol, which are known to reduce PQ in isolated thylakoids, can activate LHCII phosphorylation in the dark, whereas DCMU can inhibit light dependent LHCII phosphorylation, indicating that LHCIIs are phosphorylated when the PQ pool is reduced. Further analysis revealed that it is not the reduced PQ molecule per se but the binding of PQH_2 to Cyt b_6f complex that is critical for the activation of the kinase. *C. reinhardtii* mutants lacking Cyt b_6f complex were unable to phosphorylate the LHCII, but the other phosphoproteins in the thylakoids were unaffected (Wollman and Lemaire, 1988). An inhibitor of the stromal side of Cyt b_6f complex (Qi-site), however, did not affect the activation of the LHCII phosphorylation by PQH_2. Furthermore, a lumenal side (Qo-site) inhibitor of the Cyt b_6f complex inhibited LHCII phosphorylation, suggesting the binding of the PQH_2 to the Qo-site of Cyt b_6f activates the LHCII kinase(s) (Fig. 16). Whilst the LHCII kinase(s) is activated by the reduced PQ pool, it is deactivated by the reduced thioredoxin pool that is downstream of the PSI in the stroma of the chloroplasts.

3.4 Molecular remodelling of photosystems

The PSII supercomplex is subjected to large-scale reorganisation during the transition from state 1 to state 2, and a large number of the LHCIIs forming the peripheral antenna for PSII are undocked upon their phosphorylation. Vascular plants without PsaH, PsaL, and PsaO had an impairment in their state transitions. As these small PSI subunits were located on the opposite side of the LHCI belt (Suga and Shen, 2020;

Fig. 3), they were hypothesised to constitute a binding site for the mobile LHCII(s). This hypothesis was proven using the Cryo-EM structure of a maize PSI-LHCI-LHCII supercomplex, where a phosphorylated LHCII trimer formed close contacts with PsaL, PsaH, and PsaO (Pan et al., 2018). The Cryo-EM structure also showed that PsaO mediates excitation energy transfer from the mobile LHCII to the PSI core through two Chls bound in-between PsaO and LHCII (Fig. 15). A similar structure of the PSI-LHCI-LHCII supercomplex was reported in a green alga *C. reinhardtii*, where two LHCII trimers were present (Pan et al., 2021). One LHCII trimer (LHCII-1) was located at a position similar to the LHCII trimer in the maize PSI–LHCI–LHCII supercomplex, but shows a rotational shift (Fig. 17). The other LHCII trimer (LHCII-2) was specific to the algal PSI–LHCI–LHCII supercomplex, and bridges the gap between the LHCII-1 trimer and Lhca2. LHCII-1 is composed of LhcbM1 (LHCII type IV), LhcbM2/7 (LHCII type III), and LhcbM3/4 (LHCII type I), whereas LHCII-2 is composed of LhcbM5 (LHCII type II), LhcbM2/7 (LHCII type III), and LhcbM3/4 (LHCII type I) (Fig. 17). In the algal PSI–LHCI–LHCII structure, two LhcbM proteins face the PSI core, where LhcbM1 in LHCII-1 and LhcbM5 in LHCII-2 have a phosphorylated threonine at their N-terminal tails, suggesting that they may serve as grips for the association with the PSI core, like the phosphorylated threonine in Lhcb2 in higher plants (Pan et al., 2018). Furthermore, biochemical and functional studies on mutant strains lacking either LhcbM1 or LhcbM5 indicate that only LhcbM5 is indispensable in the PSI–LHCI–LHCII supercomplex formation (Pan et al., 2021).

COLUMN 3. Are they all shuttled?

Whilst it is known that a fraction of LHCIIs migrate between the two photosystems and their identities and binding sites on PSI are becoming clear, studies employing in vivo measurements

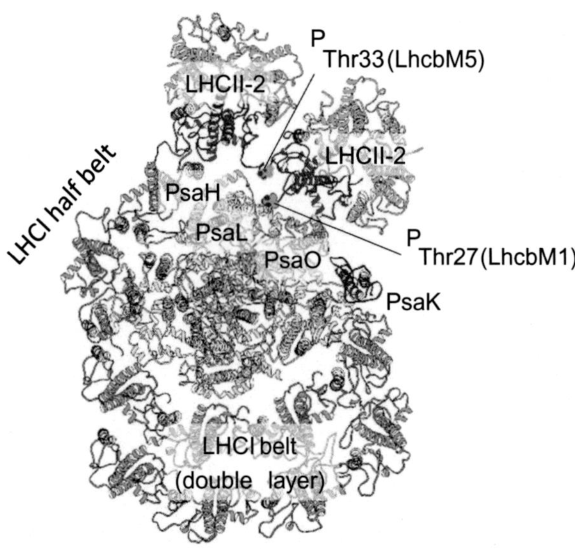

FIG. 17 **LhcbM5-mediated state transitions in green algae.** LHCII-1 is associated with the PsaO *(orange)*, PsaH *(light blue)*, PsaL *(yellow)*, and LHCII-2. LHCII-1 is associated with Lhca2 *(green)*, PsaH *(light blue)*, PsaL *(yellow)*, and LHCII-1. The phosphorylated Thr27 in LhcbM1 and the phosphorylated Thr33 in LhcbM5 have multiple interactions with the amino acid residues in the PSI subunits. View from the stromal side.

have presented an alternative view of state transitions. From the circular dichroism spectra of living *C. reinhardtii* cells, ~80% of the PSII–LHCII supercomplex was found to be preserved upon a state transition, implying that some of the energy absorbed by those LHCIIs that remained to be associated with PSII was dissipated as heat (Nagy et al., 2014). Another report showed that both the amplitude and fluorescence lifetime increased in state 2, due to PSI, but this increase was far smaller than that reported in the literature, indicating that only a fraction of LHCII was recoupled with PSI in State 2 (Ünlü et al., 2014). A comparable increase in PSI antenna size (~35%) and a decrease in PSII antenna size (~38%), were reported based on the measurements of the electrochromic shift, indicating that most of the mobile LHCII were recoupled to PSI (Nawrocki et al., 2016). Thus, to what extent the PSII and PSI antenna sizes change, and how many of the LHCIIs actually 'move' during state transitions, is still controversial.

References

Ahn, T.K., Avenson, T.J., Ballottari, M., Cheng, Y.-C., Niyogi, K.K., Bassi, R., Fleming, G.R., 2008. Architecture of a charge-transfer state regulating light harvesting in a plant antenna protein. Science 320, 794–797.

Aihara, Y., Fujimura-Kamada, K., Yamasaki, T., Minagawa, J., 2019. Algal photoprotection is regulated by the E3 ligase CUL4-DDB1^{DET1}. Nat. Plants 5, 34–40.

Allorent, G., Lefebvre-Legendre, L., Chappuis, R., Kuntz, M., Truong, T.B., Niyogi, K.K., Ulm, R., Goldschmidt-Clermont, M., 2016. UV-B photoreceptor-mediated protection of the photosynthetic machinery in *Chlamydomonas reinhardtii*. Proc. Natl. Acad. Sci. USA 113, 14864–14869.

Bennett, J., 1977. Phosphorylation of chloroplast membrane polypeptides. Nature 269, 344–346.

Bonaventura, C., Myers, J., 1969. Fluorescence and oxygen evolution from *Chlorella pyrenoidosa*. Biochim. Biophys. Acta 189, 366–383.

Boulay, C., Wilson, A., D'Haene, S., Kirilovsky, D., 2010. Identification of a protein required for recovery of full antenna capacity in OCP-related photoprotective mechanism in cyanobacteria. Proc. Natl. Acad. Sci. USA 107, 11620–11625.

Brocks, J.J., Logan, G.A., Buick, R., Summons, R.E., 1999. Archean molecular fossils and the early rise of eukaryotes. Science 285, 1033–1036.

Caffarri, S., Kouril, R., Kereiche, S., Boekema, E.J., Croce, R., 2009. Functional architecture of higher plant photosystem II supercomplexes. EMBO J. 28, 3052–3063.

Cao, P., Pan, X., Su, X., Liu, Z., Li, M., 2020. Assembly of eukaryotic photosystem II with diverse light-harvesting antennas. Curr. Opin. Struct. Biol. 63, 49–57.

Cariti, F., Chazaux, M., Lefebvre-Legendre, L., Longoni, P., Ghysels, B., Johnson, X., Goldschmidt-Clermont, M., 2020. Regulation of light harvesting in *Chlamydomonas reinhardtii* two protein phosphatases are involved in state transitions. Plant Physiol. 183, 1749–1764.

Correa-Galvis, V., Poschmann, G., Melzer, M., Stühler, K., Jahns, P., 2016. PsbS interactions involved in the activation of energy dissipation in *Arabidopsis*. Nat. Plants 2, 15225.

Daskalakis, V., Papadatos, S., Kleinekathofer, U., 2019. Fine tuning of the photosystem II major antenna mobility within the thylakoid membrane of higher plants. Biochim. Biophys. Acta 1861, 183059.

Demmig-Adams, B., Ebbert, V., Mellman, D.L., Mueh, K.E., Schaffer, L., Funk, C., Zarter, C.R., Adamska, I., Jansson, S., Adams, W.W., 2006. Modulation of PsbS and flexible vs sustained energy dissipation by light environment in different species. Physiol. Plant. 127, 670–680.

Dinc, E., Tian, L., Roy, L.M., Roth, R., Goodenough, U., Croce, R., 2016. LHCSR1 induces a fast and reversible pH-dependent fluorescence quenching in LHCII in *Chlamydomonas reinhardtii* cells. Proc. Natl. Acad. Sci. USA 113, 7673–7678.

Fan, M., Li, M., Liu, Z., Cao, P., Pan, X., Zhang, H., Zhao, X., Zhang, J., Chang, W., 2015. Crystal structures of the PsbS protein essential for photoprotection in plants. Nat. Struct. Mol. Biol. 22, 729–735.

Goral, T.K., Johnson, M.P., Duffy, C.D., Brain, A.P., Ruban, A.V., Mullineaux, C.W., 2012. Light-harvesting antenna composition controls the macrostructure and dynamics of thylakoid membranes in *Arabidopsis*. Plant J. 69, 289–301.

Goss, R., Ann Pinto, E., Wilhelm, C., Richter, M., 2006. The importance of a highly active and ΔpH-regulated diatoxanthin epoxidase for the regulation of the PS II antenna function in diadinoxanthin cycle containing algae. J. Plant Physiol. 163, 1008–1021.

Horton, P., Ruban, A.V., Rees, D., Pascal, A.A., Noctor, G., Young, A.J., 1991. Control of the light-harvesting function of chloroplast membranes by aggregation of the LHCII chlorophyll-protein complex. FEBS Lett. 292, 1–4.

Ilioaia, C., Johnson, M.P., Horton, P., Ruban, A.V., 2008. Induction of efficient energy dissipation in the isolated light-harvesting complex of photosystem II in the absence of protein aggregation. J. Biol. Chem. 283, 29505–29512.

Iwai, M., Grob, P., Iavarone, A.T., Nogales, E., Niyogi, K.K., 2018. A unique supramolecular organization of photosystem I in the moss *Physcomitrella patens*. Nat. Plants 4, 904–909.

Johnson, M.P., Goral, T.K., Duffy, C.D., Brain, A.P., Mullineaux, C.W., Ruban, A.V., 2011. Photoprotective energy dissipation involves the reorganization of photosystem II light-harvesting complexes in the grana membranes of spinach chloroplasts. Plant Cell 23, 1468–1479.

Kato, H., Tokutsu, R., Kubota-Kawai, H., Burton-Smith, R.N., Kim, E., Minagawa, J., 2020. Characterization of a giant PSI supercomplex in the symbiotic dinoflagellate symbiodiniaceae. Plant Physiol. 183, 1725–1734.

Kim, E., Kawakami, K., Sato, R., Ishii, A., Minagawa, J., 2020. Photoprotective capabilities of light-harvesting complex II trimers in the green alga *Chlamydomonas reinhardtii*. J. Phys. Chem. Lett. 11, 7755–7761.

Kirilovsky, D., Kerfeld, C.A., 2012. The orange carotenoid protein in photoprotection of photosystem II in cyanobacteria. Biochim. Biophys. Acta 1817, 158–166.

Kiss, A.Z., Ruban, A.V., Horton, P., 2008. The PsbS protein controls the organization of the photosystem II antenna in higher plant thylakoid membranes. J. Biol. Chem. 283, 3972–3978.

Kosuge, K., Tokutsu, R., Kim, E., Akimoto, S., Yokono, M., Ueno, Y., Minagawa, J., 2018. LHCSR1-dependent fluorescence quenching is mediated by excitation energy transfer from LHCII to photosystem I in *Chlamydomonas reinhardtii*. Proc. Natl. Acad. Sci. USA 115, 3722–3727.

Kouřil, R., Nosek, L., Bartos, J., Boekema, E.J., Ilik, P., 2016. Evolutionary loss of light-harvesting proteins Lhcb6 and Lhcb3 in major land plant groups—break-up of current dogma. New Phytol. 210, 808–814.

Kramer, D.M., Sacksteder, C.A., Cruz, J.A., 1999. How acidic is the lumen? Photosynth. Res. 60, 151–163.

Kromdijk, J., Glowacka, K., Leonelli, L., Gabilly, S.T., Iwai, M., Niyogi, K.K., Long, S.P., 2016. Improving photosynthesis and crop productivity by accelerating recovery from photoprotection. Science 354, 857–861.

Li, H., Wang, Y., Ye, M., Li, S., Li, D., Ren, H., Wang, M., Du, L., Li, H., Veglia, G., Gao, J., Weng, Y., 2020. Dynamical and allosteric regulation of photoprotection in light harvesting complex II. Sci. China Chem. 63, 1121–1133.

Liguori, N., Roy, L.M., Opacic, M., Durand, G., Croce, R., 2013. Regulation of light harvesting in the green alga *Chlamydomonas reinhardtii*: the C-terminus of LHCSR is the knob of a dimmer switch. J. Am. Chem. Soc. 135, 18339–18342.

Longoni, P., Douchi, D., Cariti, F., Fucile, G., Goldschmidt-Clermont, M., 2015. Phosphorylation of the light-harvesting complex II isoform Lhcb2 is central to state transitions. Plant Physiol. 169, 2874–2883.

Minagawa, J., 2011. State transitions—the molecular remodeling of photosynthetic supercomplexes that controls energy flow in the chloroplast. Biochim. Biophys. Acta 1807, 897–905.

Moseley, J.L., Allinger, T., Herzog, S., Hoerth, P., Wehinger, E., Merchant, S., Hippler, M., 2002. Adaptation to Fe-deficiency requires remodeling of the photosynthetic apparatus. EMBO J. 21, 6709–6720.

Murata, N., 1969. Control of excitation transfer in photosynthesis. I. Light-induced change of chlorophyll *a* fluoresence in *Porphyridium cruentum*. Biochim. Biophys. Acta 172, 242–251.

Nagy, G., Unnep, R., Zsiros, O., Tokutsu, R., Takizawa, K., Porcar, L., Moyet, L., Petroutsos, D., Garab, G., Finazzi, G., Minagawa, J., 2014. Chloroplast remodeling during state transitions in *Chlamydomonas reinhardtii* as revealed by noninvasive techniques in vivo. Proc. Natl. Acad. Sci. USA 111, 5042–5047.

Nawrocki, W.J., Santabarbara, S., Mosebach, L., Wollman, F.-A., Rappaport, F., 2016. State transitions redistribute rather than dissipate energy between the two photosystems in *Chlamydomonas*. Nat. Plants 2, 16031.

Nixon, P.J., Michoux, F., Yu, J., Boehm, M., Komenda, J., 2010. Recent advances in understanding the assembly and repair of photosystem II. Ann. Bot. 106, 1–16.

Niyogi, K.K., 1999. Photoprotection revisited: genetic and molecular approaches. Annu. Rev. Plant Physiol. Plant Mol. Biol. 50, 333–359.

Niyogi, K.K., Truong, T.B., 2013. Evolution of flexible non-photochemical quenching mechanisms that regulate light harvesting in oxygenic photosynthesis. Curr. Opin. Plant Biol. 16, 307–314.

Olaizola, M., La Roche, J., Kolber, Z., Falkowski, P.G., 1994. Non-photochemical fluorescence quenching and the diadinoxanthin cycle in a marine diatom. Photosynth. Res. 41, 357–370.

Pan, X., Ma, J., Su, X., Cao, P., Chang, W., Liu, Z., Zhang, X., Li, M., 2018. Structure of the maize photosystem I supercomplex with light-harvesting complexes I and II. Science 360, 1109–1113.

Pan, X., Tokutsu, R., Li, A., Takizawa, K., Song, C., Murata, K., Yamasaki, T., Liu, Z., Minagawa, J., Li, M., 2021. Structural basis of LhcbM5-mediated state transitions in green algae. Nat. Plants. https://doi.org/10.1038/s41477-021-00960-8.

Petroutsos, D., Tokutsu, R., Maruyama, S., Flori, S., Greiner, A., Magneschi, L., Cusant, L., Kottke, T., Mittag, M., Hegemann, P., Finazzi, G., Minagawa, J., 2016. A blue-light photoreceptor mediates the feedback regulation of photosynthesis. Nature 537, 563–566.

Pietrzykowska, M., Suorsa, M., Semchonok, D.A., Tikkanen, M., Boekema, E.J., Aro, E.M., Jansson, S., 2014. The light-harvesting chlorophyll a/b binding proteins Lhcb1 and Lhcb2 play complementary roles during state transitions in Arabidopsis. Plant Cell 26, 3646–3660.

Pribil, M., Pesaresi, P., Hertle, A., Barbato, R., Leister, D., 2010. Role of plastid protein phosphatase TAP38 in

LHCII dephosphorylation and thylakoid electron flow. PLoS Biol. 8, e1000288.

Puthiyaveetil, S., Woodiwiss, T., Knoerdel, R., Zia, A., Wood, M., Hoehner, R., Kirchhoff, H., 2014. Significance of the photosystem II core phosphatase PBCP for plant viability and protein repair in thylakoid membranes. Plant Cell Physiol. 55, 1245–1254.

Rochaix, J.D., 2007. Role of thylakoid protein kinases in photosynthetic acclimation. FEBS Lett. 581, 2768–2775.

Ruban, A., 2012. The Photosynthetic Membrane: Molecular Mechanisms and Biophysics of Light Harvesting. Wiley-Blackwell, Chichester.

Ruban, A.V., 2016. Nonphotochemical chlorophyll fluorescence quenching: mechanism and effectiveness in protecting plants from photodamage. Plant Physiol. 170, 1903–1916.

Ruban, A., Lavaud, J., Rousseau, B., Guglielmi, G., Horton, P., Etienne, A.L., 2004. The super-excess energy dissipation in diatom algae: comparative analysis with higher plants. Photosynth. Res. 82, 165–175.

Ruban, A.V., Berera, R., Ilioaia, C., van Stokkum, I.H., Kennis, J.T., Pascal, A.A., van Amerongen, H., Robert, B., Horton, P., van Grondelle, R., 2007. Identification of a mechanism of photoprotective energy dissipation in higher plants. Nature 450, 575–578.

Ruban, A.V., Johnson, M.P., Duffy, C.D., 2012. The photoprotective molecular switch in the photosystem II antenna. Biochim. Biophys. Acta 1817, 167–181.

Sacharz, J., Giovagnetti, V., Ungerer, P., Mastroianni, G., Ruban, A.V., 2017. The xanthophyll cycle affects reversible interactions between PsbS and light-harvesting complex II to control non-photochemical quenching. Nat. Plants 3, 16225.

Samol, I., Shapiguzov, A., Ingelsson, B., Fucile, G., Crevecoeur, M., Vener, A.V., Rochaix, J.D., Goldschmidt-Clermont, M., 2012. Identification of a photosystem II phosphatase involved in light acclimation in *Arabidopsis*. Plant Cell 24, 2596–2609.

Schreiber, U., Schliwa, U., Bilger, W., 1986. Continuous recording of photochemical and non-photochemical chlorophyll fluorescence quenching with a new type of modulation fluorometer. Photosynth. Res. 10, 51–62.

Shapiguzov, A., Ingelsson, B., Samol, I., Andres, C., Kessler, F., Rochaix, J.D., Vener, A.V., Goldschmidt-Clermont, M., 2010. The PPH1 phosphatase is specifically involved in LHCII dephosphorylation and state transitions in *Arabidopsis*. Proc. Natl. Acad. Sci. USA 107, 4782–4787.

Suga, M., Shen, J.R., 2020. Structural variations of photosystem I-antenna supercomplex in response to adaptations to different light environments. Curr. Opin. Struct. Biol. 63, 10–17.

Summons, R.E., Jahnke, L.L., Hope, J.M., Logan, G.A., 1999. 2-Methylhopanoids as biomarkers for cyanobacterial oxygenic photosynthesis. Nature 400, 554–557.

Sylak-Glassman, E.J., Malnoe, A., De Re, E., Brooks, M.D., Fischer, A.L., Niyogi, K.K., Fleming, G.R., 2014. Distinct roles of the photosystem II protein PsbS and zeaxanthin in the regulation of light harvesting in plants revealed by fluorescence lifetime snapshots. Proc. Natl. Acad. Sci. USA 111, 17498–17503.

Taddei, L., Stella, G.R., Rogato, A., Bailleul, B., Fortunato, A.-E., Annunziata, R., Sanges, R., Thaler, M., Lepetit, B., Lavaud, J., Jaubert, M., Finazzi, G., Bouly, J.P., Falciatore, A., 2016. Multisignal control of expression of the LHCX protein family in the marine diatom *Phaeodactylum tricornutum*. J. Exp. Bot. 67, 3939–3951.

Takahashi, H., Iwai, M., Takahashi, Y., Minagawa, J., 2006. Identification of the mobile light-harvesting complex II polypeptides for state transitions in *Chlamydomonas reinhardtii*. Proc. Natl. Acad. Sci. USA 103, 477–482.

Tokutsu, R., Minagawa, J., 2013. Energy-dissipative supercomplex of photosystem II associated with LHCSR3 in *Chlamydomonas reinhardtii*. Proc. Natl. Acad. Sci. USA 110, 10016–10021.

Tokutsu, R., Fujimura-Kamada, K., Matsuo, T., Yamasaki, T., Minagawa, J., 2019. The CONSTANS flowering complex controls the protective response of photosynthesis in the green alga *Chlamydomonas*. Nat. Commun. 10, 4099.

Townsend, A.J., Saccon, F., Giovagnetti, V., Wilson, S., Ungerer, P., Ruban, A.V., 2018. The causes of altered chlorophyll fluorescence quenching induction in the *Arabidopsis* mutant lacking all minor antenna complexes. Biochim. Biophys. Acta 1859, 666–675.

Ünlü, C., Drop, B., Croce, R., van Amerongen, H., 2014. State transitions in *Chlamydomonas reinhardtii* strongly modulate the functional size of photosystem II but not of photosystem I. Proc. Natl. Acad. Sci. USA 111, 3460–3465.

Wang, W., Yu, L.J., Xu, C., Tomizaki, T., Zhao, S., Umena, Y., Chen, X., Qin, X., Xin, Y., Suga, M., Han, G., Kuang, T., Shen, J.R., 2019. Structural basis for blue-green light harvesting and energy dissipation in diatoms. Science 363, eaav0365.

Wollman, F.-A., Lemaire, C., 1988. Studies on kinase-controlled state transitions in photosystem II and b_6f mutants from *Chlamydomonas reinhardtii* which lack quinone-binding proteins. Biochim. Biophys. Acta 933, 85–94.

Wraight, C.A., Crofts, A.R., 1970. Energy-dependent quenching of chlorophyll alpha fluorescence in isolated chloroplasts. Eur. J. Biochem. 17, 319–327.

Yamamoto, H.Y., Nakayama, T.O., Chichester, C.O., 1962. Studies on the light and dark interconversions of leaf xanthophylls. Arch. Biochem. Biophys. 97, 168–173.

CHAPTER

5

Abiotic stress and adaptation of electron transport: Regulation of the production and processing of ROS signals in chloroplasts

Christine H. Foyer[a] *and Guy Hanke*[b]

[a]School of Biosciences, College of Life and Environmental Sciences, University of Birmingham, Edgbaston, United Kingdom [b]School of Biological and Chemical Sciences, Queen Mary University of London, London, United Kingdom

1 Introduction

Chloroplasts are responsible for photosynthesis and the production of a wide range of metabolites and hormones. The photosynthetic processes are localised in chloroplasts that are bound by a double membrane, encapsulating a soluble inner stroma containing a naked circular genome, and the thylakoid membrane system that is essentially a single membrane comprising grana stacks linked by lamellar membranes. This structural organisation facilitates the specialised functions of chloroplasts, exemplified by the compartmentalization of the proteins and compounds required for light harvesting, electron transport, and photophosphorylation within thylakoids, and those for carbon assimilation within the soluble stroma.

All biomolecules by definition can engage in electron transfer (**redox**) processes because their atomic constituents can either act as electron donors (**red**uctants) or acceptors (**ox**idants). Redox activity depends on the chemical properties of the set of molecules with which they interact, i.e., the redox environment or interactome. Within the context of the chloroplast redox biology, emphasis is placed on ROS (particularly singlet oxygen, superoxide, and hydrogen peroxide), antioxidants such as ascorbate and a-tocopherol and thiol compounds, such as thioredoxin, peroxiredoxins, and glutathione. However, there is a vast population of redox

Photosynthesis in Action
https://doi.org/10.1016/B978-0-12-823781-6.00010-1

active molecules in chloroplasts that are often able to cross-react, forming a single synchronised network of electron-exchanging entities. Redox chemistry, therefore, encompasses chains of reactions that extend beyond the first redox couple with each reaction having the potential to generate new redox-active molecules and so on in a cascade, forming the redox interactome. Through evolution, this complex redox network acquired signalling functions and novel functional capabilities. The concept of redox homeostasis is hence not defined by the static relationships within a given system, but rather keeps cellular and organelle redox potentials within a certain range by eliminating excess reductants or oxidants in order to achieve and/or maintain a better balance. By their very nature, redox reactions imply differences in redox potential between reactants, some of which can be transported elsewhere. Hence, the current concept is that living organisms accommodate their redox environment, in which a changing redox landscape may represent a driving force as well as a potential threat. Oxidative signalling is often depicted in the form of 'excess production of ROS' overwhelming the antioxidant defence, but in reality, redox activities are coupled because the same redox agents engage simultaneously in both processes, as well as being interconnected through a common redox interactome that pervades all metabolic and functional reactions.

Energy conservation in the photosynthetic electron transport (PETC) is a paradigm of redox chemistry that integrates many different types of redox molecules. The redox reactions in the PETC are coupled to downstream electron acceptors, one of which is molecular oxygen leading to ROS production, particularly when electrons are backed up in the chain. Since the first demonstration of the Mehler reaction (Mehler, 1951) in which isolated chloroplasts were shown to reduce oxygen to form superoxide ($O_2^{\bullet-}$) and hydrogen peroxide (H_2O_2), this process has been inextricably linked to the concept of oxidative stress. Similarly, light-induced photo-damage has also been historically associated with the oxidative stress concept. ROS production by the PETC is often considered to be an unavoidable consequence of the operation of electron transport carriers in an aerobic environment (Khorobrykh et al., 2020). However, recent evidence has demonstrated that efficient regulation restricts energy and electron flow to oxygen at various sites in the PETC. It has long been recognised that the best way to protect against any collateral ROS damage is to limit or avoid their production. The second way to police ROS activity is to accumulate antioxidants, i.e., substances that interact with ROS without themselves generating highly reactive forms, in order to prevent uncontrolled oxidation. However, if we assume that all redox reactions, including ROS production, converge towards a single purpose, i.e., maintaining the integrity of the cell or organism, the advantages of the redox network more than offset the disadvantages associated with a measure of collateral damage.

Current concepts consider that the photosynthetic complexes have evolved to control and minimise electron transport to oxygen. For example, comparison of redox potentials between the electron transport chains of oxygenic and nonoxygenic species indicates redox-tuning of cofactors to minimise electron donation to oxygen (Rutherford et al., 2012). The tolerance of plants to abiotic stresses can be increased by overexpressing antioxidants and other components of the chloroplast protection network (see, for example, Badawi et al., 2004; Tang et al., 2006; Sun et al., 2010a,b; Chang et al., 2016; Wang et al., 2018, 2020). However, it is clear that such protective systems are not designed to eliminate ROS but merely to prevent excessive and uncontrolled oxidation. A key question that will be addressed in this chapter is, therefore, why oxygen metabolism has evolved to become inextricably linked to photosynthesis.

2 Redox reactions and the photosynthetic electron transport chain

As the 1937 Nobel Prize winner for the discovery of Vitamin C, Albert Szent-Györgyi said *Life is nothing but an electron looking for a place to rest*. In photosynthesis, electrons in chlorophyll are excited by light to a high energy state, and a very negative redox potential (below −0.6 V, Rutherford, 1981). This means they have a very high capacity to donate an electron to an acceptor. This excitation can be dissipated in several ways and would normally be lost as the electron returned to ground state, releasing heat or light as fluorescence (a process that takes place in the order of nanoseconds 10^{-9} s). However, the generation of photosynthetic power depends on transfer of this electron at an even faster rate (in the order of picoseconds 10^{-12} s) away from a special chlorophyll pair in the reaction centre of photosystem II (PSII) to an acceptor before it can drop back to the ground state. The final electron acceptor of PSII is plastoquinone, and this electron then passes through a series of redox centres, generally moving from more negative to more positive redox potentials (lower energy states). The energy released by this process can then be used to power metabolism. This leaves the chlorophyll in an oxidised state (lacking an electron). Earlier in evolution, photosynthetic organisms replaced this electron by exploiting chemical sources of electrons (reducing agents), such as Fe^{2+} ions or H_2S, but cyanobacteria exploited the extraordinarily positive redox potential (capacity to accept electrons) of oxidised chlorophyll (around +1.2 V, Rappaport et al., 2002) to extract electrons from an unlimited supply of H_2O, releasing O_2 as a by-product.

Molecular oxygen is highly reactive, capable of changing the structure of proteins and lipids by stripping away electrons and by forming free radicals. The oxidative power of living systems has been considered an important driving force in a reducing environment, with antireductants (Becker, 2016) rather than antioxidants essential to protecting cellular integrity (Santolini et al., 2019). Amongst the first adaptations that allowed cyanobacteria to exploit water as an electron source must therefore have been the emergence of antireductants and antioxidants that allowed the establishment and maintenance of a redox equilibrium within an ever-changing redox environment, whilst providing defence to counter-reductive or oxidative stress, respectively. These mechanisms and adaptations have evolved over time, not only in the cyanobacteria but also in the chloroplasts of algae and plants, which are their endosymbiotic descendants.

The dynamic regulation of the PETC allows continuous adjustments to accommodate changes in the availability of light and CO_2. Photosynthetic regulation is designed to maximise protection, whilst maintaining the balance between energy-producing and energy-consuming processes. Since light is a potentially dangerous energy source, and its supply is erratic, its use must be carefully managed, not least because PSII is extremely sensitive to light-induced oxidative inactivation. Optimising the efficiency of photosynthesis is a priority when light is limiting, allowing effective use of solar power and ensuring that metabolism is not limited by energy supply. In contrast, energy dissipation is progressively increased with increasing irradiance to prevent light-induced inhibition (Ruban, 2016), particularly at low temperatures or under high light. Within this context, the first question to consider is whether oxygen is an essential electron acceptor for photosynthesis? Multiple mechanisms have evolved in oxygenic photosynthesis to limit ROS generation and to prevent uncontrolled oxidation (Noctor and Foyer, 1998; Rutherford et al., 2012). However, it must be remembered that ROS production is critical for plant growth and development (Considine and Foyer, 2021). Plants, like other organisms, have numerous enzymes that produce $O_2^{\bullet -}$ and H_2O_2. For example, the NADPH oxidase called respiratory burst homologue (RBOH)D is a primary player in the ROS production, which is critical for the successful activation of immune responses to pathogens (Lee et al., 2020).

3 ROS production in the chloroplast stroma

There are two basic mechanisms by which ROS are formed in photosynthesis. First, transfer of an excited electron spin state from chlorophyll to O_2 (normally in the triplet state) to form the highly reactive singlet oxygen ($^1O_2$) (see Fig. 1). This occurs when excited chlorophyll (originally in the singlet state) drops down to the lower energy triplet state or during charge recombination following the charge recombination reaction at PSII, due to the lack of an electron acceptor (Krieger-Liszkay, 2005). Triplet chlorophyll has an even longer half-life (microsecond, 10^{-6} s) than excited singlet state chlorophyll (nanosecond, 10^{-9}), which gives ample time for excitation transfer to triplet oxygen. The great majority of $^1O_2$ generated in the photosynthetic membrane has its origin at the PSII reaction centre, but it is reported that small amounts may also form at photosystem I (PSI). The PSI reaction centre is buried deep in the protein, and it is thought that this screens it from O_2, preventing excitation transfer and the formation of $^1O_2$ (Setif et al., 1981). It has been reported that the low rates of $^1O_2$ formation actually detected at PSI originate from excitation transfer in the light harvesting complexes surrounding the PSI core (Cazzaniga et al., 2016).

Second, electrons can also be transferred directly to O_2 forming $O_2^{\bullet-}$. The mid-point redox potential for $O_2/O_2^{\bullet-}$ in aqueous solution is around −160 mV (Wardman, 1989), meaning that it is energetically favourable for O_2 to gain an electron from many reduced components of the photosynthetic electron transport chain. The predominant factors determining whether this occurs are the lifetime of the reduced cofactor, and its accessibility to O_2. It should therefore be of no surprise that PSI, the point at which electrons are transferred out of the membrane environment to soluble acceptors, is the predominant site of $O_2^{\bullet-}$ generation in chloroplasts. This reaction may be exacerbated when there is a lack of available electron acceptors, such as

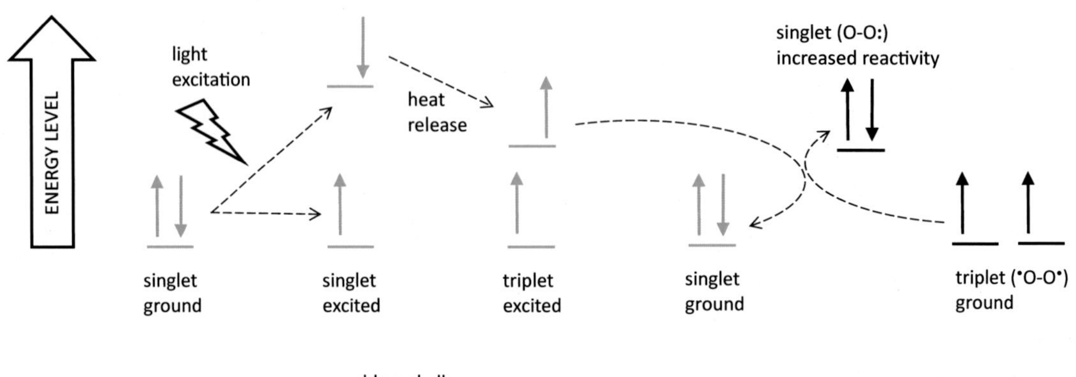

FIG. 1 Basic scheme to illustrate the energy levels (not to scale) and spin states of electrons in the outer orbitals of chlorophyll and oxygen molecules during chlorophyll excitation and relaxation, and excitation transfer from triplet chlorophyll to form of $^1O_2$. *Arrows* indicate spin states of single electrons at different energy levels (equating to different orbitals in chlorophyll). *Dotted arrows* indicate change in energy state. *Green arrows* = chlorophyll, *black arrows* = O_2.

when CO_2 concentrations limit the consumption of electrons by the Calvin–Benson–Basham cycle due to stomatal closure. However, it is likely that photorespiration becomes an effective electron sink in these circumstances (Fernie and Bauwe, 2020). Indeed, photorespiration, which produces huge amounts of H_2O_2, has been considered to be a stepping stone towards the evolution of oxygenic photosynthesis in the first instance and later C_4 prosthesis (Fernie and Bauwe, 2020).

$O_2^{\bullet -}$ production is possible at multiple points in the electron transport chain, depending on metabolic status. The reduction of molecular oxygen at PSI in the Mehler reaction initiates the pathway of pseudocyclic electron flow. This is classically considered to be an unavoidable consequence of electron transport in an aerobic environment. In this process, $O_2^{\bullet -}$ radicals are generated at the stromal surface of the thylakoid membrane and rapidly converted to H_2O_2 by the action of the thylakoid superoxide dismutases (SODs). H_2O_2 can then be reduced to water by chloroplast ascorbate peroxidases (APXs) and 2-Cys peroxiredoxins (PRXs) (Awad et al., 2015). The overall process became known as the water–water cycle because two electrons are used to produce H_2O_2 and two more are required to metabolise H_2O_2 to water, allowing the dissipation of excess excitation energy and electrons (Fig. 2; Asada, 2000). However, this process makes only a minor contribution to thylakoid acidification and the control of PSII activity because it is saturated at relatively low irradiances (Driever and Baker, 2011).

During photosynthesis, ROS are predominantly produced at the PSII reaction centre ($^1O_2$) and PSI acceptor side ($O_2^{\bullet -}$), but there are several other possible sites of free radical generation in the light harvesting and electron transport apparatus. In some cases, these could contribute a significant proportion of ROS generated (Kozuleva and Ivanov, 2016), in others, ROS generation has been proven for isolated components in vitro (Baniulis et al., 2013), but remains to be detected in vivo, and in others ROS generation is only theoretically proposed. The high reactivity of ROS means that their downstream impact through signalling cascades may depend on secondary interaction products, which in turn are dependent on the molecules in their environment when they are first generated. We will therefore set out ROS generation by the thylakoid membrane in terms of the location in the chloroplast into which they are released, and these pathways are summarised in Fig. 3.

All the cofactors on the reducing side of PSI are able to reduce O_2 in the stroma. In PSI, excited electrons pass from the reaction-centre special pair through other chlorophyll molecules and a phylloquinone (PhQ) to a chain of three 4Fe4S clusters at the stromal side of the complex (Mazor et al., 2017). The terminal FeS clusters (FA/FB) are oxidised by O_2 when the soluble electron carrier protein ferredoxin (Fd) is not available in an oxidised state. Moreover, the PhQ molecules at the A1-sites may also reduce oxygen (Kozuleva et al., 2014). Although the FB cluster is the direct electron donor to Fd (Diaz-Quintana et al., 1998), electrons reside largely on the FA site (further from the stromal surface). This is because FB has a more negative redox potential than FA (Heathcote et al., 1978), making electron transfer in the final step of the cascade through PSI 'uphill', and therefore limiting flux to O_2 in the absence of Fd. Hence, the reactivity of FeS components, such as FB and Fd, with oxygen is decreased by redox tuning and protein shielding. Although the 2Fe2S cluster containing Fd is a relatively poor electron donor to O_2 in comparison to FA/FB (Kozuleva and Ivanov, 2016), it is still sometimes described as a source of electrons for $O_2^{\bullet -}$ generation (see, for example, Allen, 1975). This finding is usually based on increased $O_2^{\bullet -}$ generation following addition of Fd to active PSI preparations or thylakoids. In this case, the unphysiologically high Fd:PSI ratios might give the false impression

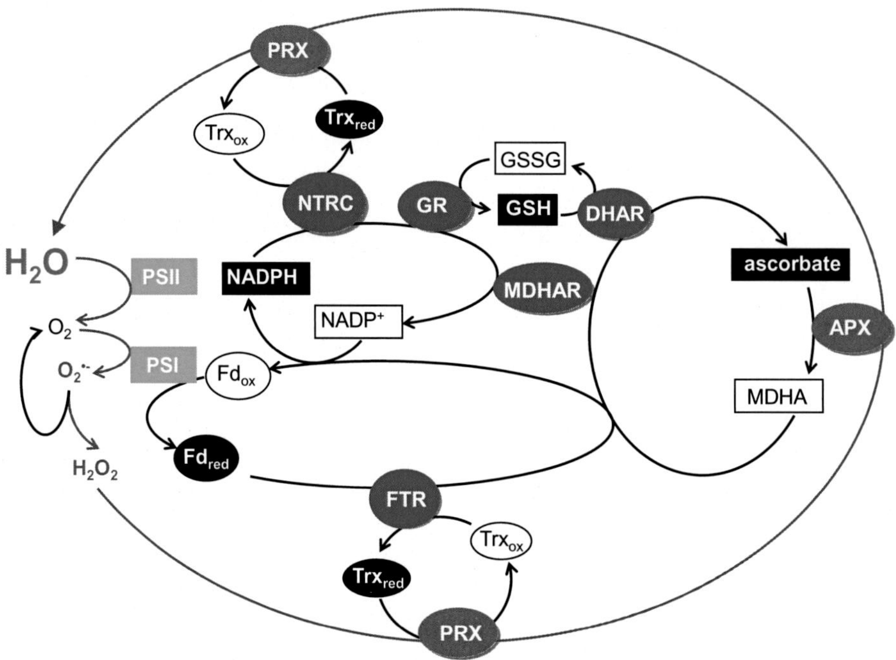

FIG. 2 Schematic model of the water/water cycle in photosynthesis. During photosynthetic electron transport, the reduction of molecular oxygen in the Mehler reaction produces superoxide anion radicals, $O_2^{\bullet -}$. $O_2^{\bullet -}$ is then either directly reduced to hydrogen peroxide (H_2O_2) by ascorbate, or this radical can participate in a dismutation reaction to produce H_2O_2, a reaction that is catalysed in chloroplasts by the thylakoid and stromal superoxide dismutases. H_2O_2 is then reduced to water either by the action of chloroplast ascorbate peroxidases (APXs) or the chloroplast 2-Cys peroxiredoxins (PRX). The APX reaction is the first step in the ascorbate/glutathione (GSH) cycle, which serves to regenerate ascorbate and maintain the ascorbate pool in its reduced form, using NADPH produced by the photosynthetic electron transport chain. These reactions start with the splitting of water in the photosystem (PS) II reaction centre and end with the reduction of $O_2^{\bullet -}$ and H_2O_2 to water in the stroma using reducing power provided by the photosynthetic electron transport chain. The overall process, which involves successive oxidations and re-reductions of ascorbate, PRXs, and thioredoxins (TRX), is often referred to as the water–water cycle of photosynthesis. This cycle is crucial in preventing the accumulation of H_2O_2 in the stroma to levels that would inhibit the thiol-modulated enzymes of the Calvin cycle and other pathways.

that Fd is a more effective O_2 reductant than the phylloquinone and FA/FB clusters of PSI. The two electron reduction of $NADP^+$ to NADPH is catalysed by Fd:NADP(H) reductase (FNR) using a flavin (FAD) cofactor. Discrepancies between supply of reduced Fd (a 1 electron carrier) and $NADP^+$ (a 2 electron acceptor) can leave this cofactor in a single electron reduced (semiquinone) state, which also readily reduces O_2 (Kozuleva et al., 2016).

In addition to the proven $O_2^{\bullet -}$ generation at PSI, there are other potential sources of $O_2^{\bullet -}$ on the stromal side of the membrane. The structure of the cyanobacterial NDH complex (more recently termed photosynthetic complex I, or PCI) has recently been solved (Laughlin et al., 2019; Schuller et al., 2019). This complex is also present in the chloroplasts of angiosperms (Sazanov et al., 1998; Corneille et al., 1998) and is a homologue of respiratory complex I, a

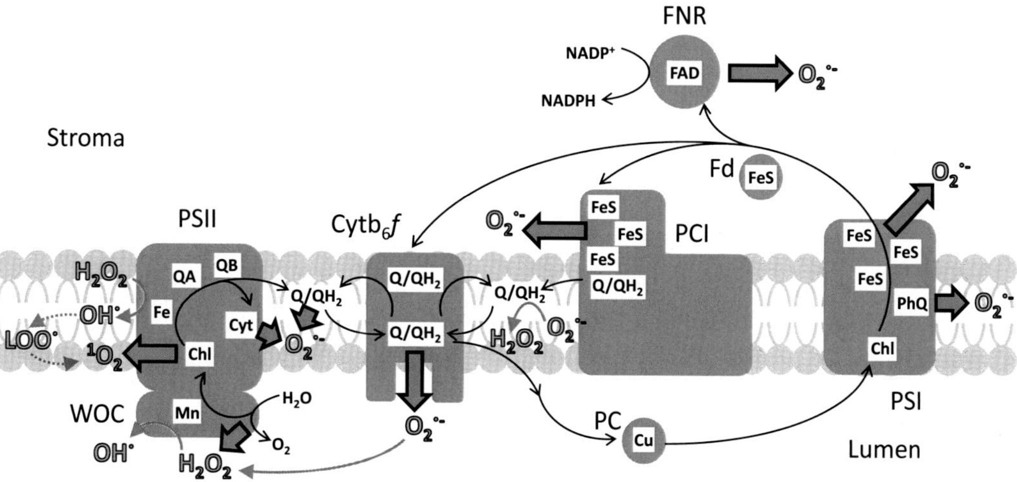

FIG. 3 Reported and proposed sites of free radical generation in the photosynthetic light harvesting and electron transport apparatus. Light excitation of chlorophyll at photosystem II (PSII) and photosystem I (PSI) drives electron transport, but in the absence of acceptors, excitation may be transferred at the PSII reaction centre to O_2 forming the singlet oxygen radical ($^1O_2$). During electron transport, donation of a single electron to the two electron carrier plastoquinone (Q) at the cytochrome b_6f (Cyt b_6f) results in formation of the semiquinone radical, which reduces oxygen to superoxide ($O_2^{\bullet -}$). This process also happens during electron transfer from the single electron carrier ferredoxin (Fd) to $NADP^+$ by Fd:NADP(H) reductase (FNR), and might occur at photosynthetic complex I (PCI). Reduction of O_2 can also occur by electron donation from: the cytochrome b_{559} coactor (Cyt) of PSII; reduced plastoquinone (QH_2); and at PSI from phyloquinone (PhQ) to O_2 in the membrane phase, and from the FeS clusters to O_2 in the stroma. In solution, $O_2^{\bullet -}$ is readily dismutated to H_2O_2, whilst in the membrane phase electron donation from QH_2 catalyses this reaction. H_2O_2 is also produced directly during incomplete oxidation of water to O_2 at the water oxidation complex (WOC). In the presence of metal centres, such as Fe or (much less efficiently) Mn, H_2O_2 is converted to hydroxyl radicals ($OH^\bullet$), which readily generate carbonyl radicals, of which only lipid peroxide radicals ($LOO^\bullet$) are shown here. *Dotted lines* indicate several intermediate steps. Steps involving excitation transfer are shown in *blue*, steps involving electron transfer are shown in *orange*. Several light harvesting and cyclic electron flow components are omitted for clarity. Note the different compartments in which radicals are produced by different cofactors.

notorious generator of $O_2^{\bullet -}$ on the *n*-side of the membrane (Hernansanz-Agustín et al., 2017; Pryde and Hirst, 2011) (stromal side in chloroplasts). The terminal FeS cofactor in PCI is very close to the solvent (Schuller et al., 2019), raising the possibility that this complex could make a large—so far uncharacterised contribution to ROS generation in some chloroplasts. Although this complex is only present at low abundance in most species (Shikanai and Yamamoto, 2017), specific cell types with high ATP demand (e.g., in C4 plants) can contain much higher levels (Darie et al., 2006; Munekage et al., 2010; Ishikawa et al., 2016), and so there may be situations where NDH/P-CI makes a significant contribution to free radical generation.

4 Lipid phase ROS production

$^1O_2$ is predominantly generated by direct energy transfer from the triplet excited state of chlorophyll to molecular oxygen (Apel and Hirt, 2004) and is so reactive that diffusion from the PSII reaction centre is unlikely given its very short half-life (nanosecond, 10^{-9}). The occurrence of photosensitizers such as triplet excited chlorophylls occurs mainly in the PSII reaction centre (Pospisil, 2016; D'Alessandro and Havaux, 2019). It is thought that $^1O_2$ is inevitably produced by photosynthesis even in low light, albeit at greatly decreased rates in comparison to high light (Fufezan et al., 2002). In contrast to H_2O_2, $^1O_2$ is a powerful oxidant that reacts rapidly with macromolecules in its vicinity, resulting in oxidation that is often referred to as 'damage', or perhaps more accurately 'collateral damage' (Apel and Hirt, 2004; Santolini et al., 2019). The vulnerability of the D1 and D2 proteins to oxidative modifications means that each reaction centre has to be rebuilt several times an hour, even under optimal irradiances (Malnoë et al., 2014). The oxygen-evolving complex is also prone to $^1O_2$-mediated damage/modification (Henmi et al., 2004).

Recently, production of significant quantities of $O_2^{\bullet -}$ within the thylakoid membrane has been described (Kozuleva and Ivanov, 2016). This is possible because the concentration of O_2 in the lipid bilayer is high, although it has a midpoint potential here of −550–600 mV (Afanas'ev, 1989), limiting the pool of potential reductants. Within the thylakoid membrane lipid phase, O_2 reduction by reduced plastoquinone (PQ), i.e., plastoquinol (PQH_2) may be important (Borisova-Mubarakshina et al., 2019; Vetoshkina et al., 2017). Whilst it has been reported that PQH_2 and the plasto-semiquinone are relatively stable in the presence of oxygen (Hasegawa et al., 2017), PQH_2 was found to reduce oxygen directly to H_2O_2, by further reducing $O_2^{\bullet -}$ (Borisova-Mubarakshina et al., 2018). PSII is also a potential stromal source of $O_2^{\bullet -}$ radicals when the plastoquinone acceptor pool is fully reduced, for example, when electrons suddenly flood the system during exposure to sun-flecks. Of the PSII cofactors that function in electron transfer from the reaction centre to plastoquinone, only pheophytin (Pheo-) has a redox potential capable of reducing O_2 in the lipid environment, but it has such a short lifetime it is unlikely to contribute (Rappaport et al., 2002). Another potential mechanism by which O_2 could be reduced in the lipid phase at PSII is via the low potential form of cytochrome *b*559 at PSII (Khorobrykh, 2019), but this is very unlikely unless PSII functions are severely perturbed.

5 Lumen side ROS reduction

Oxygen reduction at the lumenal side occurs at Q_P (the quinone binding site on the positive side of the membrane) of the cytochrome b_6f (Cyt b_6f) complex (Taylor et al., 2018), either by electron donation to O_2 from the FeS or *b*L heme cofactors. The frequency of these reactions is thought to be decreased by the dimer organisation of the b_6f complex, which provides alternative electron transport pathways between monomers (Rutherford et al., 2012). $O_2^{\bullet -}$ formation at the Cyt b_6f appears to be much more active than in the homologous respiratory complex III, which is also a dimer (Baniulis et al., 2013). Although this is a relatively small amount of ROS compared to that generated at PSI, it is presumed to be on the opposite side of the membrane and therefore might induce alternative signalling responses (see later discussion).

On the lumenal side of the membrane, direct generation of H_2O_2 is also possible at the water oxidation complex of PSII, when only 2 electrons are extracted from water. This appears to be related to loss of chloride from close to the Mn_4O_5Ca cluster, which normally helps regulate H_2O access to the metal centre (Fine and Frasch, 1992). Single electron reduction of H_2O_2 forms the more dangerous $OH^{\bullet}$ radical in the Fenton reaction, which can be catalysed by reduced Fe or (much more slowly) Mn cofactors within the electron transport chain (Pospisil et al., 2004; Šnyrychová et al., 2006). These highly reactive ROS rapidly react with lipids to form peroxidation products such as lipid hydroperoxides, which can decompose to a range of radical species, and some long lived intermediates such as lipid hydroperoxides (Pospisil, 2016), that are potential signalling molecules. Moreover, some of the high energy intermediates are capable of forming triplet carbonyls, which might be capable of energy transfer to O_2, forming $^1O_2$ (Yadav and Pospisil, 2012). The contribution of this pathway to $^1O_2$ production in vivo remains to be quantified.

6 Thioredoxin-dependent control of photosynthetic electron transport

Thioredoxins (TRX) are small ubiquitous redox proteins that contain an active site C-X-X-C motif that can form a secondary structure with βαβαββα topology. The chloroplast thioredoxin (TRX) system regulates many of the photosynthetic processes including light capture, electron transport to oxygen, induction of protection mechanisms in the thylakoid membranes, the activities of stomal enzymes, plastome transcription and translation, and chlorophyll synthesis. In addition, numerous chloroplast elongation factors, chaperones, and kinases are subject to redox modulation to provide essential control over gene expression, protein conformation, and posttranslational modification in the processes of photosystem II (PSII) repair. This regulation ensures that there is an appropriate balance between the supply and demand for essential photosynthetic components and prevents the potentially dangerous accumulation of intermediates, such as the photodynamic metabolites in tetrapyrrole and chlorophyll biosynthesis pathways. There are many examples of redox regulation of chloroplast translation such as the formation of the Nac2–RBP40 complex, which is a chloroplast mRNA translation factor that is required for the synthesis of PSII D2 protein in green algae such as *Chlamydomonas reinhardtii* (Sun and Zerges, 2015). This high-molecular-weight complex contains the RNA stabilisation factor Nac2 and the translational activator RBP40, which are linked by an intermolecular disulphide bridge that is regulated by the NADPH-dependent TRX reductase (NTRC) system (Sun and Zerges, 2015).

The redox state of chloroplast proteins is regulated by the two versatile ferredoxin/thioredoxin systems (Fig. 4). In the first system, TRX is reduced by ferredoxin in a reaction that is catalysed by the enzyme ferredoxin/TRX reductase (FTR). The second system involves reduction of TRX by chloroplast NADPH-Trx reductase (NTRC). Together, the FTR and NTRC systems reduce multiple isoforms of chloroplast TRXs (f, m, x, y, z) as well as a number of thioredoxin-like proteins (Balsera et al., 2014; Geigenberger et al., 2017; Nikkanen and Rintamäki, 2019). The TRX systems directly link photosynthetic electron flow to stromal metabolism, most importantly through reductive activation of the enzymes of the Calvin–Benson–Basham cycle enzymes such as fructose-1,6-bisphosphatase (FBPase; Buchanan, 2016). FBPase and other Calvin–Benson cycle enzymes are predominantly regulated by f-type Trxs (Fig. 4). Other important stromal enzymes such as NADP-dependent malate dehydrogenase (MDH), which is involved in the export of reducing equivalents from the stroma via the malate valve and the γ-subunit of the thylakoid ATP synthase that is involved in ATP synthesis, are regulated by both m- and f-type Trx isoforms (Geigenberger et al., 2017). The reductive activation of chloroplast proteins is reversed by the 2-CysPrx system, which serves as a rapid electron sink for the thiol network (Vaseghi et al., 2018).

Plastid TRX isoforms belonging to the x, y, and z types serve a range of functions such as reduction of peroxiredoxins (Prx), thiol-peroxidases, and methionine sulphoxide reductases. In particular, Trx Z is required for chloroplast transcription and is a key regulator of gene expression, itself regulated by the plastid-encoded RNA polymerase (PEP). Chloroplast gene transcription is facilitated by two distinct types of RNA polymerases in the chloroplasts

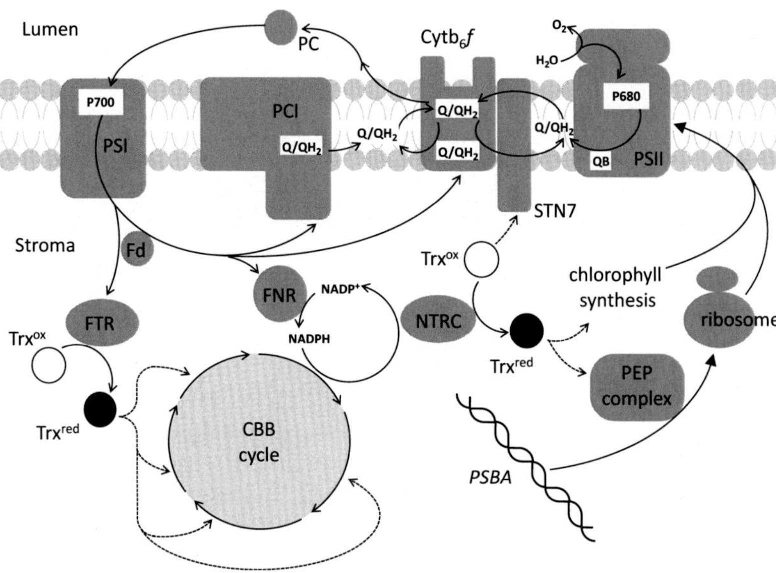

FIG. 4 Schematic model of the regulation of photosynthetic electron transport by the chloroplast thioredoxin system. The photosynthetic electron transport chain drives electrons from photosystem (PS)II through the cytochrome b_6/f (Cyt b_6/f) complex to PSI. PSI uses these electrons to convert oxidised (Fd_{ox}) ferredoxin to its reduced form (Fd_{red}). Fd_{red} can be used either to reduce NADP to NADPH, a reaction that is catalysed by ferredoxin:NADP reductase (FNR), or to reduce oxidised thioredoxin (TRX_{ox}) to its reduced form (TRX_{red}) in a reaction catalysed by the ferredoxin:NADP reductase (FTR), or to transfer electrons back to either the Cyt b_6f complex or the chloroplast NADH dehydrogenase-like complex (NDH). TRX_{ox} can also be converted to TRX_{red} by an NADPH-dependent thioredoxin reductase (NTRC) pathway. Both the FTR and the NTRC pathways can also activate certain carbon assimilation enzymes but under rather different light conditions. The NTRC pathway is responsible for regulation of chlorophyll synthesis, the expression of chloroplast genes (such as *PsbA* that encodes the D1 protein) through the plastid-encoded polymerase (PEP) complex and chloroplast translation, and D1 excision and repair pathways. NTRC is also required for the regeneration of the 2-Cys peroxiredoxins and the transfer of electrons to the thylakoid lumen via membrane-anchored CcdA/HCF164 system, in order to regulate enzymes such as the state transition (STN)7 kinase.

of higher plants: PEP and a nucleus-encoded RNA polymerase (NEP). NEP is encoded by two nuclear genes, rpoTp and rpoTmp in *Arabidopsis thaliana*, whilst PEP is a bacteria-type RNA polymerase composed of four core subunits and a promoter-recognising subunit (σ factor). The genes encoding σ factors provide the necessary promoter specificity to PEP, allowing the nucleus to regulate chloroplast gene transcription in response to environmental and developmental cues. PEP and a set of polymerase-associated proteins (PAPs) are involved in the regulation of DNA and RNA metabolism in chloroplasts. Trx Z and the plastid fructokinase-like proteins, FLN1 and FLN2, are intrinsic subunits of the PEP enzyme of chloroplasts (Schröter et al., 2010). These and other PAPS, such as PAP6/FLN1, PAP10/TrxZ, and PAP12/pTAC7, have been shown to have roles in the redox regulation originating from photosynthesis, whilst others, such as PAP4/FSD3 and PAP9/FSD2, have other ROS-related functions (Zhang et al., 2020).

The NTRC and FTR systems fulfil different roles in the regulation of photosynthesis (Fig. 4). NTRC regulates photosynthetic reactions during dark/light transitions and under low light conditions, whilst the FTR system is responsible for redox-regulation under higher light intensities that do not limit photosynthesis (Nikkanen et al., 2018; Guinea Diaz et al., 2019). Moreover, NTRC activates cyclic electron flow through modification of a redox state of the chloroplast NAD(P)H dehydrogenase-like complex (NDH) and through the regulation of state transitions (Nikkanen et al., 2018). Overexpression of NTRC was shown to increase photosynthetic activity (Guinea Diaz et al., 2019; Nikkanen et al., 2016) and improve light use efficiency particularly under shade conditions (Nikkanen et al., 2016, 2018; Guinea Diaz et al., 2019). Thiol-dependent regulation of the ferredoxin binding site on the NDH complex is thought to allow dynamic control of the ferredoxin: plastoquinone oxidoreductase activity of the complex in response to changing light conditions, together with adjustments of the protonmotive force and nonphotochemical quenching.

NTRC is also important in chloroplast biogenesis by exerting redox control over the activities of enzymes catalysing chlorophyll biosynthesis (Richter et al., 2018). For example, the NTRC dependent 2-Cys PRX activity was shown to be important in the regulation of Mg-protoporphyrin monomethyl ester cyclase activity (Stenbaek et al., 2008). Moreover, NTRC regulates the formation of stromules in response to light-dependent changes in the redox state of the chloroplast (Brunkard et al., 2015). As discussed later, the generation of stromules, which are stroma-filled tubular extensions of the chloroplast envelope containing stromal proteins, provides a potential mechanism that facilitates signal transduction between organelles, such as the nucleus, plasma membrane, endoplasmic reticulum, and other plastids (Hanson and Hines, 2018).

The reducing power of the TRX system is transferred to the thylakoid lumen via the membrane-anchored CcdA/HCF164 system, which is a trans-thylakoid thiol reducing pathway that inactivates thylakoid proteins such as the serine/threonine kinase called the state transition (STN)7 kinase under high light (Ancin et al., 2019). STN7 phosphorylates the mobile population of the light-harvesting chlorophyll *a*/*b* binding complex (LHC)II, which captures light and transfers energy towards the reaction centres of the photosystems. This kinase, which is activated through its interaction with the Cyt b_6f complex, responds to alterations of the balance between the oxidised (PQ) and reduced (PQH_2) forms of plastoquinone (Rochaix, 2014), i.e., STN7 is activated when PQH_2 binds to the Qo site of Cyt b_6f complex. This process occurs under conditions in which PSII receives more excitation energy than PSI (a situation called state 1; Bellafiore et al., 2005). Following phosphorylation, LHCII is able to migrate from PSII to PSI, thus increasing the PSI antenna size and rebalancing the excitation energy between PSII and PSI (a condition called state 2; Shapiguzov et al., 2016). STN7 has a transmembrane helix linking the kinase catalytic domain on the stromal side with its amino terminus containing two conserved cysteine residues in the lumenal domain (Bergner et al., 2015). The lumen thioredoxin called lumen thiol oxidoreductase1 (LTO1) interacts with STN7 through lumenal domains (Wu et al., 2021). The conserved cysteines in the lumenal domains of STN7 and LTO1 are important in maintaining the oxidised state of STN7 that is required for kinase activity during state transitions. LTO1 is part of the thiol-oxidising pathway which catalyses disulphide bond formation (Du et al., 2015). It contains a TRX-like domain with a similar activity to DsbA and the vitamin K epoxide reductase, VKOR, which is functionally equivalent to DsbB and transfers electrons from TRX to the final acceptor (vitamin K epoxide) in PSI (Onda, 2013). In summary, when PSI is reduced relative to PSII (State 1), STN7 is inactivated by extraction of electrons to form a disulphide bond in the lumenal domain and sequential transfer of these electrons to the TRX-like domain, then to the VKOR domain of LTO1 (Wu et al., 2021).

7 ROS as chloroplast signals

The above discussion has focused on the dynamic regulation of the photosynthetic machinery that serves to balance energy transducing reactions and to regulate ROS production. This regulation must be viewed within the context of the important functions of ROS as powerful signals that regulate supply and demand in energy metabolism (Foyer et al., 2017; Noctor and Foyer, 2016). Chloroplasts are a major source of these oxidative signals that transmit information to the nucleus to regulate gene expression (Leister, 2019). $^1O_2$, H_2O_2, and $O_2^{\bullet -}$ are products of photosynthetic electron transport that are characterised by their relative reactivities and lifetimes. As such, they are likely to function as discrete redox signals that trigger different pathways to regulate the expression of specific suites of nuclear genes. $O_2^{\bullet -}$ has a low level of reactivity, but it can react with ascorbate, plastocyanin, and nitric oxide (NO). Crucially, $O_2^{\bullet -}$ radicals can inhibit the activities of Fe—S [4 Fe—4 S] cluster-containing proteins by releasing iron. The chemical or enzymatic dismutation of $O_2^{\bullet -}$ generates H_2O_2, which is removed by enzymes such as APX and PRX. However, these enzymes do not serve to completely eliminate H_2O_2 from the chloroplasts, but rather to prevent excessive accumulation of this oxidant. H_2O_2 signalling occurs primarily though the oxidation of protein cysteine (Cys) residues. The oxidation of the redox-sensitive Cys thiols on proteins leads to the reversible formation of sulfenic acid (—SOH) that is stabilised by forming a disulphide bond (S—S) with a nearby thiol or a mixed disulphide bond with reduced glutathione (GSH). ROS-mediated Cys oxidation occurs on proteins that are located close to the sites of ROS production. ROS can react with diverse protein side chains and they have no specificity for reactive Cys. Post translational modifications (PTMs) of Cys residues may occur either through dithiol-disulphide exchange reactions or through reactions in which particular protein Cys are oxidised by ROS, reactive nitrogen species, or reactive sulphur species (Zaffagnini et al., 2019). The reversible redox regulation of Cys residues of chloroplast proteins facilitates precise and specific H_2O_2 signalling, as well as metabolic and functional regulation. Reversible redox PTMs can act as a regulatory switch that can alter the interactome, enzyme activity, conformational integrity, signalling functions, and protein stability in response to cellular redox state changes (Foyer et al., 2020b). Whilst H_2O_2 can pass from the stroma to the cytosol, recent evidence suggests that H_2O_2 can be directly transferred from chloroplasts to the nucleus through contact sites under high light (Exposito-Rodriguez et al., 2017; Mullineaux et al., 2020). The accumulation of H_2O_2 can also induce stromule formation (Hanson and Hines, 2018), allowing the direct transfer of H_2O_2 from chloroplasts to the nucleus (Caplan et al., 2015).

8 Regulation of $^1O_2$ production and signalling

The generation of $^1O_2$ by PSII is regulated by the binding of bicarbonate (Brinkert et al., 2016). Bicarbonate binding has significant effects on PSII activity. Bicarbonate binds both to the electron donor side of PSII to regulate water oxidation and to the acceptor side of PSII, which accepts the electron from the charge separation at P680. Bicarbonate binds to nonheme iron, which is situated between the quinone cofactors called QA and QB that are located on the acceptor side of PSII. When electron acceptors are in short supply, the QA quinone becomes reduced, releasing bicarbonate. The loss of bicarbonate makes the redox potential of QA^- more positive, increasing the thermodynamic barrier between QA^- and Pheo, reducing the energy gap between QA^- and QB and slowing electron transfer from QA^- to Pheo, thus regulating the probability of $^1O_2$ production (Brinkert et al., 2016). When electron transfer components become able to re-oxidise, QA^- bicarbonate binds once more, restoring rapid rates of PS II turnover (Brinkert et al., 2016).

$^1O_2$ is a powerful signalling molecule that changes the expression of nuclear genes, leading to programmed cell death or to stress acclimation, depending on the stress intensity and $^1O_2$ production (Ramel et al., 2013; Kim, 2020). Two distinct $^1O_2$-triggered chloroplast-to-nucleus signalling pathways have been characterised. In the first pathway, $^1O_2$ modifies gene expression through the oxidation of β-carotene and the generation of carotenoid breakdown products such as β-cyclocitral (β-CC), β-ionone, or dihydroactinidiolide (dhA), some of which are volatile (Ramel et al., 2012a). β-CC and dhA act as signalling molecules that elicit a genetic response, leading to a marked increase in plant stress tolerance (Ramel et al., 2012b; Shumbe et al., 2014; Fig. 3). A general feature of the transcriptomic response to β-CC is an induction of genes related to cellular defence against stress and a downregulation of genes related to cell growth and development. The β-CC pathway induces various detoxification mechanisms, including glutathione-*S*-transferases (GST) and UDP-glycosyltransferases (Ramel et al., 2012b). These participate in cellular xenobiotic detoxifying processes, which are also activated by the reactive carbonyl species that are generated by the spontaneous decomposition of lipid peroxides. β-CC induces the SCARECROW LIKE 14 (SCL14)-controlled xenobiotic detoxification pathway (D'Alessandro et al., 2019), in which SCL14 and TGAII-type transcription factors modulate the expression of the chloroplast-localised ANAC102, which in turn controls the expression of the ANAC002, ANAC031, and ANAC081 transcription factors that regulate the redox enzymes involved in the first phase of the detoxification response (D'Alessandro et al., 2019). Although crosstalk between the β-CC and PAP-dependent pathways has been proposed (D'Alessandro and Havaux, 2019) the β-CC signalling pathway is independent of other chloroplast-to-nucleus retrograde signalling pathways such as the tetrapyrolle pathway and the EXECUTER (EX)-dependent pathway (Ramel et al., 2013; Shumbe et al., 2016). For example, the OXI1-mediated pathway of programmed cell death in the *ch1* mutant is independent of the EX proteins (Shumbe et al., 2016). Moreover, the β-CC-induced and EX1-FtsH2-dependent pathways share only a small number of genes (Dogra et al., 2017). However, β-CC signalling pathway may have several branches, at least one of which involves the small zinc finger proteins called methylene blue sensitivity (MBS) proteins 1 and 2 (Shumbe et al., 2017).

The EX1 and EX2 proteins play an important role in the transmission of $^1O_2$ signals (Apel and Hirt, 2004; Wang et al., 2016). In contrast to the β-CC pathway, which operates in the grana core (Ramel et al., 2013), EX1 is localised in the non-appressed lamellar regions of the thylakoid membranes. Thus, EX1 transmits signals from $^1O_2$ produced in the margins of the grana stacks (Wang et al., 2016), where the repair of damaged/inactivated PSII reaction centres is localised. EX1 is associated with PSII reaction centre proteins, including D1 and D2, and FtsH2 proteases, together with protein-elongation factors and chlorophyll biosynthesis enzymes (Wang et al., 2016).

9 Conclusions and perspectives

Oxygenic photosynthesis is a flexible process that produces and uses energy though redox reactions. The production and utilisation of oxidants and reductants provides a mechanism for dynamic regulation of ROS signal generation that allows adaptation to a constantly changing environment. Photosynthetic redox chemistry coordinates and synchronises metabolic demand and environmental conditions. Redox molecules are not stable and exist only for short periods of time. They are continuously interacting with each other and antioxidants in the redox interactome of the chloroplasts. Chloroplasts are rich in low moecular weight antioxidants, as well as antioxidant enzymes and redox-active proteins. For example, the chloroplast stoma contains high concentrations of vitamin C (ascorbic acid), which is a general antioxidant that can detoxify $^1O_2$ as well as $O_2^{\bullet-}$ and H_2O_2. It can also regenerate lipid-soluble antioxidants, such as tocopherols and tocotrienols (vitamin E). It functions as a reducing substrate for the conversion of violaxanthin to zeaxanthin in the violaxanthin de-epoxidase (VDE) reaction in the thylakoid lumen. Ascorbate can act as an electron donor to the electron transport chain as well as an electron acceptor (Foyer et al., 2020a) The long-term benefits of the complex redox interactome of chloroplasts and their associated signalling may more than offset the disadvantages associated with the inherent reactivity of ROS that leads to the collateral damage that has often been highlighted in the literature.

The chloroplast TRXs and TRX-like proteins play a key role in this regulation, using the reducing power of the PETC to control ROS production at key electron transport components via ferredoxin/FTR and NADPH/NTRC, as well as regulating the activities of the redox-sensitive stromal enzymes in response to light. The NTRC and FTR systems coordinate stromal metabolism and electron transport activity, as well as fine-tuning ROS generation in response to changing environmental factors such as light intensity. This regulation also involves other redox components such as PRXs that control the reducing activity of chloroplast TRXs and facilitating rapid oxidation of stromal enzymes in the dark. In this discussion, we have largely considered chloroplast ROS production in isolation, only touching on how chloroplast ROS signals interact with other parts of the cellular redox network and influence other cellular compartments. However, ROS have other well established signalling roles, for example, in the reguation of plant defences to biotic and abiotic stresses (Mielecki et al., 2020). Within this context, ROS and the chloroplast redox system play a central role in plantimmunity, regulating the synthesis of phytohormones, such as jasmonic acid and salicylic acid, as well as secondary metabolites and defence compounds.

References

Afanas'ev, I.B., 1989. Superoxide Ion: Chemistry and Biological Implications. CRC Press, Boca Raton, FL.

Allen, J.F., 1975. A two-step mechanism for the photosynthetic reduction of oxygen by ferredoxin. Biochem. Biophys. Res. Commun. 66, 36–43.

Ancin, M., Fernandez-San Millan, A., Larraya, L., Morales, F., Veramendi, J., Aranjuelo, I., Farran, I., 2019. Overexpression of thioredoxin m in tobacco chloroplasts inhibits the protein kinase STN7 and alters photosynthetic performance. J. Exp. Bot. 70, 1005–1016.

Apel, K., Hirt, H., 2004. Reactive oxygen species: metabolism, oxidative stress, and signal transduction. Annu. Rev. Plant Biol. 55, 373–399.

Asada, K., 2000. The water–water cycle as alternative photon and electron sinks. Philos. Trans. R. Soc. Lond. B 355, 1419–1431.

Awad, J., Stotz, H.U., Fekete, A., Krischke, M., Engert, C., Havaux, M., Berger, S., Mueller, M.J., 2015. 2-Cysteine peroxiredoxins and thylakoid ascorbate peroxidase create a water-water cycle that is essential to protect the photosynthetic apparatus under high light stress conditions. Plant Physiol. 167, 1592–1603.

Badawi, G.H., Kawano, N., Yamauchi, Y., Shimada, E., Sasaki, R., Kubo, A., Tanaka, K., 2004. Over-expression of ascorbate peroxidase in tobacco chloroplasts enhances the tolerance to salt stress and water deficit. Physiol. Plant. 121, 231–238.

Balsera, M., Uberegui, Schürmann, P., Buchanan, B.B., 2014. Evolutionary development of redox regulation in chloroplasts. Antioxid. Redox Signal. 21, 1327–1355.

Baniulis, D., Hasan, S.S., Stofleth, J.T., Cramer, W.A., 2013. Mechanism of enhanced superoxide production in the cytochrome b(6)f complex of oxygenic photosynthesis. Biochemistry 52, 8975–8983.

Becker, P.M., 2016. Antireduction: an ancient strategy fit for future. Biosci. Rep. 36, e00367.

Bellafiore, S., Barneche, F., Peltier, G., Rochaix, J.D., 2005. State transitions and light adaptation require chloroplast thylakoid protein kinase STN7. Nature 433, 892–895.

Bergner, S.V., Scholz, M., Trompelt, K., Barth, J., Gabelein, P., Steinbeck, J., Xue, H., Clowez, S., Fucile, G., Goldschmidt-Clermont, M., Fufezan, C., Hippler, M., 2015. STATE TRANSITION7-dependent phosphorylation is modulated by changing environmental conditions, and its absence triggers remodeling of photosynthetic protein complexes. Plant Physiol. 168, 615–634.

Borisova-Mubarakshina, M.M., Naydov, I.A., Ivanov, B.N., 2018. Oxidation of the plastoquinone pool in chloroplast thylakoid membranes by superoxide anion radicals. FEBS Lett. 592, 3221–3228.

Borisova-Mubarakshina, M.M., Vetoshkina, D.V., Ivanov, B.-N., 2019. Antioxidant and signaling functions of the plastoquinone pool in higher plants. Physiol. Plant 166, 181–198.

Brinkert, K., De Causmaecker, S., Krieger-Liszkay, A., Fantuzzi, A., Rutherford, B., 2016. Proc. Natl Acad. Sci. USA 113, 12144–12149.

Brunkard, J.O., Runkel, A.M., Zambryski, P.C., 2015. Chloroplasts extend stromules independently and in response to internal redox signals. Proc. Natl Acad. Sci. USA 112, 10044–10049.

Buchanan, B.B., 2016. The path to thioredoxin and redox regulation in chloroplasts. Annu. Rev. Plant Biol. 67, 1–24.

Caplan, J.L., Kumar, A.S., Park, E., Padmanabhan, M.S., Hoban, K., Modla, S., Czymmek, K., Dinesh-Kumar, S.P., 2015. Chloroplast stromules function during innate immunity. Dev. Cell 34, 45–57.

Cazzaniga, S., Bressan, M., Carbonera, D., Agostini, A., Dall'Osto, L., 2016. Differential roles of carotenes and xanthophylls in photosystem I photoprotection. Biochemistry 55, 3636–3649.

Chang, L., Sun, H., Yang, H., Wang, X., Su, Z., Chen, F., Wei, W., 2016. Over-expression of dehydroascorbate reductase enhances oxidative stress tolerance in tobacco. Electron. J. Biotechnol. 25, 1–8.

Considine, M.J., Foyer, C.H., 2021. Oxygen and reactive oxygen species (ROS)-dependent regulation of plant growth and development. Plant Physiol. 186, 79–92.

Corneille, S., Cournac, L., Guedeney, G., Havaux, M., Peltier, G., 1998. Reduction of the plastoquinone pool by exogenous NADH and NADPH in higher plant chloroplasts. Characterization of a NAD(P)H-plastoquinone oxidoreductase activity. Biochim. Biophys. Acta 1363, 59–69.

D'Alessandro, S., Havaux, M., 2019. Sensing b-carotene oxidation in photosystem II to master plant stress tolerance. New Phytol. 223, 1776–1783.

D'Alessandro, S., Mizokami, Y., Légeret, B., Havaux, M., 2019. The apocarotenoid β-cyclocitric acid elicits drought tolerance in plants. iScience 19, 461–473.

Darie, C.C., De Pascalis, L., Mutschler, B., Haehnel, W., 2006. Studies of the Ndh complex and photosystem II from mesophyll and bundle sheath chloroplasts of the C4-type plant Zea mays. J. Plant Physiol. 163, 800–808.

Diaz-Quintana, A., Leibl, W., Bottin, H., Setif, P., 1998. Electron transfer in photosystem I reaction centers follows a linear pathway in which iron-sulfur cluster FB is the immediate electron donor to soluble ferredoxin. Biochemistry 37, 3429–3439.

Dogra, V., Duan, J., Lee, K.P., Lv, S., Liu, R., Kim, C., 2017. FtsH2-dependent proteolysis of EXECUTER1 is essential in mediating singlet oxygen-triggered retrograde signaling in Arabidopsis thaliana. Front. Plant Sci. 8, 1145.

Driever, S.M., Baker, N.R., 2011. The water-water cycle in leaves is not a major alternative electron sink for dissipation of excess excitation energy when CO_2 assimilation is restricted. Plant Cell Environ. 34, 837–846.

Du, J.-J., Zhan, C.-Y., Lu, Y., Cui, H.-R., Wang, X.-Y., 2015. The conservative cysteines in transmembrane domain of AtVKOR/LTO1 are critical for photosynthetic growth and photosystem II activity in Arabidopsis. Front. Plant Sci. 6, 238.

Exposito-Rodriguez, M., Laissue, P.P., Yvon-Durocher, G., Smirnoff, N., Mullineaux, P.M., 2017. Photosynthesis-dependent H_2O_2 transfer from chloroplasts to nuclei provides a high-light signalling mechanism. Nat. Commun. 8, 49.

Fernie, A.R., Bauwe, H., 2020. Wasteful, essential, evolutionary stepping stone? The multiple personalities of the photorespiratory pathway. Plant J. 102, 666–677.

Fine, P.L., Frasch, W.D., 1992. The oxygen-evolving complex requires chloride to prevent hydrogen peroxide formation. Biochemistry 31, 12204–12210.

Foyer, C.H., Ruban, A.V., Noctor, G., 2017. Viewing oxidative stress through the lens of oxidative signalling rather than damage. Biochem. J. 474, 877–883.

Foyer, C.H., Baker, A., Wright, M., Sparkes, I., Mhamdi, A., Schippers, J.H.M., Van Breusegem, F., 2020a. On the move: redox-dependent protein relocation. J. Exp. Bot. 71, 620–631.

Foyer, C.H., Kyndt, T., Hancock, R.D., 2020b. Vitamin C in plants: novel concepts, new perspectives and outstanding issues. Antioxid. Redox Signal. 32, 463–485.

Fufezan, C., Rutherford, A.W., Krieger-Liszkay, A., 2002. Singlet oxygen production in herbicide-treated photosystem II. FEBS Lett. 532, 407–410.

Geigenberger, P., Thormählen, I., Daloso, D.M., Fernie, A.R., 2017. The unprecedented versatility of the plant thioredoxin system. Trends Plant Sci. 22, 249–262.

Guinea Diaz, M., Nikkenen, L., Himmanen, K., Toivola, J., Rintamäki, E., 2019. Two chloroplast thioredoxin systems differentially modulate photosynthesis in Arabidopsis depending on light intensity and leaf age. Plant J. 104, 718–734.

Hanson, M.R., Hines, K.M., 2018. Stromules: probing formation and function. Plant Physiol. 176, 128–137.

Hasegawa, R., Saito, K., Takaoka, T., Ishikita, H., 2017. pKa of ubiquinone, menaquinone, phylloquinone, plastoquinone, and rhodoquinone in aqueous solution. Photosynth. Res. 133, 297–304.

Heathcote, P., Williams-Smith, D.L., Evans, M.C., 1978. Quantitative electron-paramagnetic-resonance measurements of the electron-transfer components of the photosystem-I reaction centre. The reaction-centre chlorophyll (P700), the primary electron acceptor X and bound iron-Sulphur Centre A. Biochem. J. 170, 373–378.

Henmi, T., Miyao, M., Yamamoto, Y., 2004. Release and reactive-oxygen-mediated damage of the oxygen-evolving complex subunits of PSII during photoinhibition. Plant Cell Physiol. 45, 243–250.

Hernansanz-Agustín, P., Ramos, E., Navarro, E., Parada, E., Sánchez-López, N., Peláez-Aguado, L., Cabrera-García, J.D., Tello, D., Buendia, I., Marina, A., Egea, J., López, M.G., Bogdanova, A., Martínez-Ruiz, A., 2017. Mitochondrial complex I deactivation is related to superoxide production in acute hypoxia. Redox Biol., 1040–1051.

Ishikawa, N., Takabayashi, A., Noguchi, K., Tazoe, Y., Yamamoto, H., von Caemmerer, S., Sato, F., Endo, T., 2016. NDH-mediated cyclic electron flow around photosystem I is crucial for C4 photosynthesis. Plant Cell Physiol. 57, 2020–2028.

Khorobrykh, A., 2019. Hydrogen peroxide and superoxide anion radical photoproduction in PSII preparations at various modifications of the water-oxidizing complex. Plants 8, 329.

Khorobrykh, S., Havurinne, V., Mattila, H., Tyystjärvi, E., 2020. Oxygen and ROS in photosynthesis. Plan. Theory 9, 91.

Kim, C., 2020. ROS-driven oxidative modification: its impact on chloroplasts-nucleus communication. Front. Plant Sci. 10, 1729.

Kozuleva, M.A., Ivanov, B.N., 2016. The mechanisms of oxygen reduction in the terminal reducing segment of the chloroplast photosynthetic electron transport chain. Plant Cell Physiol. 57, 1397–1404.

Kozuleva, M., Goss, T., Twachtmann, M., Rudi, K., Trapka, J., Selinski, J., Ivanov, B., Garapati, P., Steinhoff, H.J., Hase, T., Scheibe, R., Klare, J.P., Hanke, G.T., 2016. Ferredoxin:NADP(H) oxidoreductase abundance and location influences redox poise and stress tolerance. Plant Physiol. 172, 1480–1493.

Kozuleva, M.A., Petrova, A.A., Mamedov, M.D., Semenov, A.Y., Ivanov, B.N., 2014. O_2 reduction by photosystem I involves phylloquinone under steady-state illumination. FEBS Lett. 588, 4364–4368.

Krieger-Liszkay, A., 2005. Singlet oxygen production in photosynthesis. J. Exp. Bot. 56, 337–346.

Laughlin, T.G., Bayne, A.N., Trempe, J.F., Savage, D.F., Davies, K.M., 2019. Structure of the complex I-like molecule NDH of oxygenic photosynthesis. Nature 566, 411–414.

Lee, D.H., Lal, N.K., Lin, Z.-J.D., Ma, S., Liu, J., Castro, B., Toruño, T., Dinesh-Kumar, S.P., Coaker, G., 2020. Regulation of reactive oxygen species during plant immunity through phosphorylation and ubiquitination of RBOHD. Nat. Commun. 11, 1838.

Leister, D., 2019. Piecing the puzzle together: the central role of reactive oxygen species and redox hubs in chloroplast retrograde signaling. Antioxid. Redox Signal. 30, 1206–1219.

Malnoë, A., Wang, F., Girard-Bascou, J., Wollman, F.A., de Vitry, C., 2014. Thylakoid FtsH protease contributes to photosystem II and cytochrome b6f remodeling in *Chlamydomonas reinhardtii* under stress conditions. Plant Cell 26, 373–390.

Mazor, Y., Borovikova, A., Caspy, I., Nelson, N., 2017. Structure of the plant photosystem I supercomplex at 2.6 A resolution. Nat. Plants 3, 17014.

Mehler, A.H., 1951. Studies on reactions of illuminated chloroplasts. I. Mechanism of the reduction of oxygen and other Hill reagents. Arch. Biochem. Biophys. 33, 65–77.

Mielecki, J., Gawronski, P., Karpinski, S., 2020. Retrograde signaling: understanding the communication between organelles. Int. J. Mol. Sci. 21, 6173.

Mullineaux, P.M., Exposito-Rodriguez, M., Laissue, P.P., Smirnoff, N., Park, E., 2020. Spatial chloroplast-to-nucleus signalling involving plastid–nuclear complexes and stromules. Philos. Trans. R. Soc. B 375, 20190405.

Munekage, Y.N., Eymery, F., Rumeau, D., Cuine, S., Oguri, M., Nakamura, N., Yokota, A., Genty, B., Peltier, G., 2010. Elevated expression of PGR5 and NDH-H in bundle sheath chloroplasts in C4 flaveria species. Plant Cell Physiol. 51, 664–668.

Nikkanen, L., Rintamäki, E., 2019. Chloroplast thioredoxin systems dynamically regulate photosynthesis in plants. Biochem. J. 476, 1159–1172.

Nikkanen, L., Toivola, J., Rintamäki, E., 2016. Crosstalk between chloroplast thioredoxin systems in regulation of photosynthesis. Plant Cell Environ. 39, 1691–1705.

Nikkanen, L., Toivola, J., Trotta, A., Diaz, M.G., Tikkanen, M., Aro, E.V., Rintamäki, E., 2018. Regulation of cyclic electron flow by chloroplast NADPH-dependent thioredoxin system. Plant Direct 2, 1–24.

Noctor, G., Foyer, C.H., 1998. Ascorbate and glutathione: keeping active oxygen under control. Annu. Rev. Plant Physiol. Plant Mol. Biol. 49, 249–279.

Noctor, G., Foyer, C.H., 2016. Intracellular redox compartmentation and ROS-related communication in regulation and signaling. Plant Physiol. 171, 1581–1592.

Onda, Y., 2013. Oxidative protein-folding systems in plant cells. Int. J. Cell Biol. 2013, 585431.

Pospisil, P., 2016. Production of reactive oxygen species by photosystem II as a response to light and temperature stress. Front. Plant Sci. 7, 1950.

Pospisil, P., Arato, A., Krieger-Liszkay, A., Rutherford, A.W., 2004. Hydroxyl radical generation by photosystem II. Biochemistry 43, 6783–6792.

Pryde, K.R., Hirst, J., 2011. Superoxide is produced by the reduced flavin in mitochondrial complex I: a single, unified mechanism that applies during both forward and reverse electron transfer. J. Biol. Chem. 286, 18056–18065.

Ramel, F., Birtic, S., Cuine, S., Triantaphylides, C., Ravanat, J.-L., Havaux, M., 2012a. Chemical quenching of singlet oxygen by carotenoids in plants. Plant Physiol. 158, 1267–1278.

Ramel, F., Birtic, S., Ginies, C., Soubigou-Taconnat, L., Triantaphylides, C., Havaux, M., 2012b. Carotenoid oxidation products are stress signals that mediate gene responses to singlet oxygen in plants. Proc. Natl. Acad. Sci. USA 109, 5535–5540.

Ramel, F., Ksas, B., Akkari, E., Mialoundama, A.S., Monnet, F., Krieger-Liszkay, A., et al., 2013. Light induced acclimation of the Arabidopsis chlorina1 mutant to singlet oxygen. Plant Cell 25, 1445–1715.

Rappaport, F., Guergova-Kuras, M., Nixon, P.J., Diner, B.A., Lavergne, J., 2002. Kinetics and pathways of charge recombination in photosystem II. Biochemistry 41, 8518–8527.

Richter, A., Pérez-Ruiz, J.M., Cejudo, F.J., Grimm, B., 2018. Redox-control of chlorophyll biosynthesis mainly depends on thioredoxins. FEBS Lett. 292, 3111–3115.

Rochaix, J.-D., 2014. Regulation and dynamics of the light-harvesting system. Annu. Rev. Plant Biol. 65, 287–309.

Ruban, A.V., 2016. Nonphotochemical chlorophyll fluorescence quenching: mechanism and effectiveness in protecting plants from photodamage. Plant Physiol. 170, 1903–1916.

Rutherford, A.W., 1981. EPR evidence for an acceptor functioning in photosystem II when the pheophytin acceptor is reduced. Biochem. Biophys. Res. Commun. 102, 1065–1070.

Rutherford, A.W., Osyczka, A., Rappaport, F., 2012. Back-reactions, short-circuits, leaks and other energy wasteful reactions in biological electron transfer: redox tuning to survive life in O(2). FEBS Lett. 586, 603–616.

Santolini, J., Wootton, S.A., Jackson, A.A., Feelish, M., 2019. The redox architecture of physiological function. Curr. Opin. Physio. 9, 34–47.

Sazanov, L.A., Burrows, P.A., Nixon, P.J., 1998. The plastid ndh genes code for an NADH-specific dehydrogenase: isolation of a complex I analogue from pea thylakoid membranes. Proc. Natl. Acad. Sci. USA 95, 1319–1324.

Schröter, Y., Steiner, S., Matthai, K., Pfannschmidt, T., 2010. Analysis of oligomeric protein complexes in the chloroplast sub-proteome of nucleic acid-binding proteins from mustard reveals potential redox regulators of plastid gene expression. Proteomics 10, 2191–2204.

Schuller, J.M., Birrell, J.A., Tanaka, H., Konuma, T., Wulfhorst, H., Cox, N., Schuller, S.K., Thiemann, J., Lubitz, W., Sétif, P., et al., 2019. Structural adaptations of photosynthetic complex I enable ferredoxin-dependent electron transfer. Science 363, 257–260.

Setif, P., Hervoa, G., Mathisa, P., 1981. Flash-induced absorption changes in photosystem I, radical pair or triplet state formation? Biochim. Biophys. Acta Bioenerg. 638, 257–267.

Shapiguzov, A., Chai, X., Fucile, G., Longoni, P., Zhang, L., Rochaix, J.D., 2016. Activation of the Stt7/STN7 kinase through dynamic interactions with the cytochrome b6f complex. Plant Physiol. 171, 82–92.

Shikanai, T., Yamamoto, H., 2017. Contribution of cyclic and pseudo-cyclic electron transport to the formation of proton motive force in chloroplasts. Mol. Plant 10, 20–29.

Shumbe, L., Bott, R., Havaux, M., 2014. Dihydroactinidiolide, a high light-induced beta-carotene derivative that can regulate gene expression and photoacclimation in Arabidopsis. Mol. Plant 7, 1248–1251.

Shumbe, L., Chevalier, A., Legeret, B., Taconnat, L., Monnet, F., Havaux, M., 2016. Singlet oxygen-induced cell death in Arabidopsis under high-light stress is controlled by OXI1 kinase. Plant Physiol. 170, 1757–1771.

Shumbe, L., D'Alessandro, S., Shao, N., Chevalier, A., Ksas, B., Bock, R., Havaux, M., 2017. METHYLENE BLUE SENSITIVITY 1 (MBS1) is required for acclimation of Arabidopsis to singlet oxygen and acts downstream of beta-cyclocitral. Plant Cell Environ. 40, 216–226.

Šnyrychová, I., Pospíšil, P., Nauš, J., 2006. Reaction pathways involved in the production of hydroxyl radicals in thylakoid membrane: EPR spin-trapping study. Photochem. Photobiol. Sci. 5, 472–476.

Stenbaek, A., Hansson, A., Wulff, R.P., Hansson, M., Dietz, K.J., Jensen, P.E., 2008. NADPH-dependent thioredoxin reductase and 2-Cys peroxiredoxins are needed for the protection of Mg-protoporphyrin monomethyl ester cyclase. FEBS Lett. 582, 2773–2778.

Sun, Y., Zerges, W., 2015. Translational regulation in chloroplasts for development and homeostasis. Biochim. Biophys. Acta 1847, 809–882.

Sun, W.H., Duan, M., Shu, D.F., Yang, S., Meng, Q.W., 2010a. Over-expression of StAPX in tobacco improves seed germination and increases early seedling tolerance to salinity and osmotic stresses. Plant Cell Rep. 29, 917–926.

Sun, W.H., Duan, M., Li, F., Shu, D.F., Yang, S., Meng, Q.W., 2010b. Overexpression of tomato tAPX gene in tobacco improves tolerance to high or low temperature stress. Biol. Plant. 54, 614–620.

Tang, L., Kwon, S.Y., Kim, S.H., Kim, J.S., Choi, J.S., Cho, K.Y., Sung, C.K., Kwak, S.S., Lee, H.S., 2006. Enhanced tolerance of transgenic potato plants expressing both superoxide dismutase and ascorbate peroxidase in chloroplasts against oxidative stress and high temperature. Plant Cell Rep. 25, 1380–1386.

Taylor, R.M., Sallans, L., Frankel, L.K., Bricker, T.M., 2018. Natively oxidized amino acid residues in the spinach cytochrome b6f complex. Photosynth. Res. 137, 141–151.

Vaseghi, M.-J., Chibani, K., Telman, W., Liebthal, M.F., Gerken, M., Schnitzer, H., Mueller, S.M., Dietz, K.J.,

2018. The chloroplast 2-cysteine peroxiredoxin functions as thioredoxin oxidase in redox regulation of chloroplast metabolism. elife 7, e38194.

Vetoshkina, D.V., Ivanov, B.N., Khorobrykh, S.A., Proskuryakov, I.I., Borisova-Mubarakshina, M.M., 2017. Involvement of the chloroplast plastoquinone pool in the Mehler reaction. Physiol. Plant 161, 45–55.

Wang, L., Kim, C., Xu, X., Piskurewicz, U., Dogra, V., Singh, S., Mahler, H., Apel, K., 2016. Singlet oxygen- and EXECUTER1-mediated signaling is initiated in grana margins and depends on the protease FtsH2. Proc. Natl. Acad. Sci. USA 113, 3792–3800.

Wang, B., Li, Z., Ran, Q., Li, P., Peng, Z., Zhang, J., 2018. ZmNF-YB16 overexpression improves drought resistance and yield by enhancing photosynthesis and the antioxidant capacity of maize plants. Front. Plant Sci. 9, 709.

Wang, W.R., Liang, J.H., Wang, G.F., Sun, M.X., Peng, F.T., Xiao, Y.S., 2020. Overexpression of PpSnRK1alpha in tomato enhanced salt tolerance by regulating ABA signaling pathway and reactive oxygen metabolism. BMC Plant Biol. 20, 128.

Wardman, J., 1989. Reduction potentials of one-electron couples involving free radicals in aqueous solution. J. Phys. Chem. Ref. Data Monogr. 18, 1637.

Wu, J., Rong, L., Lin, W., Kong, L., Wei, D., Zhang, L., Rochaix, J.-D., Xu, X., 2021. Functional redox link between LTO1 and STN7 in arabidopsis. Plant Physiol. 186, 964–976.

Yadav, D.K., Pospisil, P., 2012. Evidence on the formation of singlet oxygen in the donor side photoinhibition of photosystem II: EPR spin-trapping study. PLoS One 7, e45883.

Zaffagnini, M., Fermani, S., Marchand, C.H., Costa, A., Sparla, F., Rouhier, N., Geigenberger, P., Lemaire, S.D., Trost, P., 2019. Redox homeostasis in photosynthetic organisms: novel and established thiol-based molecular mechanisms. Antioxid. Redox Signal. 31, 155–210.

Zhang, Y., Zhang, A., Li, X., Lu, C., 2020. The role of chloroplast gene expression in plant responses to environmental stress. Int. J. Mol. Sci. 21, 6082.

CHAPTER

6

Abiotic stress, acclimation, and adaptation in carbon fixation processes

Erik H. Murchie, Lorna McAusland, and Alexandra J. Burgess

School of Biosciences, University of Nottingham, Sutton Bonington, Loughborough, United Kingdom

1 Introduction

The physical environment frequently places constraints on photosynthesis when plants move beyond optimal conditions of light, temperature, water, and humidity. Soil conditions providing low or excessive nutrients or the presence of salt exacerbates this. Understanding the sensitivity of photosynthesis to the environment is not always straightforward because it is a multicomponent process including physics (gas diffusion, light absorption, energy transfer) and metabolism (electron transport, Calvin–Benson cycle (CBC), sucrose synthesis). All components are sensitive to variation in the physical environment. Moreover, photosynthesis is embedded into the plant form, supported by whole plant mechanisms of water transport and biomechanics of growth and responding to internal signals sensed locally and from other plant parts (and even other plants). It follows that in the natural environment, any response to a physical limitation will involve both intrinsic and extrinsic factors operating together. For example, we may consider that high temperatures will directly limit activity of the CBC via oxygenation, allosteric regulation, or inhibition of Rubisco, but it is also likely to be combined with stomatal closure and down regulation of photosystem II (PSII), further limiting CO_2 uptake and assimilation. These processes must be disentangled in any study of plant stress.

In this chapter, we identify the sites of limitation to CO_2 fixation as gas uptake and availability within the leaf, largely via stomata, activity and photorespiratory flux. The basics of how these processes function is covered in the previous chapters. While Chapter 4 laid out the molecular fundamentals of the response of light harvesting and photoprotection to abiotic stress, Chapter 5 covered the role of electron transport and oxidative stress. This chapter extends this discussion by focusing on the physical factors determining limitations to the uptake and

Photosynthesis in Action
https://doi.org/10.1016/B978-0-12-823781-6.00011-3

assimilation of CO_2 in chloroplasts, leaves, cells, and canopies in suboptimal situations. We identify key abiotic factors, the sites of limitation, and adaptations that have occurred within nature to optimise CO_2 fixation.

1.1 What is stress? The physical environment and optimality

A broad framework to define 'stress' could be as follows. Generally, a well-adapted plant within an environment (temperature, light, humidity, soil moisture, nutrients) that promotes high growth can be said to be within an optimal range of conditions. This optimum is often described as a bell curve and can vary substantially between and within species (Fig. 1). Optimum environments promote the efficient conversion of solar energy to biomass through photosynthesis. Furthermore, if we think of plants as energy storage and transduction systems then the concept of homeostasis is useful, i.-e., equilibrium is maintained between energy absorbed vs growth achieved. A disruption to homeostatic stability due to *a change* in environmental conditions pushes the system away from optimum. In such circumstances, absorbed light may not be converted to growth with such high efficiency (Fig. 1). If this shift in environment is minor, the homeostatic disruption becomes temporary. Then, through sensing/perceiving the change in conditions the plant may acclimate, producing a change in the composition or structure of the plant usually via gene expression and protein synthesis. It shifts the entire curve, moving the optimum value but may or may not reduce photosynthetic rate depending on the efficiency of acclimation or the extent of long-term damage to tissue. Acclimation to higher temperatures consists of processes such as a decline in respiration (after an initial rise) and synthesis of heat stable enzymes such as Rubisco activase (Sage and Kubien, 2007). Acclimation to lower temperatures typically consists of increased electron transport, sugar and starch synthesis, lipid desaturation, and enhanced Rubisco content. A substantial environmental change may divert the system far from the optimal range of growth or photosynthesis, beyond the range at which acclimation can re-achieve homeostasis. Both low and high temperatures inhibit enzyme activity, lowering photosynthetic capacity and risking membrane disruption. Heat waves (short periods of extreme heat) result in enzyme inactivation and pushing CBC enzymes beyond the optimum. This may involve irreversible damage. Combined high light and high temperature causes further photoinhibition and oxidative stress (Murchie and Niyogi, 2011). In each case, tissue may be stabilised by the synthesis of chaperone proteins, photoprotective proteins, and antioxidant enzymes. Under these high temperature conditions, light absorption continues and creates reduced PSII, which in nonacclimated plants can be thought of as a potential stress and loss of homeostasis (Öquist and Huner, 2003).

It follows that it may not be immediately clear whether a condition is merely suboptimal and results in rapidly reversible effects or whether there is substantial deviation from optimality, which may involve damage to cellular constituents which can be described as 'stress'. To discern this, various measurements are often employed to characterise plant tissue in each case, these can range from plant growth, measurements of CO_2 uptake, chlorophyll fluorescence, reflectance indices, gene expression, and protein content.

In this chapter, we consider acclimation but also substantial movement away from optimality that limits photosynthesis. Even though these processes strictly encompass acclimation too, we refer to the latter as abiotic stress. We define stress as climactic conditions that are suboptimal for maintaining cellular homeostasis, which

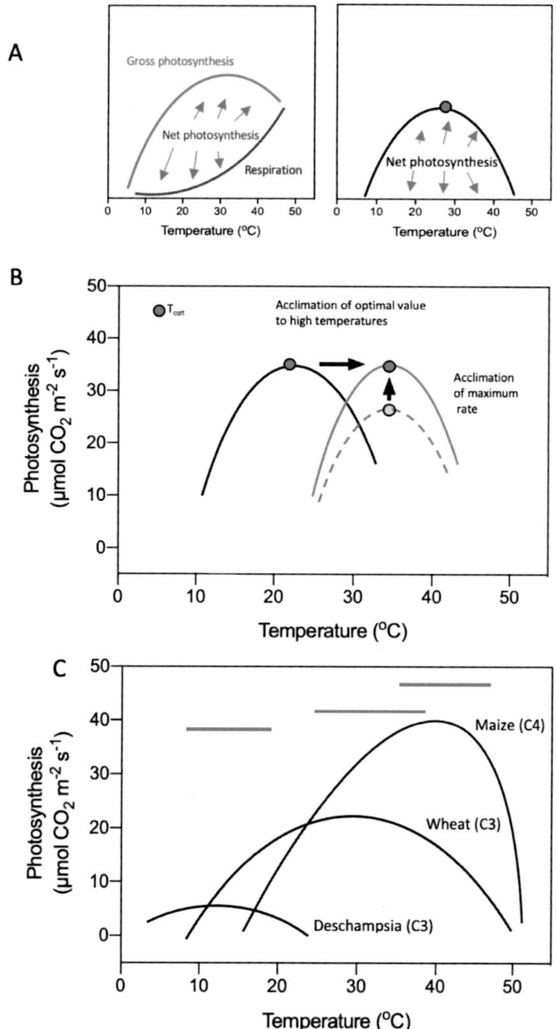

FIG. 1 Schematic figure showing (A) the concept of photosynthesis dependence on temperature and how this results from the difference between the responses of net photosynthetic CO_2 assimilation and mitochondrial respiratory CO_2 release. (B) The concept of acclimation of temperature optima following transfer to a higher temperature. The circles shift in terms of optimum but may show a greater decline in photosynthetic capacity light saturated photosynthesis (A_{max}) if the genotype is maladapted to the new environment. (C) Temperature optima for a cold tolerant grass Antarctic vascular species (*Deschampsia antarctica*), wheat (*Triticum aestivum*), and maize (*Zea mays*). *Blue bars* show the range of optima for each species. *Data used are from Xiong, F.S., Mueller, E.C., Day, T.A., 2000. Photosynthetic and respiratory acclimation and growth response of Antarctic vascular plants to contrasting temperature regimes. Am. J. Bot. 87, 700–710; Murchie, E., Mooney, S., Khalil, A. (U. Nottingham, unpublished); Yin, X., Struik, P.C., 2009.* C_3 *and* C_4 *photosynthesis models: an overview from the perspective of crop modelling. NJAS—Wagen. J. Life Sci. 57, 27–38.*

eventually impairs plant growth and fitness. Abiotic stress can then be divided into transient (short-term) or chronic (long-term) episodes (Mickelbart et al., 2015). The intensity and timing (e.g., growth stage or diurnal cycle) of these episodes determines the severity of the negative impact on plant growth and productivity or yield. Here, we adopt a process-based approach, discussing the major sites of limitation in plants and how different abiotic factors can limit their operation.

1.2 The interplay between environment, photoprotection, and electron transport determine CO_2 assimilation

The impact of environment on the CBC must also consider the processes of light harvesting, electron transport, carbohydrate synthesis, and even whole plant carbon dynamics. We refer the reader to other chapters in this book: the concept of integration of all components of photosynthesis is specifically covered in Chapter 10. The CBC can be directly affected via physical changes felt by the leaf tissue (e.g., leaf temperature) or via the interactions with other affected plant processes. Indirect regulation occurs in several ways. The electron transport chain regulates the activation state of CBC enzymes, e.g., via thioredoxin and the ATP/NADPH ratio by cyclic electron transport. Closure of stomata in response to drought or low humidity results in lowered CO_2 diffusion and a temporary reduction in internal leaf CO_2 (C_i). Photorespiratory flux and alternative electron sinks can be of greater significance during periods of low leaf conductance.

Fig. 2 shows a tool that is commonly employed to help understand the site of limitations to photosynthesis in the environment: the light response curve. Typically, in an infra-red gas analyser, leaf CO_2 assimilation and stomatal conductance are measured while light applied is stepped up (or down) between zero and saturating, moving through light-limited phases towards light saturation. This allows comparison with environmental parameters to predict rates empirically. It follows that light frequently exceeds the requirements of photosynthesis (Murchie and Niyogi, 2011), and so high light can exacerbate any factor that limits the light saturated rate of photosynthesis. Another method, described in Chapters 10 and 11, is the CO_2 response curve or A-C_i curve where CO_2 is varied instead of light. Using biochemical models, this enables the separation of phases to calculate the maximum rates of carboxylation and electron transport (V_{Cmax} and J_{max}). Both methods are important when attempting to understand limitations to CO_2 assimilation (Johnson and Murchie, 2011).

The fluctuating nature of the environment is of importance: this creates temporary imbalances in homeostasis between light harvesting, electron transport and CO_2 assimilation. The photosynthetic response to fluctuating conditions are still less well characterised than traditional steady-state measurements but are predicted to have large effects on integrated rates of carbon assimilation (Kaiser et al., 2018). A good example is that of light, the timescales for which can happen over seconds, minutes, or hours and be of an unpredictable or stochastic nature. A sudden increase in light can result in the electron transport chain temporarily becoming highly reduced if the enzymes of the CBC are not activated and/or stomatal conductance is low. In such conditions, the flexibility of electron sinks within the chloroplast (e.g. cyclic, Mehler reaction, photorespiration) and light harvesting flexibility become important in order to avoid the possibility of the over generation of reactive oxygen species (ROS). The ability of photosynthesis to rapidly activate enzymes of the CBC and open stomata is

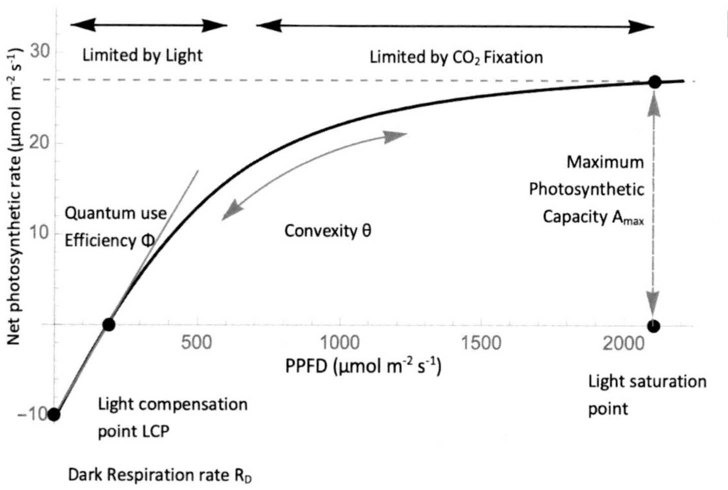

FIG. 2 Example light response curve of net photosynthetic CO_2 assimilation as denoted by the nonrectangular hyperbola indicating the shaping parameters. PPFD = photosynthetic photon flux density.

predicted to be limiting to photosynthesis in the field, as shown recently for rice and wheat (Acevedo-Siaca et al., 2020; Taylor and Long, 2017). Additionally, the slow response of stomata can either reduce instantaneous water use efficiency when conductance stays high during a high to low light transition or improve it when stomata are slow to open during a low to high light transition (Lawson and Blatt, 2014).

Photoprotective mechanisms such as nonphotochemical quenching (NPQ) help to prevent PSII inactivation and also act as a means of homeostasis, buffering PSII, and electron transport against overreduction (Chapter 4). However, such quenching momentarily reduces the quantum yield of photosynthesis, which is not a problem under high light but will limit photosynthesis when the leaf returns to low light, in a similar way that long term photoinactivation achieves. Therefore, in a fluctuating environment, the slow recovery of NPQ limits photosynthesis when integrated in a canopy in the field. This idea was impressively validated using tobacco engineered to have a rapid recovery from NPQ. Plants showed higher CO_2 assimilation during low light recovery and in the field demonstrated higher biomass and yield (Kromdijk et al., 2016).

2 The CBC under stress

Limitations in the catalytic properties of Rubisco can compromise the efficiency of photosynthesis (Parry et al., 2007). Relative to other enzymes involved in the CBC, Rubisco has a low turnover rate and is a major rate limiting step. Furthermore, due to the competing reaction with oxygen in the process of photorespiration, energy and fixed carbon can be lost from the process. Other key limiting steps in the CBC are SBPase (see Chapter 7). Here, we consider Rubisco adaptations to ensure optimal performance under varying environmental conditions.

2.1 Photosynthetic acclimation to growth temperature

As discussed earlier, the photosynthetic machinery requires optimisation towards current conditions, known as temperature acclimation. It is generally seen that the temperature which maximises photosynthetic rate (T_{opt}: Fig. 1) will increase with an increase in growth temperature to a certain limit (Berry and Bjorkman, 1980; Hikosaka et al., 2006).

Temperature acclimation requires changes to the photosynthetic rate (Fig. 1). Accordingly, the highest photosynthetic rate is generally found in the leaves that are measured under the same temperature to which they were grown (Way and Yamori, 2013), suggesting that photosynthesis adjusts to maximally exploit growth temperature. Temperature acclimation generally includes changes to both the photosynthetic thermal optimum (T_{opt}), photosynthetic rates at the growth temperature (A_{growth}), and mitochondrial respiration often by altering enzyme capacities (O'Leary et al., 2019). However, in some species such as *Plantago asiatica*, the photosynthetic rate is constant in leaves growth at 15° C or 30°C (19.0 and 19.4 $\mu mol\,m^{-2}\,s^{-1}$, respectively), confirming that temperature acclimation may represent a homeostatic response to maintain photosynthetic rate (Hikosaka, 2005).

In most plants, the light—saturated photosynthetic capacity, A_{max} will be low under extreme hot and cold conditions, and highest under intermediate conditions (Fig. 1). Changes in the temperature dependence of photosynthesis may, in part, be due to changes in the specific activity and/or total activity of photosynthetic components, or to changes in CO_2 concentration in the carboxylation site. However, the response of each factor to temperature seems to differ among species suggesting both intra- and interspecific differences in temperature acclimation between species (Berry and Bjorkman, 1980). C_3 plants generally exhibit a greater ability to adapt to a broad range of growth temperatures, whilst C_4 plants are generally restricted to warmer climates (see later). In contrast, CAM species acclimate differentially to day- and night-time temperatures (Yamori et al., 2014).

Based on the biochemical model of photosynthesis (Farquhar et al., 1980; von Caemmerer, 2000), changes in the photosynthesis–temperature curve can be attributed to four factors:

(1) intercellular CO_2 concentration (C_i);
(2) activation energy of the maximum rate of RuBP (ribulose-1,5-bisphosphate) carboxylation (V_{cmax});
(3) activation energy of the rate of RuBP regeneration (J_{max}); and
(4) the ratio of J_{max} to V_{cmax}.

Among these factors, the predominant response is through changes in V_{cmax}, with increased activation energy associated with increasing temperature (Hikosaka et al., 2006). Mesophyll conductance may also be significant and is discussed below.

Since the RuBP regeneration phase of the CBC limits photosynthesis at higher CO_2 concen trations, the optimal temperature is also high under high [CO_2] (see Chapter 11). Under lower CO_2 concentrations, the carboxylation rate is less sensitive to temperature because an increase in the Michaelis–Menten constant of Rubisco for CO_2 (K_c) partly offsets the increase in V_{cmax}. The actual relationship between temperature and photosynthetic response will also depend on temperature dependence of leaf internal [CO_2] - C_i. The optimal temperature of photosynthesis will remain low if C_i is low, regardless of an increase leaf temperature. It has also been shown that stomatal conductance and therefore C_i is more sensitive to the vapour pressure deficit (VPD) as opposed to temperature, which has been proposed to be regulated so as to maintain the ratio of C_i to C_a (the CO_2 concentration of the air) (Leuning, 1995). There is considerable variation within and between species in the effect of growth temperature on C_i (Hikosaka et al., 1999, 2006; Hendrickson et al., 2004).

In general, V_{cmax} exponentially increases from 15°C to 30°C although deactivation can occur rapidly at high temperatures. This pattern of deactivation is also seen for J_{max} possibly as a result of changes in the heat tolerance of components in RuBP regeneration (Leuning, 2002). The deactivation of V_{cmax} can be related to a decrease in the activation state of Rubisco. The activation state of Rubisco (see previous chapters) may have significant effect on temperature dependence of photosynthesis in some species with evidence indicating that temperatures both below and above the optimum can both influence activation effectiveness (Salvucci and Crafts-Brandner, 2004). Temperature dependence may also, in part, be due to the form of Rubisco present (Hikosaka et al., 2006). Whilst higher plants contain one gene encoding the large subunit, the small subunit is encoded by a multigene family containing between 2 and 12 members (Gutteridge and Gatenby, 1995); thus, different small subunits may produce Rubisco with contrasting properties (Galmés et al., 2005) including those related to temperature dependence and acclimation and this concept of 'mixing and matching' Rubisco properties according to environment has made recent progress (Sharwood, 2020; Lin et al., 2020).

Temperature dependence of photosynthesis will depend upon the rate limiting step, and thus, the balance between V_{cmax} and J_{max} is an important factor. When plants have a higher J_{max} to V_{cmax} ratio, the optimal temperature required for photosynthesis will decrease, and vice versa (Hikosaka et al., 2006). Whilst many species balance carboxylation and RuBP regeneration across different conditions (i.e., the ratio of J_{max} to V_{cmax} is constant), other species exhibit a change in the ratio dependent on growth temperature (Hikosaka et al., 2006). Under ambient CO_2 conditions and saturating light, changes in temperature dependence of photosynthesis can mostly be explained by V_{cmax} as photosynthesis is generally limited by RuBP carboxylation. However, under elevated CO_2, photosynthesis becomes limited by RuBP regeneration and so temperature dependence of J_{max} and the ratio of J_{max} to V_{cmax} will become more important.

2.2 The CBC at temperature extremes

Reactive oxygen species (ROS) increase in response to heat and cold stress in the light, risking oxidative damage to proteins found in the stroma and thylakoid, as well as inducing remodeling of thylakoid lipids. Stress leads to oxidative modification of specific residues on Rubisco, which mark the enzyme for degradation (Parry et al., 2008). Oxidative stress may also influence Rubisco abundance through reduced levels of the transcript for the small subunit, postulated to be a result of increased ethylene levels (Glick et al., 1995).

Whilst C_4 plants are generally (potentially) more efficient than C_3 in terms of light, water, and nitrogen use, C_4 plants are generally more susceptible to chilling than C_3 (Long, 1983). This has proposed to be due to the decrease in photorespiration in C_3 plants with decreasing temperature, thus reducing the net energy required to fix CO_2 under cooler conditions in addition to cold—lability of specific proteins in the C_4 pathway and general chilling sensitivity (Osborne et al., 2007). In addition, C_4-specific proteins and enzymes generally have a greater chilling sensitivity to those found in C_3; a 60%–80% reduction in Rubisco quantity on a total protein basis in C_4 plants means that under conditions where turnover is reduced such as cold temperatures, Rubisco catalysis is unable to match the CO_2 levels or RuBP regeneration (Sage, 2002). Whilst some C_4 species, such as *Miscanthus*, are relatively tolerant of cold temperatures, others such as Maize (*Zea mays*) exhibit a decrease in assimilation rate of over 60% in response to chilling (Naidu and Long, 2004; Wang et al., 2008; Long and Spence, 2013). This is due to a combination of chilling-dependent photoinhibition and impaired synthesis of PSII and light-harvesting complex proteins, resulting in nonfunctional PSII reaction centres (Long et al., 1994). Additionally, enzyme activities are reduced under chilling temperatures, photosynthetic genes are downregulated, and protein stability and solubility are altered (Ruelland et al., 2009). Despite this, there is evidence from C_4 plants such as *Miscanthus,* which have improved tolerance to cold that there is no fundamental reason why the C_4 pathway prevents acclimation to cool environments and that the advantages of C_4 photosynthesis such as improved light, nitrogen, and water use efficiency can be realised under cooler temperatures (Long and Spence, 2013).

Carbon dioxide is not the only substrate fixed by Rubisco, O_2 is also able to bind to Rubisco in a process called photorespiration. A process that reduces C_3 plant photosynthetic efficiency and water use efficiency (Chapter 3). During photorespiration, an oxygenase reaction occurs between RuBP and O_2 leading to the production of a three-carbon compound, 3-phosphoglyceric acid (3-PGA), and two molecules of the two-carbon compound, phosphoglycolate. Whilst 3-PGA can enter the CBC, phosphoglycolate cannot, effectively removing four carbons. Three of these four carbons can be recovered in a salvage pathway, but the overall effect is the net loss of fixed carbon. A decline in the CO_2/O_2 solubility with rising temperature accounts for about a third of the rise in photorespiration, while the remainder is caused by the reduction in relative specificity. The latter is largely driven by a greater increase in the Rubisco Michaelis–Menton constant for CO_2 relative to that of O_2 (von Caemmerer and Quick, 2000). At mild temperatures, Rubisco's affinity for CO_2 is approximately 80 times higher than its affinity for O_2 (Vu, 2005). However, under high temperatures, the affinity for O_2 increases. This leads to a rise in photorespiration and mitochondrial respiration, increasing the CO_2 compensation point (Brooks and Farquhar, 1985; Sharkey, 1988; Sage et al., 1990), a process which is overcome by the carbon concentrating mechanism found in C_4 species. Both temperature and the CO_2:O_2 ratio can influence the probability that Rubisco will use O_2 as a substrate (Fig. 3). Any process which limits the opening of stomata and reduces CO_2 diffusion into the leaf will therefore also increase photorespiration. Whilst the process generally leads a loss of fixed carbon, evidence suggests that photorespiration may have a photoprotective role and help maintain redox balance in cells (Eisenhut et al., 2017).

Above the T_{opt}, the rate of electron transport declines in parallel with CO_2-saturated assimilation rate in a number of species (Yamasaki et al., 2002; Wise et al., 2004; Cen and Sage, 2005). In sweet potato, increasing CO_2 concentrations above ambient levels increased the temperature sensitivity of photosynthesis due to a transition from Rubisco limitation to electron transport limitation (Cen and Sage, 2005). This reduced the breadth of the thermal response curve whilst increasing the thermal optimum. Whilst a reduction in electron transport may be responsible for reduced assimilation under high temperatures in some species, in others it can be attributed to the sensitivity of the protein Rubisco activase (Salvucci and Crafts-Brandner, 2004).

The activation state of Rubisco changes in response to temperature, light, CO_2 concen tration, and other environmental factors (Parry et al., 2013). Rubisco activase is essential for the activation and maintenance of Rubisco catalytic activity by promoting the removal of any tightly bound, inhibitory sugar phosphates from the catalytic site of both the carbamylated and decarbamylated forms of Rubisco (Parry et al., 2013). Whilst Rubisco itself is heat stable, Rubisco activase is highly sensitive to temperature (Robinson and Portis, 1989; Eckardt and Portis, 1997). As a result, Rubisco deactivates under high temperatures to a point where CO_2 assimilation is limited by its ability to consume RuBP. When compared separately, the small isoform of Rubisco activase is considerably more

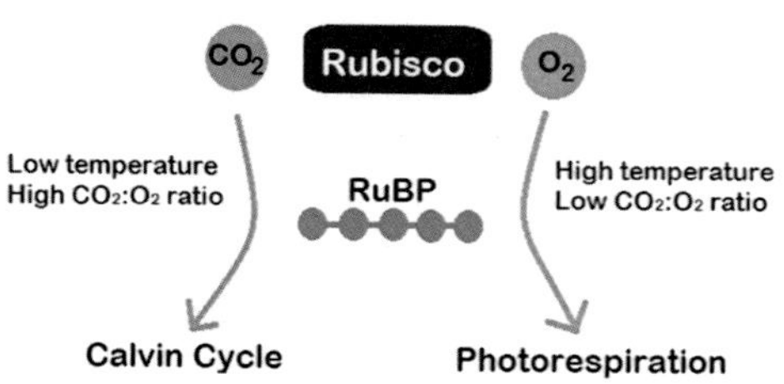

FIG. 3 Depending on the temperature and CO_2:O_2 ratio, Rubisco will either use CO_2 as a substrate in the first step of the Calvin cycle or use O_2 in the process of photorespiration.

heat labile compared to that of the large isoform, whilst mixtures containing both resemble heat sensitivity of the larger isoform (Crafts-Brandner et al., 1997). This indicates that the larger isoform can confer thermal stability of the smaller isoform. The close correlation between CO_2 assimilation and Rubisco activity between 28°C and 45°C in cotton and wheat leaves supports the role of Rubisco activase in temperature inhibition of photosynthesis and presents a potential route to improving net photosynthesis under elevated temperatures (Law and Crafts-Brandner, 1999). There is now plenty of evidence that Rubisco activase is an important component of the maintenance of CO_2 assimilation at high temperatures. It is established that heat stable activase isoforms improves growth and photosynthesis under high temperatures and this can be enhanced by over expressing both Rubisco and Rubisco activase (Perdomo et al., 2017; Qu et al., 2021). Heat also induces the differential expression of isoforms, which differ in heat sensitivity (Degen et al., 2021).

At thermal optimum, sedoheptulose 1,7-bisphosphate (SBPase) exerts control over the CBC flux. A 35% reduction in SBPase can lead to a significant decrease in photosynthesis under ambient CO_2 levels and a larger decrease under elevated CO_2 (Ölçer et al., 2001). Overexpression of SBPase seems to lead to greater thermal tolerance of the CBC (Feng et al., 2007). Similarly, transketolase and aldolase also exhibit significant control over assimilation rate (Raines, 2003). At chilling temperatures in tomato, SBPase is found in high concentrations suggesting that SBPase and fructose 1,6-bisphosphatase (FBPase) activities are limiting (Sassenrath et al., 1990). However, under chilling temperatures and low light, SBPase and FBPase do not appear to be limiting, indicating that photoinhibition of photosystem II (PSII) in high light leads to regulatory deactivation in bisphosphatase activity (Brüggemann and Linger, 1994). In spinach, low metabolite levels in the Calvin cycle above thermal optimum have been attributed to excessive sucrose synthesis, which effectively drains the Calvin cycle of intermediates, reducing its turnover (Stitt and Grosse, 1988).

2.3 Salt and nutrient stress

Salt stress effects plant growth and yield worldwide due to ionic toxicity, osmotic stress, and alterations to the soil ionic balance and water status (Acosta-Motos et al., 2017). Salinity leads to an increase in the concentration of NaCl in the cytoplasm of cyanobacteria and in chloroplasts of higher plants, affecting growth rate. In salt-susceptible plants such as potato, tomato, and pea, chlorophyll content decreases, whereas in salt-tolerant plants such as wheat, pearl millet and mustard, chlorophyll content increases under salt stress (Sudhir and Murthy, 2004; Acosta-Motos et al., 2017). Salt stress further impacts photosynthesis through stomatal limitation, stomatal closure, PSII inhibition, photosynthetic pigment loss, degradation of photosynthetic membrane proteins, membrane instability, decreased electron transport activities, a reduction in Rubisco content, and/or activity and inhibition of assimilation and enhancement of photorespiration.

Nitrogen (N) accounts for a large portion of the machinery required for plant metabolism including the proteins and enzymes of the photosynthetic apparatus and the respiratory system (Evans and Clarke, 2019). Therefore, a reduced N supply leads to a reduction in photosynthetic machinery's ability to process carbohydrates. Even under low O_2 conditions where photorespiration is eliminated, photosynthetic rate cannot be increased due to the lack of available machinery. This increases the sensitivity of Rubisco to O_2 under low N conditions and leads to a corresponding increase in starch accumulation (Paul and Driscoll, 1997). N limitation may also affect assimilation rate indirectly through limitations to growth and the subsequent accumulation of carbohydrates, thus feedback limitation on photosynthesis (Paul and Driscoll, 1997). It has been hypothesised that an increase in leaf carbohydrate levels may provide the signal to downregulate photosynthesis through modulation of photosynthetic genes, such as the small subunit of Rubisco (Paul and Foyer, 2001).

Phosphorous (P) is also integral to photosynthesis. Phosphate plays a regulatory role in the starch/sucrose biosynthesis pathway, Rubisco activation, and is used for the phosphorylation of intermediates in the CBC plus energy availability such as ATP and NADPH (Edwards and Walker, 1983; Sawada et al., 1992). Similarly to N, under low P, feedback limitation has been suggested to decrease assimilation rate (Pieters et al., 2001). However, within tomato seedlings, a corresponding decrease in starch accumulation under decreasing P indicates an alternative impact of deficiency to P on photosynthesis (De Groot et al., 2003). Contrary to N deficiency, tomato plants grown under low P and O_2 were better able to process carbohydrates gained by eliminating photorespiration than plants grown under high P. This indicates that limitation was due to a reduced production of assimilates and not due to the utilisation of photosynthates (i.e., not sink limited). This contrasts to experiments performed in tobacco and *Arabidopsis,* which indicate that during P deficiency (following complete P deprivation), photosynthesis was limited by low sink demand (Pieters et al., 2001). Decreases in the amount and activation of Rubisco have been reported in C_3 species, including spinach, soybean, and sunflower (Brooks, 1986; Jacob and Lawlor, 1992). Recently, it was concluded that photosynthetic limitation by phosphate may not be common in tropical forests as previously thought due to optimal allocation to photosynthetic tissue (Mo et al., 2019).

Measures of leaf absorbance under high irradiance indicate that reduced N leads to a reduction in light harvesting following a reduction in chlorophyll concentration whereas low P mainly affects the functioning of PSII. Under both nutrient deficiencies, the functioning of PSII and PSI remains coordinated indicating acclimation of photosynthesis to nutrient stress and consistent electron transport function (i.e., noncyclic vs cyclic) (Harbinson, 1994). The effect of nutrient stress is also dependent on growth irradiance. Under low irradiance, plants invest in a higher proportion of chlorophyll-protein complexes for light harvesting, rather than in Calvin cycle enzymes (see later). This reduction in cytosolic enzymes and structures can induce feedback limitation of photosynthesis when measured under saturating light and reduced O_2 concentrations due to the inability to process carbohydrates. Thus, leaves grown under low irradiance are more likely to be feedback limited at saturating light and low O_2 than those grown under high irradiance. Under such conditions, acclimation to low irradiance can be of greater importance than acclimation to low nutrient supply.

3 CO_2 uptake by leaves under stress

At the heart of maintaining plant growth and productivity is sustaining CO_2 uptake for assimilation in the CBC. For the majority of species, and all major crop plants, CO_2 uptake is mediated by the stomata. These small pores are located in the surface of most aerial parts of the plant, gated by a pair of specialised guard cells (GCs). Using changes in turgor, the guard cells regulate the aperture of the pore to optimise gas exchange between the environment and the intracellular spaces inside the leaf (see Chapter 3). Guard cells balance the release of water via evapotranspiration with the uptake of CO_2 for photosynthetic fixation at the mesophyll. It is important to mention mesophyll conductance (G_m), which is known to present resistance to diffusion of CO_2 inside the leaf between air and liquid phases and limits access of chloroplasts to CO_2. A number of physical chemical and anatomical properties are thought to contribute to G_m. While G_m increases with temperature from 15°C to 40°C, results vary according to species and the mechanism is not fully understood (Evans, 2021).

Closure of the stomatal pore is characteristic of the onset of abiotic stress. While stomatal closure promotes the conservation of water, it reduces or prevents CO_2 uptake and therefore limits photosynthetic CO_2 assimilation required for maintaining growth. Stomatal responses to abiotic stress can be broadly separated into two categories; an adaptive physiological response (speed and maintenance of closure) to avoid damage under transient episodes and modification of anatomical characteristics—a longer-term adaptation to prolonged adverse conditions.

3.1 Stomatal closure

Drought, water-deficit, or dehydration is symptomatic for a many abiotic stresses. While we tend to think of drought and water-deficit as the result of high irradiance and heat in the field—encouraging rapid rates of evapotranspiration from the leaf—a decrease in water availability can also be the result of other environmental stresses, including high salinity, freezing, and anaerobia (Neill et al., 2008). Closure of the stomatal pore is a preventative measure to limit H_2O loss; however, in doing so, CO_2 is also prevented from entering the leaf.

The phytohormone, abscisic acid (ABA), is one of the most studied signals known to stimulate stomatal closure in response to stress (Kollist et al., 2014). At the onset of drought, [ABA] increases, initiating a signalling cascade that results in stomatal closure through a decrease in GC turgor. For in depth reviews, see Kim et al. (2010) and Assmann and Jegla (2016). In brief, ABA is a rapid, ameliorative mechanism overseeing a plethora of whole plant responses to mediate the effects of dehydration (Neill et al., 2008). In the case of stomata, ABA stimulates GC closure, triggering the accumulation of cytosolic calcium ions (Ca^{2+}) and enhancing the sensitivity of the ion efflux channels on the GC membrane. These channels mediate a slow/sustained (S-type) or rapid/transient (R-type) anion efflux from the guard cells, further decreasing the water potential in the surrounding cytosol. The anion efflux also acts to depolarise the membrane and stimulate potassium (K^+) efflux channels. Finally, the gluconeogenic conversion of malate^{2+} into starch further removes ions from the GC cytoplasm, increasing the water potential gradient across the guard cell membrane, promoting the movement of water out of the GC, reducing turgor, and closing over the exposed stomatal pore (Fig. 4).

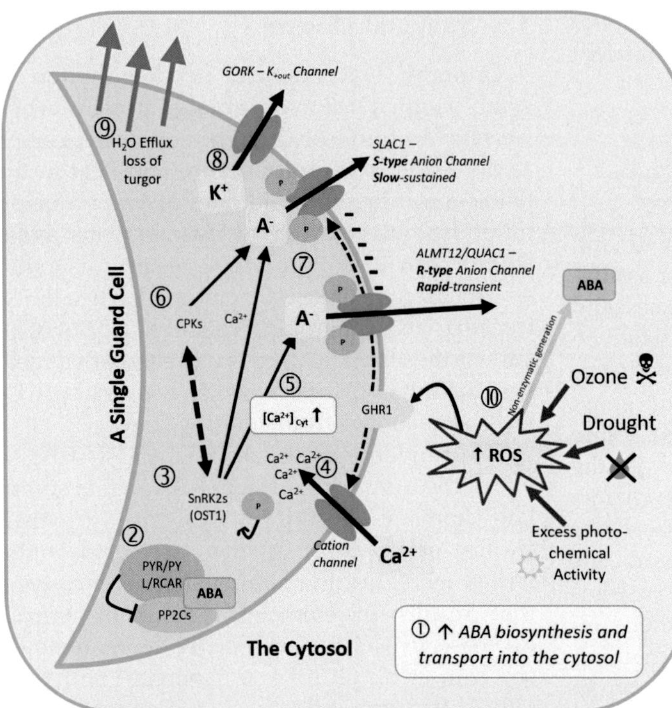

1. [ABA] builds in the cytosol or GC
2. PYR/PYL/RCAR ABA receptors inhibit PP2Cs
3. Inhibition of PP2Cs enables activation of SnRK2s
4. Ca^{2+} channels pump in Ca^{2+} from cytosol
5. $[Ca^{2+}]$ increases inside the GC
6. Increasing $[Ca^{2+}]$ activates CPKs which also stimulate phosphorylation of SnRK2s
7. SnRK2s and CPKs phosphorylate S-type to stimulate anion efflux. OST1 phosphorylates R-type anion channels. The plasma membrane depolarises.
8. Depolarisation drives K^+ efflux from the GC via the GORK cation channel
9. High concentrations of ions outside the GC drive water out of the GC – lowering turgor and closing the pore
10. ROS generated by a number of abiotic stresses. GHR1 mediates ABA activation of cation influx and S-type anion channels for closing the pore

FIG. 4 A summary of the guard cell-specific response to an increase in the concentration of the phytohormone, abscisic acid (ABA). *Adapted from Munemasa, S., Hauser, F., Park, J., Waadt, R., Brandt, B., Schroeder, J.I., 2015. Mechanisms of abscisic acid-mediated control of stomatal aperture. Curr. Opin. Plant Biol. 28, 154–162; Kollist, H., Nuhkat, M., Roelfsema, M.R.G., 2014. Closing gaps: linking elements that control stomatal movement. New Phytol.*

There is a growing body of work linking components of the ABA pathways with other abiotic stress pathways within the plant, for example, OST-1 and ICE-1 (also known as SCRM—Chinnusamy et al., 2003). Open Stomata 1 (OST1), a Ca^{2+}-independent protein kinase SNF1-related kinase 2s (SnRK2s), phosphorylates multiple downstream targets including the S and R-type anion efflux channels responsible for driving membrane depolarisation during the response to increased [ABA]. OST1 has also been shown to positively regulate the C-repeat-binding factor (CBF)-dependent cold signalling pathway (Ding et al., 2015), phosphorylating Inducer of CBF expression 1 (ICE1) a substrate of MPK3/MPK6 in the plant freezing tolerance pathway. This process stabilises ICE1 under low temperature promoting activity under cold stress (Larue and Zhang, 2019). However, a definitive role for the ICE1 CBF pathway is still under discussion (Kidokoro et al., 2020; Thomashow and Torii, 2020). Furthermore, ICE1 or SCRM has also been shown to be involved in the initiation of stomatal patterning on the leaf surface (Kanaoka et al., 2008) forming heterodimers with other stomatal transcription factors such as SPEECHLESS (SPCH; Kanaoka et al., 2008), with the mutant ice1-1 producing excessive numbers of stomata instead of pavement cells. Reflected in the complexity and wide-reaching genetic interactions, often responses to transient abiotic stress can generate changes in anatomical characteristics (such as stomatal density) to adapt to the onset of chronic abiotic stress.

Another abiotic stress, which can limit CO_2 uptake, is freezing. Ice can enter the leaf through the stomata when the guard cells are open (Pearce, 2001); however, nocturnal stomatal closure can physically defend against entry of ice. For freezing tolerance, it should be noted that the anatomy of the stomatal complex can also serve to maintain CO_2 uptake during adverse conditions. Plants from cooler regions tend to have greater densities of stomata, perhaps to maximise cell expansion through increased CO_2 and nutrient uptake under low temperatures (Körner and Larcher, 1988; Loveys et al., 2002).

Ozone (O_3), an anthropogenic pollutant responsible for increases in reactive oxygen species (ROS), is another abiotic stress that leads to significant decreases in global plant productivity (Ainsworth et al., 2012). Stomatal closure in response to O_3-induced ROS production is one of the oldest reported responses to pollutants (Koritz and Went, 1953). While transient exposure to high ozone induces a brief stomatal closure (3–6min) through the generation of GC ROS (Kollist et al., 2007; Vahisalu et al., 2010), stomata often recover within 30–40min with little impact on carbon uptake (Kollist et al., 2007). However, longer exposures can lead to a significant reduction in stomatal conductance, which is irreversible (Morgan et al., 2003), leading to a decrease in CO_2 uptake and impacting on the ability of the GCs to close in response to further stress (Barnes et al., 1990; McLaughlin et al., 2007). There have also been reports of decreased stomatal sensitivity to ABA in plants exposed to O_3 (Mills et al., 2009; Wilkinson and Davies, 2010).

3.2 Stomatal anatomical adaptation

While ABA works to mitigate transient abiotic stress—e.g., through closing stomata at midday when evaporative demand can be high: this hormone is also an important regulator of stomatal development. Plants grown under high [ABA] develop leaves with lower stomatal densities (Casson et al., 2008; Wei et al., 2021) while plants deficient in ABA biosynthesis have higher densities (Jalakas et al., 2018). In addition, the response of stomata to the presence (or absence) of ABA is systemic—with the experiences of older material shaping the development of younger leaves (Casson and Hetherington, 2010). The role of density and morphology of stomata in the context of heat and drought stress has been examined using transgenic plants of multiple species which over express or have lower expression of the epidermal patterning factor (EPF) gene family. By reducing the density of stomata in species such as rice, a reduction in stomatal conductance and transpiration was achievable with little cost to photosynthesis, enhancing leaf water use efficiency and providing greater heat tolerance through conservation of soil water (Caine et al., 2019).

4 Changing environments and acclimation

The efficiency of photosynthesis under a given set of conditions will depend upon the absorption of light by plant material, the transfer of this energy to reaction centres and its final use in CO_2 assimilation. Four aspects of light are important for driving photosynthesis and controlling plant growth and development (Geiger, 1994):

(1) Irradiance: determines the rate at which energy is delivered to the reaction centres.
(2) Duration: influences the total energy received during a given period.
(3) Spectral quality: influences the ability to drive carbon sequestration due to the probabilities of absorbing different wavelengths.
(4) Timing: determines the effectiveness of light in the regulation of various plant processes according to plant development, for example, source–sink effects.

PAR refers to the spectral range of solar radiation that can be used by plants: between 400 and 700 nm. This is often referred to as the photosynthetic photon flux density or PPFD, with each quantum of light called a photon, and quantified as μmol photons $m^{-2} s^{-1}$. The PPFD intercepted by each section of leaf, and the total amount of PPFD intercepted by the whole plant, are key determinants of the rate of CO_2 assimilation, and thus of whole plant photosynthesis.

Leaf photosynthesis responds nonlinearly to light intensity. Under highly heterogeneous light environments, light intensities will vary from limited to excessive depending on the shape of light response curve. The light response curve may be described by a nonrectangular hyperbola (Eq. 1; Fig. 2).

$$A = \phi L + (1+\alpha)A_{max} - \sqrt{(\phi L + (1+\alpha)A_{max})^2} \frac{-4\theta\phi L(1+\alpha)A_{max}}{2\theta - \alpha A_{max}} \quad (1)$$

The curve relates net photosynthetic assimilation rate, A, to PPFD, L. In the absence of light, net photosynthesis will be negative and relates to a dark respiration rate, R_D. It is assumed that the rate of dark respiration is proportional to the maximum photosynthetic according to the relationship $R_D = \alpha A_{max}$ (Niinemets and Tenhunen, 1997; Retkute et al., 2015). The light response curve can be characterised by three shaping parameters: quantum yield otherwise known as quantum use efficiency (ϕ), convexity, (θ) and maximum photosynthetic capacity (A_{max}). The quantum yield refers to the initial linear portion of the curve and describes the maximum efficiency with which light can be used to fix carbon whilst the convexity, or bending factor, describes the curvature. The net photosynthesis rate (A) rises until it reaches a maximum: the maximum photosynthetic capacity (A_{max}). The value at which photosynthesis matches respiration (where net carbon assimilation is equal to zero) is known as the light compensation point (LCP).

The shape of the light response curve, and thus the values of the shaping parameters, will depend upon the biochemical pathway employed (i.e., C_3/C_4) the light absorption properties of plant material, the relative concentration of the structures involved in light harvesting, and the current 'induction' status of the leaf (Murchie and Horton, 1997; Retkute et al., 2015). The relationship between light and photosynthesis can also be extended to a population of photosynthesising cells, for example, a leaf or a whole canopy. Each section of photosynthetic material on a plant will place somewhere along a light response curve, this can be summed up over the whole organ or plant and thus build a curve that represents the whole structure. The shape of the resultant curve will depend upon a number of factors including those mentioned above combined with architectural features of the plant and the canopy it resides in. Furthermore, the features of the light response curve is not fixed, but rather can change as a result of the environmental conditions to which the plant is exposed.

The sessile lifestyle of plants necessitates a sophisticated mechanism to optimise resource capture in a changing environment (Dietzel and Pfannschmidt, 2008). The arrangement and movement of plant leaves in three-dimensional space combined with changes in wind, solar movement, and cloud cover plus developmental changes in plant structure can lead to a highly variable pattern of light reaching individual leaf sections. As the most variable environmental driver light imposes a twofold challenge, the need to efficiently utilise as many photons as possible whilst simultaneously preventing harm caused by excess radiation. Achieving the optimal balance between these two states is critical to maximise both productivity and mitigate radiation-induced damage (Demmig-Adams et al., 2012). One process by which plants optimise photosynthesis is acclimation.

4.1 Photosynthetic acclimation to changes in light

Photosynthetic acclimation (photoacclimation) is the process whereby leaves alter their morphology and/or biochemistry to optimise photosynthetic efficiency and productivity according to long-term (i.e., days) changes in the light environment. Photoacclimation can be broadly split into two different mechanisms: developmental acclimation and dynamic acclimation. Developmental acclimation refers to changes occurring during leaf development, which are largely irreversible, whereas dynamic acclimation is the ability for fully developed leaves to change their photosynthetic capacity. These mechanisms are an integral part of photosynthetic regulation, and there is emerging evidence that any alterations may impact upon the ability of a plant to assimilate carbon over long periods of time, thus affecting biomass production (Athanasiou et al., 2010; Retkute et al., 2015).

The extent of the propensity to acclimate will depend on the plant's genotype, which will, to a greater or lesser extent, match the environment to which it is adapted through evolution. Any given acclimation state of a leaf is defined by A_{max} as well as R_D (see Fig. 5). There is substantial variation between species in their ability to acclimate, with plants from semishaded environments seemingly exhibiting the greatest plasticity in acclimation capacity (Murchie and Horton, 1997). There is also evidence for species-specific differences in the relative durations of cellular division and expansion during leaf development (Stiles and Van Volkenburgh, 2002) and biochemical differences, i.e., in chlorophyll contents and ratios, Rubisco amounts, electron transport capacity, or enzyme activity (Murchie and Horton, 1997; Carmo-Silva et al., 2015). This suggests that there are both benefits and costs associated with acclimation. At the whole canopy level, the ability for individual plant leaves to acclimate is also dependent upon leaf age and availability of nutrients (Hikosaka, 2005). This inherent plasticity enables foliage photosynthetic potentials to increase with an increasing light availability (e.g., Johnson et al., 2010). Depending upon the species, photosynthetic capacity can vary between 2- and 20-fold from the canopy top to bottom.

Variations in light intensity can occur over different spatial or temporal scales. Spatial scales include variation that can be attributed to shading effects within a plant stand, or a single plant canopy, whereas temporal scales may refer to long-term solar radiation changes (e.g., as a result of seasonal change) or short-term response such as fluctuating light enforced by sun position, cloud, or leaf movement. Therefore, the interception of light will depend on a number of different factors including leaf orientation and shape, the spatial arrangement of photosynthetic surfaces (i.e., uniform vs clumping), sun elevation, the finite width of the sun's disc, and changes in spectral distribution of PPFD within the canopy (Nobel et al., 1993). Such variable patterns will lead to periods of time where photosynthesis is fully saturated, and others where photosynthesis may be below or approaching the light compensation point.

4.1.1 Developmental acclimation

One of the simplest examples of acclimation within a spatial scale can be seen in the anatomical and physiological differences between sun and shade leaves and is a key example of developmental acclimation. Sun leaves differ from shade leaves primarily in their higher A_{max} and R_D rates (Lambers and Oliviera, 2019). Differences in anatomy (Fig. 6) are determined early in development and are considered largely irreversible in many species. These anatomical differences can constrain the potential of leaves

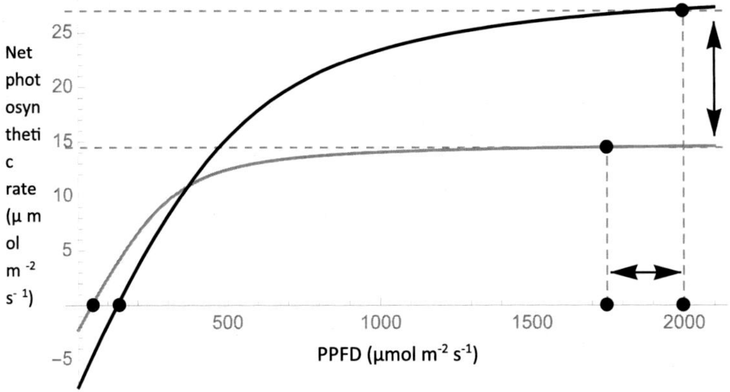

FIG. 5 Example light response curves from a sunlit (high-light acclimated; *black line*) versus shaded (low-light acclimated; *grey line*) leaf.

to [dynamically] acclimate further (Murchie et al., 2005). Sun leaves are generally thicker, with differing cellular structure, providing more space for photosynthetic components per unit leaf area and have thicker palisade parenchyma, either through an increased length or through multiple layering of cells. Contrary to this, shade leaves are often thinner with a greater surface area, requiring less investment in terms of nitrogen and carbon and a reduced stomatal conductance. Methods for 3D analysis of leaf structure and airspaces might present new opportunities for understanding the role of anatomy in light acclimation (Lehmeier et al., 2017).

Further differences can be seen in the biochemical properties of the two types of leaves; sun leaves contain a greater chlorophyll *a*:chlorophyll *b* ratio, larger amounts of Calvin-cycle enzymes, and more components of the electron transport chain (including b_6f cytochromes and ATPase). For some plants, the change in A_{max} between different acclimation states shows an almost linear relationship to an increase in the amount of photosynthetic compounds (Evans and Seemann, 1989); thus, investment in compounds that determine photosynthetic capacity translates to higher photosynthetic rate at increased irradiance levels (e.g., Murchie et al., 2002; Walters, 2005). These differences help sun leaves to exploit high irradiances more efficiently. Their ability to regenerate more ATP and NADPH to alleviate the overreduction of PSII reaction centres at high PPFD helps to minimise their risk of damage to high light intensities (i.e., photoinhibition; Chow, 1994; Baker and Oxborough, 2004). Recent dissection of the thylakoid proteome indicates a greater involvement of low abundance proteins in acclimation of photosynthesis and photoprotection (Flannery et al., 2021).

4.1.2 *Dynamic acclimation*

Dynamic acclimation is typically reversible, fluctuating over timescales of hours to days, although it is predicted that between 5 and 10 days are required for full dynamic acclimation to occur (Murchie and Horton, 1997; Retkute et al., 2015). The ability of preexisting foliage to dynamically acclimate requires a transition from high photosynthetic capacity under high irradiances to high light efficiency under low irradiances and vice versa (Hikosaka and Terashima, 1996). Dynamic acclimation is, at least in part, mechanistically different from developmental acclimation (Murchie and Horton, 1997; Athanasiou et al., 2010). Such a

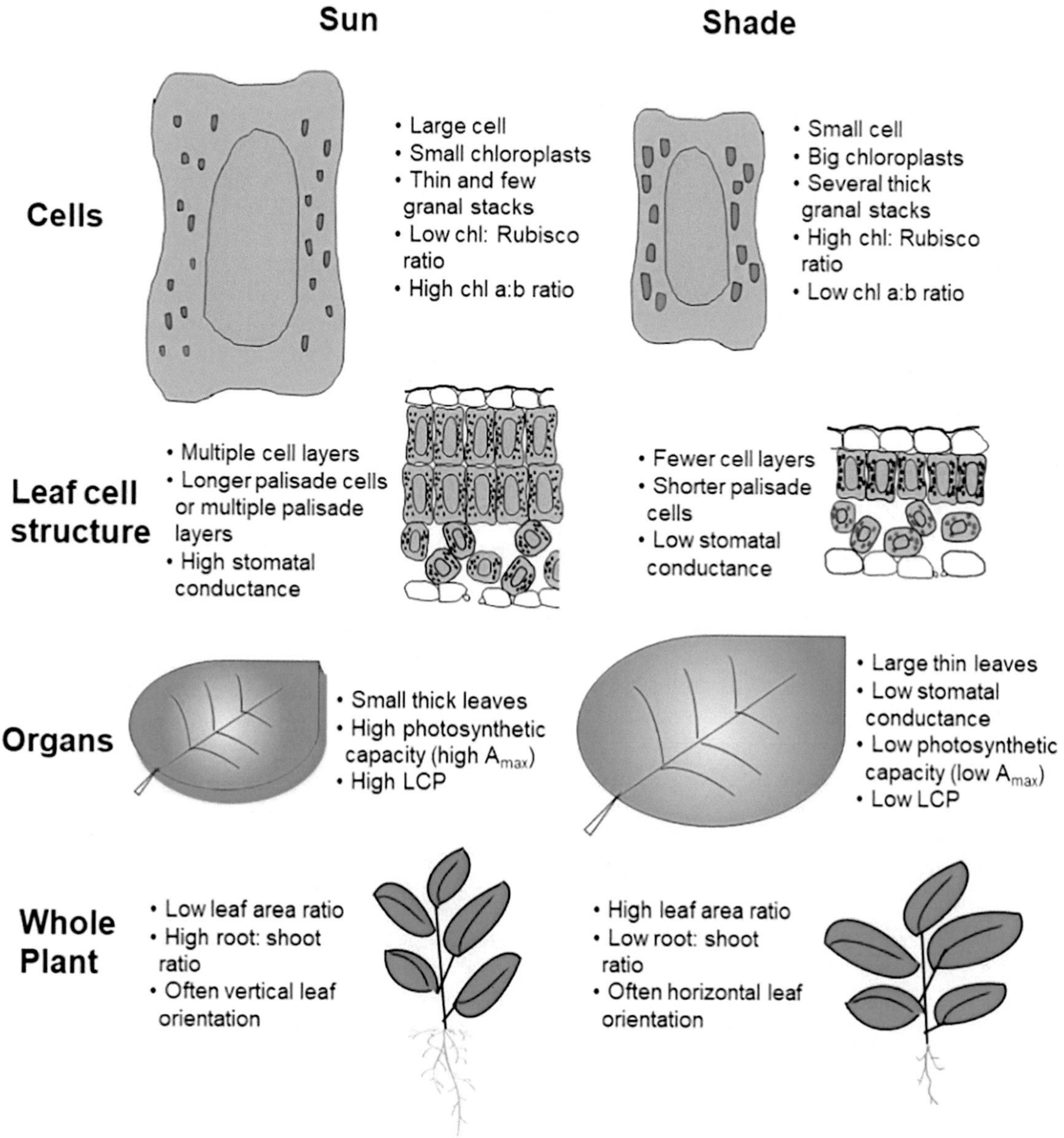

FIG. 6 Anatomical and physiological differences between sun and shade leaves, an example of developmental acclimation.

transformation will alter both the total carbon assimilation and the susceptibility to photoinhibition (Baker and Oxborough, 2004). Acclimation to an increased irradiance can include adjustments in both physiological and morphological traits to achieve an increase in amounts of photosynthetic components per unit area. The extent of these changes will depend on whether the increase in irradiance occurs before or after leaf development becomes fixed (i.e., before or after leaf expansion) (Murchie et al., 2005; Oguchi et al., 2005). Contrary to biochemical changes, morphological features are largely irreversible (Sims and Pearcy, 1992), thus may limit complete acclimation to the light environment in some cases (Oguchi et al., 2003, 2005). This is of relevance because the ability of mature leaves to acclimate to changes in irradiances is generally limited to existing chloroplasts and cells and coupled with gene expression data, requiring modification of an existing protein profile (Miller et al., 2017).

Within seconds–minutes of a change in irradiance, nuclear gene expression will be altered affecting hundreds of transcripts (Suzuki et al., 2015; Crisp et al., 2017; Schneider et al., 2019). Dynamic acclimation often includes alterations in the quantity and stoichiometry of photosynthetic components—including Rubisco, cytochrome-b/f complexes, light harvesting complexes, ATPase, and enzymes involved in carbohydrate synthesis. Under low light conditions, a reduced investment in Rubisco and other Calvin-cycle enzymes and an increased investment in chlorophyll ± protein complexes has been shown for several species (Hikosaka and Terashima, 1996; Evans and Poorter, 2001). This enables dynamic regulation of light harvesting capacity, nonphotochemical quenching, chloroplast movement, CO_2 diffusion, and enzyme plus electron transport activities (Kaiser et al., 2018; Murchie et al., 2018a,b). This results in long-term changes to leaf properties such as A_{max}, R_D, and the LCP (Walters, 2005; Athanasiou et al., 2010). Despite the variation seen within and between species, there are conserved trends in relation to changes in the shape of the photosynthetic light response curve that are useful for acclimation modelling approaches. For example, the maximum quantum yield is unaffected by (nonstressful) growth conditions (Long and Drake, 1991) and leaf absorptance is unlikely to be substantially altered during dynamic acclimation (Pearcy and Sims, 1994). This means that studies on dynamic acclimation can focus on changes to A_{max}.

4.1.3 Excess light: Photoprotection

As growth irradiance increases, absorbed photons may become in excess if they are produced quicker than they can be used in photosynthesis (Murchie and Niyogi, 2011). Due to the sensitivity of PSII, high light levels may lead to damage to the photosynthetic apparatus. Plants have an ability to regulate the amount of light they intercept through changes in leaf area, leaf angle or chloroplast movement, or on a molecular level, through acclimatory adjustments in LHC antenna size (state transitions). However, if excess energy has been absorbed, it can be dissipated via a number of different routes, broadly termed photoprotection, including nonphotochemical quenching or NPQ. Photoprotection itself can acclimate, e.-g., through changes in the pool sizes of the xanthophyll cycle (Demmig-Adams and Adams, 1992, 2006). Despite photoprotection, high light can lead to PSII photoinactivation and a reduction in quantum yield known as photoinhibition. The effect of photoinhibition on shaping parameters of the photosynthesis light response curve is already well characterised (Fig. 2). The primary effect of photoinhibition is the reduction in Φ, which is important under low light conditions (Bjorkman and Demmig, 1987). However, under conditions causing photoinhibition, a reduction in Φ is often accompanied by a similar reduction in θ (Leverenz, 1994) (Fig. 7).

Previous empirical models have looked at the effects of distorting such shaping parameters to empirically quantify values for reduction in carbon gain (Werner et al., 2001; Burgess et al., 2015). A reduction in both parameters is of particular importance under natural conditions, because light is thought to be a limiting resource to photosynthesis for the majority of time in a large number of environments (Ort and Baker, 1988). In terms of productivity, the best strategy may be to minimise photoinhibition at the biochemical level (Hubbart et al., 2018). Photoprotective mechanisms, such as the xanthophyll cycle and PsbS-dependent quenching, are

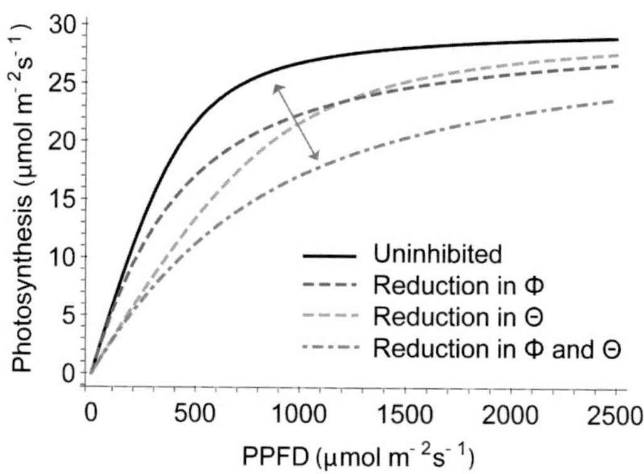

FIG. 7 Example light response curves from an uninhibited versus photoinhibited leaf dependent on changes in the shaping parameters Φ and Θ.

known to reduce the level of photoinhibition in leaves (Li et al., 2002; Niyogi et al., 2005). However, both the costs and as benefits of high levels of thylakoid-level photoprotection must be determined when assessing overall productivity (Hubbart et al., 2018; Kromdijk et al., 2016). In the latter case, the lowered quantum yield of CO_2 assimilation that occurs during NPQ can limit photosynthesis on the canopy scale when extended from transient high light periods into transient low light.

Both acclimation and photoprotection represent a subset of regulatory mechanisms used to accommodate for variations in light availability and can be effective in reducing damage due to excess excitation energy. However, the different processes will interact together and thus the actual productivity of the plant will depend upon the balance between different states. For example, exposure to excess light levels may lead to the enhancement of photoprotective mechanisms and in turn photoprotection may place an upper limit on the capacity to acclimate (e.g., Demmig-Adams et al., 2012).

4.1.4 Photosynthesis and fluctuating light

Acclimation of photosynthesis is an essential component of environmental adaptation, but assessment of its 'effectiveness' in a complex temporal and spatial environment can be difficult. Whilst it is relatively well understood how a plant responds to a change from low to high growth irradiance, or vice versa, response to fluctuating light is less well known. Daily carbon gain cannot be derived from the average values of light due to the nonlinear response of photosynthesis (Niinemets and Anten, 2009). As such, measured profiles of photosynthetic capacity in plant crowns typically do not match those of average irradiance (Buckley et al., 2013). Therefore, the optimal maximum photosynthetic capacity under fluctuating light patterns is different compared to those obtained from averaged light intensity. Understanding the response of a collection of photosynthetic cells (i.e., a whole leaf or the whole canopy) is even more complex. The adaptive significance of acclimation and photoprotection under fluctuating light patterns has been little studied but is key to understanding the limits placed on plants in natural environments. This has become increasingly important since uncovering that the short-term response of photosynthesis (seconds and minutes) is under genetic control (Cruz et al., 2016), and thus, both gene editing or exploiting natural genetic variation can provide a means to improve productivity (Kromdijk et al., 2016; Soleh et al., 2017).

Over the short term (seconds and minutes), the mechanisms that plants use to deal with these changes are relatively well understood: it is possible to invoke enzyme activation states, metabolite concentrations, and the state of energisation of the thylakoid membrane as a 'memory' of short-term past light history (Horton and Ruban, 2005; Murchie et al., 2009). Short-term responses are regulated by processes such as phosphorylation of thylakoid components, allosteric regulation of enzymes, and the physical state of the thylakoid (Tikkanen et al., 2010, 2012; Ruban et al., 2012). Two examples of processes on such short timescales are photosynthetic induction—the delay in the rise in carbon assimilation immediately following a light increase (Pearcy et al., 1997)—and thylakoid photoprotective processes (Demmig-Adams and Adams, 1992; Murchie and Niyogi, 2011). Over longer time scales, photoacclimation becomes more important. Under fluctuating light conditions, all mechanisms will interact together to determine the overall effect on photosynthesis.

Optimal plant metabolism would track current environmental changes and alter photosynthesis instantaneously (Retkute et al., 2015). However, this does not happen and there is a time lag before the leaf can fully respond to changes (Fig. 8) (Walters and Horton, 1994; Athanasiou et al., 2010). The length of the time lag will depend upon the process being evoked. For acclimation to a change in light intensity, the time lag for increasing light intensity is longer than that for a decreasing light intensity. This is thought to be due to the protein synthesis, maintenance, and investment requirements (in terms of carbon, nitrogen, and other resources) for an increased A_{max} (Athanasiou et al., 2010; Retkute et al., 2015).

It is often assumed that acclimation involves a strategy of optimisation geared towards maximum carbon gain in a given environment (Pons, 2012); however, the impact of fluctuating light on the photosynthetic membrane is yet to be fully elucidated. Experimental evidence indicates that A_{max} is dependent on the pattern of switches between high and low light intensities, as well as the relative time spent under each condition (Fig. 9) (Athanasiou et al., 2010; Retkute et al., 2015; Townsend et al., 2018). This enables the optimal A_{max} of a given leaf to be higher than that obtained for the averaged light intensity if the fraction of higher light intensity is large enough. Thus, the relative frequency distribution of irradiance is more critical in determining optimal A_{max} that the average irradiance over an entire leaf lifetime (Takenaka, 1989). The potential for this response will therefore depend on the recent conditions to which a leaf has been

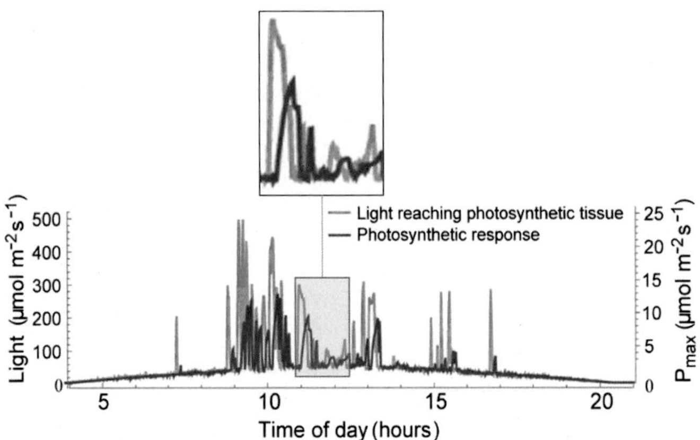

FIG. 8 Photosynthesis *(red line)* does not respond instantaneously to changes in light intensity *(grey line)*. This will lead to a time lag before the leaf can respond, and the length of this time lag will depend upon a variety of different photosynthetic processes. The inset box magnifies this lag.

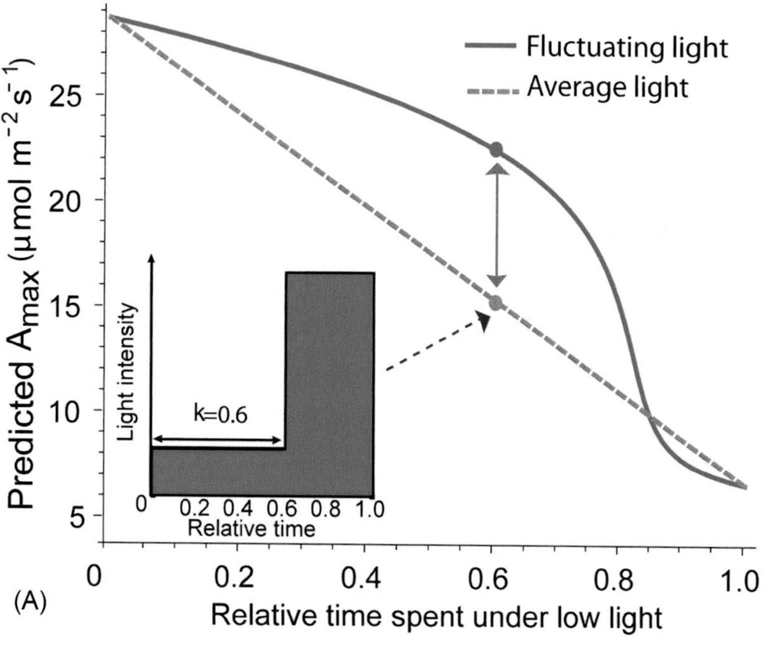

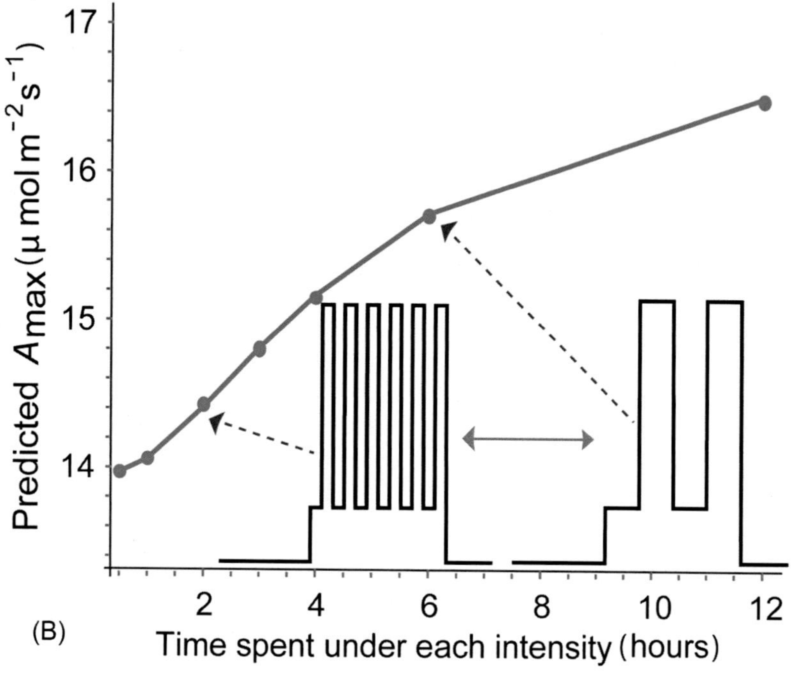

FIG. 9 The optimal P_{max} will depend on the period of time spent under low versus high light intensity (A) as well as the frequency of switches between the two (B). This can lead to a higher optimal A_{max} under fluctuating conditions than those under the same average light intensity. *Redrawn from Retkute, R., Smith-Unna, S., Smith, R., Burgess, A., Jensen, O., Johnson, G., Preston, S., Murchie, E., 2015. Exploiting heterogeneous environments: DOES photosynthetic acclimation optimize carbon gain in fluctuating light? J. Exp. Bot. 66, 2437–2447.*

exposed and may either lead to an increase or a decrease in biomass production (Athanasiou et al., 2010; Vialet-Chabrand et al., 2017; Matthews et al., 2018). To account for 'past light history', Retkute et al. (2015) applied a time-weighted average to a model of photosynthetic acclimation, which effectively mimics the notion that the photosynthetic machinery has a fading memory, enabling a greater adaptation to more recent experienced light patterns.

There are very few experimental investigations producing data that would allow us to understand the influences of light pattern on dynamic acclimation. This may be partly due to past difficulties in developing lighting systems that could cope with rapid switching between light levels of greatly differing magnitude. Recent work indicates that within field settings, light fluctuations may occur over a period of <2 s or even on the subsecond scale (Kaiser et al., 2018, 2019). However, the current shortest length of fluctuations used in experiments has been 20 s (Kaiser et al., 2018), with many using fluctuation ≥ 180 s. Recent developments with LED lighting have overcome such problems, and it is now possible to accurately replicate light dynamics from virtually any environment (Pattison et al., 2018). However, the true features of the natural light environment are poorly understood, and first we require a full characterisation of the four features of light that drive photosynthesis in a variety of settings: irradiance, duration, spectral quality, and timing.

5 Agriculture and productivity

For crop species, the maximum efficiency of utilisation of light energy is surprisingly conserved for C_3 and C_4 crop species in optimal conditions and encompassed in the measurement of radiation use efficiency or biomass produced per unit radiation intercepted (Murchie et al., 2018a,b). Away from research stations and controlled environments towards the more unpredictable situation of farmer's fields, it is common to experience conditions that shift from optimal homeostasis where temperature, water availability, and humidity act to limit CO_2 uptake and assimilation. Moreover, man-made climate change is causing an increase of terrestrial temperatures that may rise by 1–6°C, and this has already begun to compound both the severity and unpredictability of abiotic stress such that the current rate of year on year yield gains is less than the 2.4% needed to double yields by 2050 (Ray et al., 2013), and thus, adaptation is needed to prevent substantial risk to global food security (Challinor et al., 2014). Rising global temperatures already have a negative effect on crop yields by acting directly on sensitive processes of plant development and metabolism. High temperatures directly inhibit photosynthesis, especially in C_3 crop species. With each 1°C rise in temperature, yields will decrease by up to 7.4% (Zhao et al., 2017, Tigchelaar et al., 2018). Water deficit and drought restrict growth processes directly via turgor loss but also affect CO_2 assimilation through stomatal closure and by reducing transpirational cooling, increasing leaf temperature (Tardieu et al., 2018). High temperatures commonly act in combination with drought, high light intensities, and an increased water vapour deficit (Grossiord et al., 2020), meaning that the greatest impact will be felt in regions with greater socioeconomic vulnerability. There is a requirement to improve photosynthesis not only in ideal, 'yield potential' conditions but increasingly in conditions where CO_2 assimilation is limited by abiotic stress factors. It is common for multiple stress factors to occur simultaneously such as drought, high temperature, and high light (Moore et al., 2021).

When considering limitations to carbon fixation with respect to productivity in the field, it must be viewed in the context of whole plant processes. The products of the CBC in source organs such as leaves are ultimately exported to the rest of the plant in the form of sucrose, synthesised within the cytosol. Clearly, there needs to be regulation between the rate of source photosynthate and the ability of the rest of the plant to utilise it for growth. Sucrose moves between mesophyll cells towards vascular tissue where it is loaded into phloem tissue by apoplastic or symplastic means and then to sink tissues (such as grains, fruit, or growing regions) by mass transfer mechanisms (Zhang and Turgeon 2018; Milne et al., 2018). However, the capacity of sinks, their location, and activity can vary, which presents the possibility that the mobilisation and movement of carbohydrates around the plant itself can be a limiting factor for plant productivity. Such 'sink limitation' of growth has been demonstrated for important crops at key growth stages, for example, wheat shows evidence of this in the post flowering and grain filling phase (Reynolds et al., 2012). Plants with large sinks can also make greater use of high CO_2 environments. However, can long-term signalling by weak sinks limit CO_2 assimilation in source leaves? This is a complex question in specific experimental circumstances in plants with multiple sinks such as rice where there is a high capacity for leaf export and that stem, grain, and leaf can all act as multiple and substantial sinks for carbon (Morita and Nakano, 2011). An emerging role has appeared or trehalose-6-phosphate (T6P) as a signalling molecule for whole plant resource allocation via the SnRK1 protein kinase regulatory pathway and this pathway seems to be involved in the regulation of photosynthesis (Oszvald et al., 2018). Thus, TPU has also been promoted as a mechanism of regulation between source generation and consumption. In leaves, sucrose accumulation in the cytosol may allow feedback via triose-phosphate (TPU) utilisation to regulate photosynthesis under certain conditions (Fabre et al., 2019; Paul and Foyer, 2001). It seems clear that sucrose accumulation in the leaf caused by limited export will decrease TPU utilisation and result in a down regulation of photosynthesis. In principle, any part of the plant capable of accumulating energy-dense material is capable of behaving as a sink with the potential to relieve limitations and this has opened up interesting new areas (Paul and Eastmond, 2020).

Co-ordination between distal plant parts is well known, for example, the synthesis of ABA in roots in response to drying soil that, when sensed by the stomatal guard cells, results in closure (Takahashi and Shinozaki, 2019). Strigolactones, microRNAs, jasmonic acid, peptides, and cytokinins all are involved in some form of shoot—root communication (Ko and Helariutta, 2017). The signalling of a stress event, whether abiotic or pathogen/pest from one part of the plant to another, has been well established, and this normally takes the form of calcium, ROS, and hydraulic signals (Fichman and Mittler, 2020). In this way, defensive or stress mitigation responses are induced via established signalling pathways, but many questions remain unanswered. ABA is also involved in ROS/Ca^{2+} associated closure in response to very high light on both a local and long-distance scale, activating stomatal closure at a systemic level (Takahashi and Shinozaki, 2019).

Similarly, developmental and dynamic acclimation (see above) is subject to influence from systemic signals to light and CO_2 (Coupe et al., 2006; Kangasjärvi et al., 2009). Such signalling may be an efficient means of 'priming' emerging leaves to a prevailing condition rather than relying solely on sensing cues from a single plant location. However, the nature of such signals remains to be identified. Such responses demonstrate the emerging importance of whole plant signalling for the long-term regulation of CO_2 fixation processes in plants.

6 Summary

This chapter has highlighted the major processes by which the environment can influence chloroplastic CO_2 assimilation and, in particular, how plants respond and acclimate in order to achieve optimisation and alleviate stress. We have considered this to be a cell, organ and whole plant level process. We considered the environment to be a dynamic set of conditions and introduced the theme of multiple changes, fluctuations, and corresponding acclimation. These concepts (which are not new) have undergone recent exploration and validation, and we have highlighted these in the context of agricultural yield improvement.

References

Acevedo-Siaca, L.G., Coe, R., Wang, Y., Kromdijk, J., Quick, W.P., Long, S.P., 2020. Variation in photosynthetic induction between rice accessions and its potential for improving productivity. New Phytol. 227 (4), 1097–1108.

Acosta-Motos, J.R., Ortuño, M.F., Bernal-Vicente, A., Diaz-Vivancos, P., Sanchez-Blanco, M.J., Hernandez, J.A., 2017. Plant responses to salt stress: adaptive mechanisms. Agronomy 7, 18.

Ainsworth, E.A., Yendrek, C.R., Sitch, S., Collins, W.J., Emberson, L.D., 2012. The effects of tropospheric ozone on net primary productivity and implications for climate change. Annu. Rev. Plant Biol. 63, 637–661.

Assmann, S.M., Jegla, T., 2016. Guard cell sensory systems: recent insights on stomatal responses to light, abscisic acid, and CO_2. Curr. Opin. Plant Biol. 33, 157–167.

Athanasiou, K., Dyson, B., Webster, R., Johnson, G., 2010. Dynamic acclimation of photosynthesis increases plant fitness in changing environments. Plant Physiol. 152, 366–373.

Baker, N.R., Oxborough, K., 2004. Chlorophyll fluorescence as a probe of photosynthetic productivity. In: Papageorgiou, G.C., Govindjee (Eds.), Chlorophyll a Fluorescence: A Signature of Photosynthesis. Springer, pp. 65–82.

Barnes, J., Eamus, D., Davison, A., Ro-Poulsen, H., Mortensen, L., 1990. Persistent effects of ozone on needle water loss and wettability in Norway spruce. Environ. Pollut. 63 (4), 345–363.

Berry, J., Bjorkman, O., 1980. Photosynthetic response and adaptation to temperature in higher plants. Annu. Rev. Plant Physiol. 31, 491–543.

Bjorkman, O., Demmig, B., 1987. Photon yield of O_2 evolution and chlorophyll fluorescence characteristics at 77 K among vascular plants of diverse origins. Planta 170, 489–504.

Brooks, A., 1986. Effects of phosphorus nutrition on ribulose-1,5-bisphosphate carboxylase activation, photosynthetic quantum yield and amounts of some Calvin-cycle metabolites in spinach leaves. Aust. J. Plant Physiol. 13, 221–237.

Brooks, A., Farquhar, G., 1985. Effect of temperature on the CO_2/O_2 specificity of ribulose-1,5-bisphosphate carboxylase/oxygenase and the rate of respiration in the light—estimates from gas-exchange measurements on spinach. Planta 165, 397–406.

Brüggemann, W., Linger, P., 1994. Long-term chilling of young tomato plants under low light. IV. Differential responses of chlorophyll fluorescence quenching coefficients in lycopersicon species of different chilling sensitivity. Plant Cell Physiol. 35 (4), 585–591.

Buckley, T., Cescatti, A., Farquhar, G., 2013. What does optimization theory actually predict about crown profiles of photosynthetic capacity when models incorporate greater realism? Plant Cell Environ. 36, 1547–1563.

Burgess, A., Retkute, R., Pound, M., Preston, S., Pridmore, T., Foulkes, M., Jensen, O., Murchie, E., 2015. High-resolution 3D structural data quantifies the impact of photoinhibition on long term carbon gain in wheat canopies in the field. Plant Physiol. 15.00722v1-p.00722.2015.

Caine, R.S., Yin, X., Sloan, J., et al., 2019. Rice with reduced stomatal density conserves water and has improved drought tolerance under future climate conditions. New Phytol. 221, 371–384.

Carmo-Silva, E., Scales, J.C., Madgwick, P.J., Parry, M.A.J., 2015. Optimizing Rubisco and its regulation for greater resource use efficiency. Plant Cell Environ. 38, 1817–1832.

Casson, S., Hetherington, A.M., 2010. Environmental regulation of stomatal development. Curr. Opin. Plant Biol. 13, 90–95.

Casson, S., Franklin, K.A., Gray, J.E., Grierson, C.S., Whitelam, G.C., Hetherington, A.M., 2008. PhytochromeB and PIF4 regulate stomatal development in response to light quality. Curr. Biol. 19, 229–234.

Cen, Y., Sage, R., 2005. The regulation of Rubisco activity in response to variation in temperature and atmospheric CO_2 partial pressure in sweet potato. Plant Physiol. 139, 979–990.

Challinor, A.J., Watson, J., Lobell, D.B., Howden, S.M., Smith, D.R., Chhetri, N., 2014. A meta-analysis of crop yield under climate change and adaptation. Nat. Clim. Change 4, 287–291.

Chinnusamy, V., Ohta, M., Kanrar, S., Lee, B.-h., Hong, X., Agarwal, M., Zhu, J.-K., 2003. ICE1: a regulator of cold-induced transcriptome and freezing tolerance in Arabidopsis. Genes Dev. 17 (8), 1043–1054.

Chow, W.S., 1994. Photoprotection and photoinhibitory damage. Adv. Mol. Cell Biol. 10, 151–196.

Coupe, S.A., Palmer, B.G., Lake, J.A., Overy, S.A., Oxborough, K., Woodward, F.I., Gray, J.E., Quick, W.P., 2006. Systemic signalling of environmental cues in Arabidopsis leaves. J. Exp. Bot. 57, 329–341.

Crafts-Brandner, S., Van de Loo, F., Salvucci, M., 1997. The two forms of ribulose-1,5-bisphosphate carboxylase/oxygenase activase differ in sensitivity to elevated temperature. Plant Physiol. 114, 439–444.

Crisp, P., Ganguly, D., Smith, A., Murray, K., Estavillo, G., Searle, I., Ford, E., Bogdanović, O., Lister, R., Borevitz, J., et al., 2017. Rapid recovery gene downregulation during excess-light stress and recovery in arabidopsis. Plant Cell 29, 1836–1863.

Cruz, J., Savage, L., Zegarac, R., Hall, C., Satoh-Cruz, M., Davis, G., Kovac, W., Chen, J., Kramer, D., 2016. Dynamic environmental photosynthetic imaging reveals emergent phenotypes. Cell Syst. 2, 365–377.

De Groot, C., Van Den Boogaard, R., Marcelis, L., Harbinson, J., Lambers, H., 2003. Contrasting effects of N and P deprivation on the regulation of photosynthesis in tomato plants in relation to feedback limitation. J. Exp. Bot. 54, 1957–1967.

Degen, G.E., Orr, D.J., Carmo-Silva, E., 2021. Heat-induced changes in the abundance of wheat Rubisco activase isoforms. New Phytol. 229, 1298–1311.

Demmig-Adams, B., Adams, W.W., 1992. Photoprotection and other responses of plants to high light stress. Annu. Rev. Plant Physiol. Plant Mol. Biol. 43, 599–626.

Demmig-Adams, B., Adams III, W.W., 2006. Photoprotection in an ecological context: the remarkable complexity of thermal energy dissipation. New Phytol. 172, 11–21.

Demmig-Adams, B., Cohu, C., Muller, O., Adams, W., 2012. Modulation of photosynthetic energy conversion efficiency in nature: from seconds to seasons. Photosynth. Res. 113, 75–88.

Dietzel, L., Pfannschmidt, T., 2008. Photosynthetic acclimation to light gradients in plant stands comes out of shade. Plant Signal. Behav. 3, 1116–1118.

Ding, Y., Li, H., Zhang, X., Xie, Q., Gong, Z., Yang, S., 2015. OST1 kinase modulates freezing tolerance by enhancing ICE1 stability in Arabidopsis. Dev. Cell 32 (3), 278–289.

Eckardt, N., Portis, A., 1997. Heat denaturation profiles of ribulose-1,5-bisphosphate carboxylase/oxygenase (Rubisco) and Rubisco activase and the inability of Rubisco activase to restore activity of heat-denatured Rubisco. Plant Physiol. 113, 243–248.

Edwards, G., Walker, D., 1983. C3, C4: Mechanisms and Cellular and Environmental Regulation of Photosynthesis. Blackwell Scientific Publication, Oxford.

Eisenhut, M., Bräutigam, A., Timm, S., Florian, A., Tohge, T., Fernie, A., Bauwe, H., Weber, A., 2017. Photorespiration is crucial for dynamic response of photosynthetic metabolism and stomatal movement to altered CO_2 availability. Mol. Plant 10, 47–61.

Evans, J.R., 2021. Mesophyll conductance: walls, membranes and spatial complexity. New Phytol. 229 (4), 1864–1876.

Evans, J.R., Clarke, V.C., 2019. The nitrogen cost of photosynthesis. J. Exp. Bot. 70, 7–15.

Evans, J.R., Poorter, H., 2001. Photosynthetic acclimation of plants to growth irradiance: the relative importance of specific leaf area and nitrogen partitioning in maximizing carbon gain. Plant Cell Environ. 24, 755–767.

Evans, J., Seemann, J., 1989. The allocation of nitrogen in the photosynthetic apparatus: costs, consequences and control. In: Briggs, W.R. (Ed.), Photosynthesis. A.R. Liss, pp. 183–205. New York.

Fabre, D., Yin, X., Dingkuhn, M., Clément-Vidal, A., Roques, S., Rouan, L., Soutiras, A., Luquet, D., 2019. Is triose phosphate utilization involved in the feedback inhibition of photosynthesis in rice under conditions of sink limitation? J. Exp. Bot. 70 (20), 5773–5785.

Farquhar, G.D., Caemmerer, S., Berry, J.A., 1980. A biochemical model of photosynthetic CO_2 assimilation in leaves of C3 species. Planta 149, 78–90.

Feng, L., Wang, K., Li, Y., Tan, Y., Kong, J., Li, H., Li, Y., Zhu, Y., 2007. Overexpression of SBPase enhances photosynthesis against high temperature stress in transgenic rice plants. Plant Cell Rep. 26 (9), 1635–1646.

Fichman, Y., Mittler, R., 2020. Rapid systemic signaling during abiotic and biotic stresses: is the ROS wave master of all trades? Plant J. 102 (5), 887–896.

Flannery, S.E., Hepworth, C., Wood, W.H.J., Pastorelli, F., Hunter, C.N., Dickman, M.J., Jackson, P.J., Johnson, M.P., 2021. Developmental acclimation of the thylakoid proteome to light intensity in Arabidopsis. Plant J. 105, 223–244.

Galmés, J., Flexas, J., Keys, A.J., Cifre, J., Mitchell, R.A.C., Madgwick, P.J., Haslam, R.P., Medrano, H., Parry, M.A.J., 2005. Rubisco specificity factor tends to be larger in plant species from drier habitats and in species with persistent leaves. Plant Cell Environ. 28, 571–579.

Geiger, D., 1994. General lighting requirements for photosynthesis. In: International Lighting in Controlled Environments Workshop. Wisconsin University, pp. 3–18.

Glick, R., Schlagnhaufer, C., Arteca, R., Pell, E., 1995. Ozone-induced ethylene emission accelerates the loss of ribulose-1,5-bisphosphate carboxylase/oxygenase and nuclear-encoded mRNAs in senescing potato leaves. Plant Physiol. 109, 891–898.

Grossiord, C., Buckley, T.N., Cernusak, L.A., Novick, K.A., Poulter, B., Siegwolf, R.T.W., Sperry, J.S., McDowell, N.G., 2020. Plant responses to rising vapor pressure deficit. New Phytol. 226, 1550–1566.

Gutteridge, S., Gatenby, A., 1995. Rubisco synthesis, assembly, mechanism, and regulation. Plant Cell 7, 809–819.

Harbinson, J., 1994. The response of thylakoid electron transport and light utilization efficiency to sink limitation of photosynthesis. In: Baker, N., Bowers, J. (Eds.), Photoinhibition of Photosynthesis, From Molecular Mechanisms to the Field. BIOS Scientific Publishers, Oxford, pp. 273–295.

Hendrickson, L., Ball, M., Wood, J., Chow, W., Furbank, R., 2004. Low temperature effects on photosynthesis and growth of grapevine. Plant Cell Environ. 27, 795–809.

Hikosaka, K., 2005. Leaf canopy as a dynamic system: ecophysiology and optimality in leaf turnover. Ann. Bot. 95, 521–533.

Hikosaka, K., Terashima, I., 1996. Nitrogen partitioning among photosynthetic components and its consequenses in sun and shade plants. Funct. Ecol. 10, 335–343.

Hikosaka, K., Murakami, A., Hirose, T., 1999. Balancing carboxylation and regeneration of ribulose-1,5-bisphosphate in leaf photosynthesis: temperature acclimation of an evergreen tree, Quercus myrsinaefolia. Plant Cell Environ. 22, 841–849.

Hikosaka, K., Ishikawa, K., Borjigidai, A., Muller, O., Onoda, Y., 2006. Temperature acclimation of photosynthesis: mechanisms involved in the changes in temperature dependence of photosynthetic rate. J. Exp. Bot. 57, 291–302.

Horton, P., Ruban, A., 2005. Molecular design of the photosystem II light-harvesting antenna: photosynthesis and photoprotection. J. Exp. Bot. 56, 365–373.

Hubbart, S., Smillie, I., Heatley, M., Swarup, R., Foo, C., Zhao, L., Murchie, E., 2018. Enhanced thylakoid photoprotection can increase yield and canopy radiation use efficiency in rice. Commun. Biol. 1, 22.

Jacob, J., Lawlor, D., 1992. Dependence of photosynthesis of sunflower and maize leaves on phosphate supply, ribulose-1,5-bisphosphate carboxylase/oxygenase activity, and ribulose-1,5-bisphosphate pool size. Plant Physiol. 98, 801–807.

Jalakas, P., Merilo, E., Kollist, H., Brosché, M., 2018. ABA-mediated regulation of stomatal density is OST 1-independent. Plant Direct 2 (9), e00082.

Johnson, G., Murchie, E.H., 2011. Gas exchange measurements for the determination of photosynthetic efficiency in *Arabidopsis* leaves. Methods Mol. Biol. 775, 311–326.

Johnson, I.R., Thornley, J.H.M., Frantz, J.M., Bugbee, B., 2010. A model of canopy photosynthesis incorporating protein distribution through the canopy and its acclimation to light, temperature and CO_2. Ann. Bot. 106, 735–749.

Kaiser, E., Morales, A., Harbinson, J., 2018. Fluctuating light takes crop photosynthesis on a rollercoaster ride. Plant Physiol. 176, 977–989.

Kaiser, E., Galvis, V., Armbruster, U., 2019. Efficient photosynthesis in dynamic light environments: a chloroplast's perspective. Biochem. J. 476, 2725–2741.

Kanaoka, M.M., Pillitteri, L.J., Fujii, H., Yoshida, Y., Bogenschutz, N.L., Takabayashi, J., Zhu, J.-K., Torii, K.U., 2008. SCREAM/ICE1 and SCREAM2 specify three cell-state transitional steps leading to Arabidopsis stomatal differentiation. Plant Cell 20 (7), 1775–1785.

Kangasjärvi, S., Nurmi, M., Tikkanen, M., Aro, E.-M., 2009. Cell-specific mechanisms and systemic signalling as emerging themes in light acclimation of C3 plants. Plant Cell Environ. 32, 1230–1240.

Kidokoro, S., Kim, J.-S., Ishikawa, T., Suzuki, T., Shinozaki, K., Yamaguchi-Shinozaki, K., 2020. DREB1A/CBF3 is repressed by transgene-induced DNA methylation in the Arabidopsis ice1-1 mutant. Plant Cell 32 (4), 1035–1048.

Kim, T.H., Böhmer, M., Hu, H., Nishimura, N., Schroeder, J.I., 2010. Guard cell signal transduction network: advances in understanding abscisic acid, CO_2, and Ca^{2+} signaling. Annu. Rev. Plant Biol. 61, 561.

Ko, D., Helariutta, Y., 2017. Shoot–root communication in flowering plants. Curr. Biol. 27, R973–R978.

Kollist, T., Moldau, H., Rasulov, B., Oja, V., Rämma, H., Hüve, K., Jaspers, P., Kangasjärvi, J., Kollist, H., 2007. A novel device detects a rapid ozone-induced transient stomatal closure in intact Arabidopsis and its absence in abi2 mutant. Physiol. Plant. 129 (4), 796–803.

Kollist, H., Nuhkat, M., Roelfsema, M.R.G., 2014. Closing gaps: linking elements that control stomatal movement. New Phytol. 203 (1), 44–62.

Koritz, H.G., Went, F., 1953. The physiological action of smog on plants. I. Initial growth and transpiration studies. Plant Physiol. 28 (1), 50.

Körner, Ch., Larcher, W., 1988. Plant life in cold climates. In: Symposia of the Society for Experimental Biology. vol. 42, pp. 25–57.

Kromdijk, J., Głowacka, K., Leonelli, L., Gabilly, S., Iwai, M., Niyogi, K., Long, S., 2016. Improving photosynthesis and crop productivity by accelerating recovery from photoprotection. Science 354, 857–861.

Lambers, H., Oliviera, R.S., 2019. Plant Physiological Ecology. Springer Nature, Switzerland.

Larue, H., Zhang, S., 2019. SCREAM in the making of stomata. Nat. Plant 5 (7), 648–649.

Law, R., Crafts-Brandner, S., 1999. Inhibition and acclimation of photosynthesis to heat stress is closely correlated with activation of ribulose-1,5-bisphosphate carboxylase/oxygenase. Plant Physiol. 120, 173–181.

Lawson, T., Blatt, M., 2014. Stomatal size, speed, and responsiveness impact on photosynthesis and water use efficiency. Plant Physiol. 164, 1556–1570.

Lehmeier, C., Pajor, R., Lundgren, M.R., et al., 2017. Cell density and airspace patterning in the leaf can be manipulated to increase leaf photosynthetic capacity. Plant J. 92, 981–994.

Leuning, R., 1995. A critical appraisal of a combined stomatal-photosynthesis model for C3 plants. Plant Cell Environ. 18, 339–355.

Leuning, R., 2002. Temperature dependence of two parameters in a photosynthesis model. Plant Cell Environ. 25, 1205–1210.

Leverenz, J., 1994. Factors determining the nature of the light dosage response curve of leaves. In: Baker, N., Bowyer, J.

(Eds.), Photoinhibition of Photosynthesis—Molecular Mechanisms to the Field. BIOS Scientific Publishers, pp. 239–254.

Li, X.-P., Phippard, A., Pasari, J., Niyogi, K., 2002. Structure-function analysis of photosystem II subunit S (PsbS) in vivo. Funct. Plant Biol. 29, 1131–1139.

Lin, M.T., Stone, W.D., Chaudhari, V., Hanson, M.R., 2020. Small subunits can determine enzyme kinetics of tobacco Rubisco expressed in Escherichia coli. Nat. Plant 6 (10), 1289–1299.

Long, S.P., 1983. C_4 photosynthesis at low temperatures. Plant Cell Environ. 6 (4), 345–363.

Long, S.P., Drake, B.G., 1991. Effect of the long-term elevation of CO(2) concentration in the field on the quantum yield of photosynthesis of the C(3) sedge, Scirpus olneyi. Plant Physiol. 96, 221–226.

Long, S., Spence, A., 2013. Toward cool C4 crops. Annu. Rev. Plant Biol. 64, 701–722.

Long, S., Humphries, S., Falkowski, P., 1994. Photoinhibition of photosynthesis in nature. Annu. Rev. Plant Physiol. Plant Mol. Biol. 45, 633–662.

Loveys, B., Scheurwater, I., Pons, T., Fitter, A., Atkin, O., 2002. Growth temperature influences the underlying components of relative growth rate: an investigation using inherently fast-and slow-growing plant species. Plant Cell Environ. 25 (8), 975–988.

Matthews, J., Vialet-Chabrand, S., Lawson, T., 2018. Acclimation to fluctuating light impacts the rapidity and diurnal rhythm of stomatal conductance. Plant Physiol. 01809.2017.

McLaughlin, S.B., Nosal, M., Wullschleger, S.D., Sun, G., 2007. Interactive effects of ozone and climate on tree growth and water use in a southern Appalachian forest in the USA. New Phytol. 174 (1), 109–124.

Mickelbart, M.V., Hasegawa, P.M., Bailey-Serres, J., 2015. Genetic mechanisms of abiotic stress tolerance that translate to crop yield stability. Nat. Rev. Genet. 16 (4), 237.

Miller, M., O'Cualain, R., Selley, J., Knight, D., Karim, M., Hubbard, S., Johnson, G., 2017. Dynamic acclimation to high light in *Arabidopsis thaliana* involves widespread reengineering of the leaf proteome. Front. Plant Sci. 8, 1239.

Mills, G., Hayes, F., Wilkinson, S., Davies, W.J., 2009. Chronic exposure to increasing background ozone impairs stomatal functioning in grassland species. Glob. Change Biol. 15 (6), 1522–1533.

Milne, R.J., Grof, C.P., Patrick, J.W., 2018. Mechanisms of phloem unloading: shaped by cellular pathways, their conductances and sink function. Curr. Opin. Plant Biol. 43, 8–15.

Mo, Q., Li, Z., Sayer, E.J., et al., 2019. Foliar phosphorus fractions reveal how tropical plants maintain photosynthetic rates despite low soil phosphorus availability. Funct. Ecol. 33, 503–513.

Moore, C.E., Meacham-Hensold, K., Lemonnier, P., Slattery, R.A., Benjamin, C., Bernacchi, C.J., Lawson, T., Cavanagh, A.P., 2021. The effect of increasing temperature on crop photosynthesis: from enzymes to ecosystems. J. Exp. Bot. 72, 2822–2844.

Morgan, P., Ainsworth, E., Long, S.P., 2003. How does elevated ozone impact soybean? A meta-analysis of photosynthesis, growth and yield. Plant Cell Environ. 26 (8), 1317–1328.

Morita, S., Nakano, H., 2011. Nonstructural carbohydrate content in the stem at full heading contributes to high performance of ripening in heat-tolerant rice cultivar nikomaru. Crop Sci. 51, 818–828.

Murchie, E.H., Horton, P., 1997. Acclimation of photosynthesis to irradiance and spectral quality in British plant species: chlorophyll content, photosynthetic capacity and habitat preference. Plant Cell Environ. 20, 438–448.

Murchie, E.H., Niyogi, K.K., 2011. Manipulation of photoprotection to improve plant photosynthesis. Plant Physiol. 155, 86–92.

Murchie, E.H., Hubbart, S., Chen, Y., Peng, S., Horton, P., 2002. Acclimation of rice photosynthesis to irradiance under field conditions. Plant Physiol. 130, 1999–2010.

Murchie, E.H., Hubbart, S., Peng, S., Horton, P., 2005. Acclimation of photosynthesis to high irradiance in rice: gene expression and interactions with leaf development. J. Exp. Bot. 56, 449–460.

Murchie, E.H., Pinto, M., Horton, P., 2009. Agriculture and the new challenges for photosynthesis research. New Phytol. 181 (3), 532–552.

Murchie, E.H., Burgess, A., Reynolds, M., 2018a. Crop radiation capture and use efficiency. In: Encyclopedia of Sustainability Science and Technology. Springer International Publishing, New York, pp. 2615–2638.

Murchie, E.H., Kefauver, S., Araus, J., Muller, O., Rascher, U., Flood, P., Lawson, T., 2018b. Measuring the dynamic photosynthome. Ann. Bot. 122, 207–220.

Naidu, S., Long, S., 2004. Potential mechanisms of low-temperature tolerance of C4 photosynthesis in Miscanthus x giganteus: an in vivo analysis. Planta 220, 145–155.

Neill, S., Barros, R., Bright, J., Desikan, R., Hancock, J., Harrison, J., Morris, P., Ribeiro, D., Wilson, I., 2008. Nitric oxide, stomatal closure, and abiotic stress. J. Exp. Bot. 59 (2), 165–176.

Niinemets, Ü., Anten, N., 2009. Packing the photosynthetic machinery: from leaf to canopy. In: Laisk, A., Nedbal, L., Govindjee (Eds.), Photosynthesis in Silico: Understanding Complexity From Molecules to Ecosystems. Springer, Dordrecht, pp. 363–399.

Niinemets, U., Tenhunen, J.D., 1997. A model separating leaf structural and physiological effects on carbon gain along light gradients for the shade-tolerant species Acer saccharum. Plant Cell Environ. 20, 845–866.

Niyogi, K.K., Li, X.P., Rosenberg, V., Jung, H.S., 2005. Is PsbS the site of non-photochemical quenching in photosynthesis? J. Exp. Bot. 56, 375–382.

Nobel, P.S., Forseth, I.N., Long, S.P., 1993. Canopy structure and light interception. In: Hall, D.O., Scurlock, J.M.O, Bolhàr-Nordenkampf, H.R., Leegood, R.C., Long, S.P. (Eds.), Photosynthesis and Production in a Changing Environment. Springer, Dordrecht, pp. 79–90.

O'Leary, B.M., Asao, S., Millar, A.H., Atkin, O.K., 2019. Core principles which explain variation in respiration across biological scales. New Phytol. 222, 670–686.

Oguchi, R., Hikosaka, K., Hirose, T., 2003. Does the photosynthetic light-acclimation need change in leaf anatomy? Plant Cell Environ. 26, 505–512.

Oguchi, R., Hikosaka, K., Hirose, T., 2005. Leaf anatomy as a constraint for photosynthetic acclimation: differential responses in leaf anatomy to increasing growth irradiance among three deciduous trees. Plant Cell Environ. 28, 916–927.

Ölçer, H., Lloyd, J., Raines, C., 2001. Photosynthetic capacity is differentially affected by reductions in sedoheptulose-1,7-bisphosphatase activity during leaf development in transgenic tobacco plants. Plant Physiol. 125, 982–989.

Öquist, G., Huner, N.P.A., 2003. Photosynthesis of overwintering vergreen plants. Annu. Rev. Plant Biol. 54, 329–355.

Ort, D., Baker, N.R., 1988. Consideration of photosynthetic efficiency at low light as a major determinant of crop photosynthetic performance. Plant Physiol. Biochem. 26, 555–565.

Osborne, C., Wythe, E., Ibrahim, D., Gilbert, M., Ripley, B., 2007. Low temperature effects on leaf physiology and survivorship in the C 3 and C4 subspecies of Alloteropsis semialata. J. Exp. Bot. 59, 1743–1754.

Oszvald, M., Primavesi, L.F., Griffiths, C.A., Cohn, J., Basu, S.S., Nuccio, M.L., Paul, M.J., 2018. Trehalose 6-phosphate regulates photosynthesis and assimilate partitioning in reproductive tissue. Plant Physiol. 176, 2623–2638.

Parry, M., Madgwick, P., Carvalho, J., Andralojc, P., 2007. Prospects for increasing photosynthesis by overcoming the limitations of Rubisco. J. Agric. Sci. 145, 31–43.

Parry, M., Keys, A., Madgwick, P., Carmo-Silva, A., Andralojc, P., 2008. Rubisco regulation: a role for inhibitors. J. Exp. Bot. 59, 1569–1580.

Parry, M.A.J., Andralojc, P.J., Scales, J.C., Salvucci, M.E., Carmo-Silva, E.A., Alonso, H., Whitney, S.M., 2013. Rubisco activity and regulation as targets for crop improvement. J. Exp. Bot. 64, 717–730.

Pattison, P., Tsao, J., Brainard, G., Bugbee, B., 2018. LEDs for photons, physiology and food. Nature 563, 493–500.

Paul, M., Driscoll, S., 1997. Sugar repression of photosynthesis: the role of carbohydrates in signalling nitrogen deficiency through source:sink imbalance. Plant Cell Environ. 20, 110–116.

Paul, M.J., Eastmond, P.J., 2020. Turning sugar into oil: making photosynthesis blind to feedback inhibition. J. Exp. Bot. 71, 2216–2218.

Paul, M., Foyer, C., 2001. Sink regulation of photosynthesis. J. Exp. Bot. 52, 1383–1400.

Pearce, R.S., 2001. Plant freezing and damage. Ann. Bot. 87 (4), 417–424.

Pearcy, R., Sims, D.A., 1994. Photosynthetic acclimation to changing light environments: scaling from the leaf to the whole plant. In: Caldwell, M., Pearcy, R.W. (Eds.), Exploitation of Environmental Heterogeneity by Plants: Ecophysiological Processes Above and Below Ground. Academic Press, London, pp. 145–170.

Pearcy, R., Gross, L., He, D., 1997. An improved dynamic model of photosynthesis for estimation of carbon gain in sunfleck light regimes. Plant Cell Environ. 20, 411–424.

Perdomo, J.A., Capó-Bauçà, S., Carmo-Silva, E., Galmés, J., 2017. Rubisco and Rubisco activase play an important role in the biochemical limitations of photosynthesis in rice, wheat, and maize under high temperature and water deficit. Front. Plant Sci. 8, 490.

Pieters, A., Paul, M., Lawlor, D., 2001. Low sink demand limits photosynthesis under Pi deficiency. J. Exp. Bot. 52, 1083–1091.

Pons, T., 2012. Interaction of temperature and irradiance effects on photosynthetic acclimation in two accessions of *Arabidopsis thaliana*. Photosynth. Res. 113, 207–219.

Qu, Y., Sakoda, K., Fukayama, H., Kondo, E., Suzuki, Y., Makino, A., Terashima, I., Yamori, W., 2021. Overexpression of both Rubisco and Rubisco activase rescues rice photosynthesis and biomass under heat stress. Plant Cell Environ. 44, 2308–2320.

Raines, C., 2003. The Calvin cycle revisited. Photosynth. Res. 75, 1–10.

Ray, D.K., Mueller, N.D., West, P.C., Foley, J.A., 2013. Yield trends are insufficient to double global crop production by 2050. PLoS One 8, e66428.

Retkute, R., Smith-Unna, S., Smith, R., Burgess, A., Jensen, O., Johnson, G., Preston, S., Murchie, E.H., 2015. Exploiting heterogeneous environments: does photosynthetic acclimation optimize carbon gain in fluctuating light? J. Exp. Bot. 66, 2437–2447.

Reynolds, M., Foulkes, J., Furbank, R., Griffiths, S., King, J., Murchie, E., Parry, M., Slafer, G., 2012. Achieving yield gains in wheat. Plant Cell Environ. 35, 1799–1823.

Robinson, S., Portis, A., 1989. Adenosine triphosphate hydrolysis by purified Rubisco activase. Arch. Biochem. Biophys. 268, 93–99.

Ruban, A., Johnson, M., Duffy, C., 2012. The photoprotective molecular switch in the photosystem II antenna. Biochim. Biophys. Acta Bioenerg. 1817, 167–181.

Ruelland, E., Vaultier, M., Zachowski, A., Hurry, V., 2009. Chapter 2 Cold signalling and cold acclimation in plants.

In: Kader, J., Delseny, M. (Eds.), Advances in Botanical Research. Academic Press, Cambridge, MA, pp. 35–150.

Sage, R., 2002. Variation in the kcat of Rubisco in C3 and C4 plants and some implications for photosynthetic performance at high and low temperature. J. Exp. Bot. 53, 609–620.

Sage, R.F., Kubien, D.S., 2007. The temperature response of C3 and C4 photosynthesis. Plant Cell Environ. 30, 1086–1106.

Sage, R., Sharkey, T., Pearcy, R., 1990. The effect of leaf nitrogen and temperature on the CO_2 response of photosynthesis in the C3 *dicot Chenopodium* album L. Aust. J. Plant Physiol. 17, 135–148.

Salvucci, M., Crafts-Brandner, S., 2004. Inhibition of photosynthesis by heat stress: the activation state of Rubisco as a limiting factor in photosynthesis. Physiol. Plant. 120, 179–186.

Sassenrath, G., Ort, D., Portis, A., 1990. Impaired reductive activation of stromal bisphosphatases in tomato leaves following low-temperature exposure at high light. Arch. Biochem. Biophys. 282, 302–308.

Sawada, S., Usuda, H., Tsukui, T., 1992. Participation of inorganic orthophosphate in regulation of the ribulose-1,5-bisphosphate carboxylase activity in response to changes in the photosynthetic source-sink balance. Plant Cell Physiol. 33, 943–949.

Schneider, T., Bolger, A., Zeier, J., Preiskowski, S., Benes, V., Trenkamp, S., Usadel, B., Farré, E., Matsubara, S., 2019. Fluctuating light interacts with time of day and leaf development stage to reprogram gene expression. Plant Physiol. 179, 1632–1657.

Sharkey, T., 1988. Estimating the rate of photorespiration in leaves. Physiol. Plant. 73, 147–152.

Sharwood, R.E., 2020. Mix-and-match Rubisco subunits. Nat. Plant 6 (10), 1199–1200.

Sims, D.A., Pearcy, R.W., 1992. Response of leaf anatomy and photosynthetic capacity in *Alocasia macrorrhiza* (Araceae) to a transfer from low to high light. Am. J. Bot. 79, 449.

Soleh, M., Tanaka, Y., Kim, S., Huber, S., Sakoda, K., Shiraiwa, T., 2017. Identification of large variation in the photosynthetic induction response among 37 *soybean* [Glycine max (L.) Merr.] genotypes that is not correlated with steady-state photosynthetic capacity. Photosynth. Res. 131, 305–315.

Stiles, K.A., Van Volkenburgh, E., 2002. Light-regulated leaf expansion in two Populus species: dependence on developmentally controlled ion transport. J. Exp. Bot. 53, 1651–1657.

Stitt, M., Grosse, H., 1988. Interactions between sucrose synthesis and CO_2 fixation IV. Temperature-dependent adjustment of the relation between sucrose synthesis and CO_2 fixation. J. Plant Physiol. 133, 392–400.

Sudhir, P., Murthy, S., 2004. Effects of salt stress on basic processes of photosynthesis. Photosynthetica 42, 481–486.

Suzuki, N., Devireddy, A., Inupakutika, M., Baxter, A., Miller, G., Song, L., Shulaev, E., Azad, R., Shulaev, V., Mittler, R., 2015. Ultra-fast alterations in mRNA levels uncover multiple players in light stress acclimation in plants. Plant J. 84, 760–772.

Takahashi, F., Shinozaki, K., 2019. Long-distance signaling in plant stress response. Curr. Opin. Plant Biol. 47, 106–111.

Takenaka, A., 1989. Optimal leaf photosynthetic capacity in terms of utilizing a natural light environment. J. Theor. Biol. 139, 517–529.

Tardieu, F., Simonneau, T., Muller, B., 2018. The physiological basis of drought tolerance in crop plants: a scenario-dependent probabilistic approach. Annu. Rev. Plant Biol. 69, 733–759.

Taylor, S.H., Long, S.P., 2017. Slow induction of photosynthesis on shade to sun transitions in wheat may cost at least 21% of productivity. Philos. Trans. R. Soc. B 372, 20160543.

Thomashow, M.F., Torii, K.U., 2020. SCREAMing twist on the role of ICE1 in freezing tolerance. Am. Soc. Plant Biol. 32 (4), 816–819.

Tigchelaar, M., Battisti, D.S., Naylor, R.L., Ray, D.K., 2018. Future warming increases probability of globally synchronized maize production shocks. Proc. Natl. Acad. Sci. U. S. A. 115, 6644–6649.

Tikkanen, M., Grieco, M., Kangasjärvi, S., Aro, E., 2010. Thylakoid protein phosphorylation in higher plant chloroplasts optimizes electron transfer under fluctuating light. Plant Physiol. 152, 723–735.

Tikkanen, M., Grieco, M., Nurmi, M., Rantala, M., Suorsa, M., Aro, E., 2012. Regulation of the photosynthetic apparatus under fluctuating growth light. Philos. Trans. R. Soc., B 367, 3486–3493.

Townsend, A., Retkute, R., Chinnathambi, K., Randall, J., Foulkes, J., Carmo-Silva, E., Murchie, E.H., 2018. Suboptimal acclimation of photosynthesis to light in wheat canopies. Plant Physiol. 176, 1233–1246.

Vahisalu, T., Puzõrjova, I., Brosché, M., Valk, E., Lepiku, M., Moldau, H., Pechter, P., Wang, Y.S., Lindgren, O., Saljärvi, J., 2010. Ozone-triggered rapid stomatal response involves the production of reactive oxygen species, and is controlled by SLAC1 and OST1. Plant J. 62 (3), 442–453.

Vialet-Chabrand, S., Matthews, J., Simkin, A., Raines, C., Lawson, T., 2017. Importance of fluctuations in light on plant photosynthetic acclimation. Plant Physiol. 173, 2163–2179.

von Caemmerer, S., 2000. Biochemical Models of Leaf Photosynthesis. CSIRO Publishing.

von Caemmerer, S., Quick, W.P., 2000. Rubisco: physiology in vivo. In: Leegood, R.C., Sharkey, T.D., von Caemmerer, S. (Eds.), Photosynthesis. Advances in Photosynthesis and Respiration. vol. 9. Springer, Dordrecht, pp. 85–113.

Vu, J.C.V., 2005. Rising atmospheric CO_2 and C_4 photosynthesis. Handbook of Photogynthesis, second ed. CRC

Press, Taylor & Francis Group, Boca Raton, FL, pp. 315–326.

Walters, R., 2005. Towards an understanding of photosynthetic acclimation. J. Exp. Bot. 56, 435–447.

Walters, R., Horton, P., 1994. Acclimation of Arabidopsis thaliana to the light environment—changes in composition of the photosynthetic apparatus. Planta 195, 248–256.

Wang, D., Naidu, S., Portis, A., Moose, S., Long, S., 2008. Can the cold tolerance of C_4 photosynthesis in Miscanthus x giganteus relative to Zea mays be explained by differences in activities and thermal properties of Rubisco? J. Exp. Bot. 59, 1779–1787.

Way, D., Yamori, W., 2013. Thermal acclimation of photosynthesis: on the importance of adjusting our definitions and accounting for thermal acclimation of respiration. Photosynth. Res. 119, 89–100.

Wei, H., Jing, Y., Zhang, L., Kong, D., 2021. Phytohormones and their cross-talk in regulating stomatal development and patterning. J. Exp. Bot. 72 (7), 2356–2370.

Werner, C., Ryel, R.J., Correia, O., Beyschlag, W., 2001. Effects of photoinhibition on whole-plant carbon gain assessed with a photosynthesis model. Plant Cell Environ. 24, 27–40.

Wilkinson, S., Davies, W.J., 2010. Drought, ozone, ABA and ethylene: new insights from cell to plant to community. Plant Cell Environ. 33 (4), 510–525.

Wise, R., Olson, A., Schrader, S., Sharkey, T., 2004. Electron transport is the functional limitation of photosynthesis in field-grown Pima cotton plants at high temperature. Plant Cell Environ. 27, 717–724.

Yamasaki, T., Yamakawa, T., Yamane, Y., Koike, H., Satoh, K., Katoh, S., 2002. Temperature acclimation of photosynthesis and related changes in photosystem II electron transport in winter wheat. Plant Physiol. 128, 1087–1097.

Yamori, W., Hikosaka, K., Way, D., 2014. Temperature response of photosynthesis in C_3, C_4, and CAM plants: temperature acclimation and temperature adaptation. Photosynth. Res. 119, 101–117.

Zhang, C., Turgeon, R., 2018. Mechanisms of phloem loading. Curr. Opin. Plant Biol. 43, 71–75.

Zhao, C., Liu, B., Piao, S., Wang, X., Lobell, D.B., Huang, Y., Huang, M., Yao, Y., Bassu, S., Ciais, P., et al., 2017. Temperature increase reduces global yields of major crops in four independent estimates. Proc. Natl. Acad. Sci. U. S. A. 114, 9326–9331.

PART III

Action

CHAPTER

7

Improving light harvesting

Zeno Guardini, Rodrigo L. Gomez, and Luca Dall'Osto

Department of Biotechnology, University of Verona, Verona, Italy

1 Functional architecture and molecular physiology of light harvesting in plants and green algae

'Net primary production' is the difference between carbon (CO_2) fixed into sugars during photosynthesis and carbon released by respiration (Clark et al., 2001), that is, it represents the energy stored in plant material that becomes available as a source of food, feed, and fibre. Sunlight, the driving force of photosynthesis, is a diffuse energy source that must be concentrated before conversion into chemical form can occur. So light harvesting is a fundamental step in net primary production; indeed, it has been identified as a target for synthetic improvement of biomass and bioenergy production in plants and algae.

In each photosystem (PS), sunlight is absorbed by chlorophylls (Chl) and excitation energy is rapidly transferred to a reaction centre (RC). There, a charge separation event fuels the electron transport chain and promotes water oxidation and $NADP^+$ reduction, generating a trans-membrane proton motive force used for ATP synthesis (Nelson and Ben-Shem, 2004). Each PS is made up of a core complex that houses the RC and an array of membrane-embedded light-harvesting complexes (LHCs), shaping a modular antenna system that surrounds the core. Together, these elements form a so-called supercomplex (Fig. 1) (Croce and van Amerongen, 2020).

Evolution generated a wide group of photoautotrophs, ranging from cyanobacteria to higher plants. These optimised the photosynthesis process, making it possible to occupy the most diverse ecological niches, which occurred along with the enlargement of the LHC superfamily (Dall'Osto et al., 2015), in contrast with the high level of conservation in the composition of core complexes (Neilson and Durnford, 2010). Antenna subunits comprise about 40% of the protein content in the thylakoid membrane. They all share common structural motifs, with membrane-spanning regions hosting many Chl-binding residues (Pan et al., 2011).

In addition to enlarging the absorption capacity of the supercomplexes, antenna systems regulate light-use efficiency and provide an enhanced level of photoprotection. Indeed, while the efficiency of energy conversion is maximal under moderate irradiance levels, photosynthesis is hampered under excess light (EL) conditions. In these cases, the concentration of Chl singlet excited states ($^1Chl^*$) in the photosynthetic machinery exceeds the capacity to use this energy, a phenomenon known as

Photosynthesis in Action
https://doi.org/10.1016/B978-0-12-823781-6.00005-8

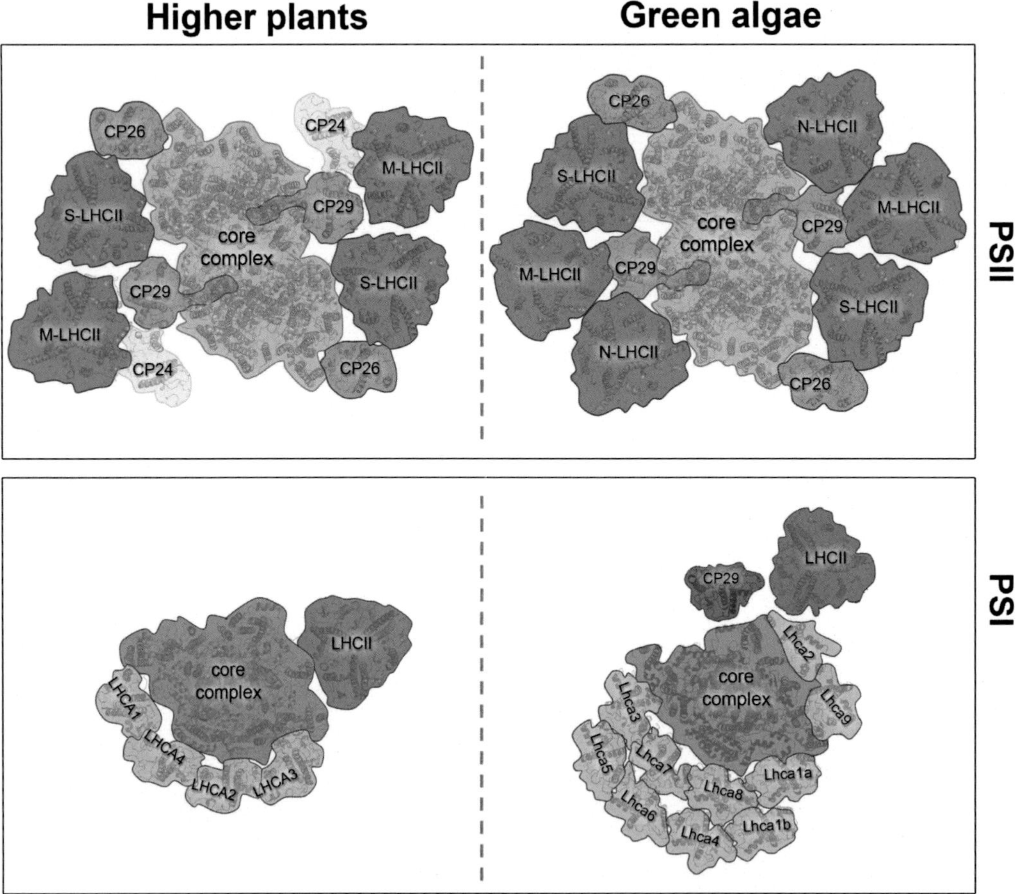

FIG. 1 Models of the supramolecular organisation of photosystems in land plants and green microalga *Chlamydomonas reinhardtii*. (Upper part) PSII supercomplex of plants (left panel) and green algae (right panel). Models were assembled based on the 3D structures of *Pisum sativum* $C_2S_2M_2$ (Su et al., 2017) (PDB 5XNM) and *C. reinhardtii* $C_2S_2M_2N_2$ (Shen et al., 2019) (PDB 6KAF). (Lower part) PSI supercomplexes of land plants (left panel) and *C. reinhardtii* (right panel). Models were assembled based on the 3D structures of *Zea mays* PSI-LHCI-LHCII (Pan et al., 2018) (PDB 5ZJI), *C. reinhardtii* PSI-LHCI (Suga et al., 2019) (PDB 6JO5), and the projection map of algal PSI upon state transitions (Drop et al., 2014). The structures of CP29 and LHCII (Shen et al., 2019) (PDB 6KAF) were used.

overexcitation. Under these conditions, the probability of the spontaneous formation of Chl triplets (^{3}Chl*) increases, leading to the release of singlet oxygen ($^1O_2$) (Fig. 2), a reactive oxygen species (ROS) that might act as damaging agent as well as signalling molecule (Dogra and Kim, 2020).

Molecular safety valves are built-in LHC proteins that catalyse detoxification of $^1O_2$ or prevent its formation by quenching ^{1}Chl* excited states (see above) (Li et al., 2009). The evolutionary selection of LHCs that are more efficient in their photoprotective responses was probably crucial during transition from an aquatic to a land environment, in which a concomitant increase in O_2 concentration led to a higher risk of $^1O_2$ formation and photodamage (Triantaphylides et al., 2008). The evolution of

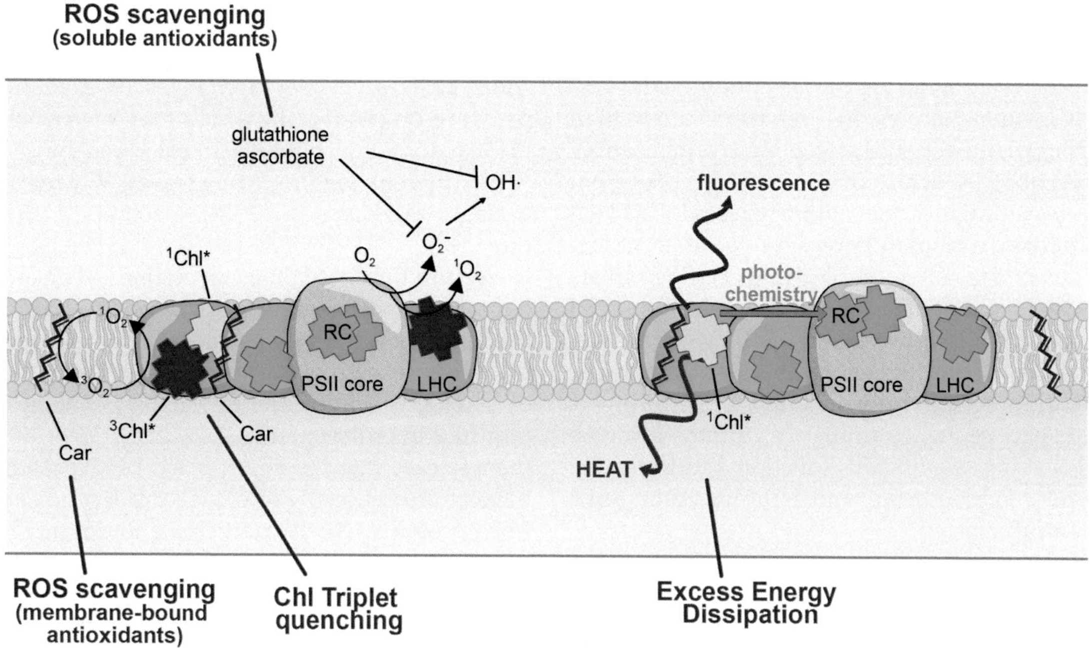

FIG. 2 Photoprotective mechanisms of PSII. (Left) In pigment-binding complexes, the transition ^{1}Chl* → ^{3}Chl* leads to $^1O_2$ release, and possibly the univalent reduction of O_2 produces O_2^- and OH•. Carotenoid (Car) scavenge ROS and modulate ^{3}Chl* yield. Other antioxidant systems include the soluble compounds glutathione and ascorbate. (Right) NPQ mechanism regulates ^{1}Chl* lifetime. Light absorbed by the PSII can be used for driving photosynthetic electron transport or can be lost as fluorescence or heat. Since these processes are in competition with each other, an increased efficiency of a process leads to a decreased efficiency of the two others. The mechanism of NPQ accounts for light-induced excitation decay by heat loss.

photosynthesis, therefore, occurred together with the diversification of LHC on multiple isoforms, which gave rise to a genetic complexity within a conserved supramolecular assembly. These events were important during evolution and led to a number of gene products tuned to a different balance between light harvesting and photoprotection capacity.

LHCII, the major antenna of higher plants (Liu et al., 2004), has three membrane-spanning helices, connected by both stroma- and lumen-exposed loops, and two amphipathic helices exposed to the lumen. The complex binds four carotenoids, eight Chl *a*, and six Chl *b* molecules. More recently, a cryo-electron microscopy structure resolved at 2.7 Å showed that the PSII supercomplex from *Pisum sativum* is comprised of a dimeric core-complex (C_2), and each half of the core binds the so-called trimeric antenna LHCII-S (strongly bound, named for their resistance to detachment by detergents) and one copy of each monomeric antenna proteins Lhcb4 (CP29), Lhcb5 (CP26), and Lhcb6 (CP24). The latter coordinates additional LHCII-L (loosely bound) trimers that accumulate in LL conditions. The whole supercomplex is thus labelled $C_2S_2M_2L_2$ (Su et al., 2017). In the green alga *Chlamydomonas reinhardtii*, an additional trimeric LHCII (LHCII-N) occupies the position of CP24, which is present in the higher plants PSII–LHCII but absent in this species (Fig. 1) (Shen et al., 2019).

The core complex of PSI is also endowed with a peripheral antenna system called LHCI (light-harvesting complex of PSI), which, in plants, consists of four antenna proteins (Lhca1–4), one copy each per supercomplex (Fig. 1) (Pan et al., 2018). The study of the PSI-LHCI supercomplex in organisms other than higher plants found differences in organisation: a recent 3D structure of PSI supercomplex from the green alga *C. reinhardtii* revealed 10 LHCI subunits distributed between two different locations of the supercomplex (Fig. 1) (Suga et al., 2019). LHCII is also known to move from PSII and bind to PSI, increasing its light-harvesting capacity. Beside the more abundant members, the LHC superfamily includes other proteins that share sequence similarity with the formers, yet having significant functional differences, namely PsbS, LhcSR, and light-harvesting-like (LIL) proteins.

PsbS is a four-helix protein that is essential for both the photoprotective mechanism of excess energy dissipation NPQ (nonphotochemical quenching, see above) and the EL-dependent reorganisation of the antenna system within PSII in plants. On the other hand, PsbS protein is transiently expressed in *C. reinhardtii* under EL acclimation (Redekop et al., 2020; Tibiletti et al., 2016).

LhcSR is also essential for NPQ but in green algae and mosses, since plants lack orthologue of *lhcsr* genes. In *C. reinhardtii*, LhcSR was first described as a stress-related protein, the transcripts of which accumulate in response to EL conditions (Tokutsu et al., 2019).

LIL can be found in both plants and algae. This class includes proteins that differ in their number of transmembrane segments: three-helix early light-inducible proteins (ELIPs), one-helix proteins (OHPs), and stress-enhanced proteins (SEPs) (Rochaix and Bassi, 2019). LILs are probably involved in photoprotection rather than light harvesting (Li et al., 2019).

Why do photosystems need to equip themselves with an outer antenna system? In addition to light harvesting, the specific properties of the LHCs have the ability (i) to actively regulate PSII light-use efficiency and (ii) to catalyse photoprotective reactions. Indeed, fluctuations of light quality and quantity on a daily as well as seasonal basis give rise to changes of excitation pressure on plant PSII, which may overwhelm the capacity for photochemical quenching of ^{1}Chl*, thereby leading to increased $^1O_2$ release (Krieger-Liszkay et al., 2008; Roach et al., 2020) (Fig. 2).

We can think of PSII as an engine: when sunlight is too abundant, it is like an increase in pressure that turns on a 'safety valve' and gets rid of the excess excitation energy. Activation of photoprotective safety systems is therefore required in order to either detoxify ROS or limit their release into the chloroplast. LHC proteins have an irreplaceable role in all these processes. Lhcb proteins actively participate in downregulation of the ^{1}Chl* lifetime through the process of NPQ, which is rapidly activated (timescale in seconds) to preserve PSII activity (Fig. 2) (Ruban, 2016). NPQ allows excess excitation energy to be safely dissipated as heat and is located within the PSII antenna system: indeed, the depletion of LHC proteins in plant mutants, such as *ch1* in the model plants *Arabidopsis thaliana*, leads to impaired NPQ activation and a dramatic increase in photodamage as well as a decline in the photosynthetic yield (Havaux et al., 2004). Additional regulation mechanisms under natural fluctuating light environments include the reversible migration of phosphorylated LHCII trimers between PSII and PSI (timescale in minutes), triggered by photosynthetic electron chain overreduction, thereby balancing excitation distribution of PSs through the so-called state I–state II transition (ST) (Pesaresi et al., 2011). In the green alga *C. reinhardtii*, the amplitude of ST is far larger than in higher plants: the migration of CP26 and CP29 to PSI, in addition to that of LHCII, results in the dissociation of PSII supercomplexes and the enlargement of the PSI antenna (Tokutsu et al., 2009). Additional photoprotective mechanisms include long-term (timescale in days) photoacclimatory responses through the re-programming of gene transcription and protein turnover, leading to stoichiometric reduction of the trimeric LHCII complement (Flannery et al., 2021), thus relieving chronic overexcitation on the PSII.

Analysis of isolated LHC has been largely used as an in vitro system for studying NPQ and incorporates a multitude of different models. However, data derived mainly from in vitro measurements might not reflect the in vivo situation. To gain a deeper understanding, *A. thaliana* plants lacking one or more Lhcbs have been produced, using a reverse genetic approach, and then characterised (Pietrzykowska et al., 2014; Dall'Osto et al., 2017). While findings supporting a reversible, conformational switch between unquenched and quenched states of LHCII units have been provided, there was no common agreement on the molecular events that lead to NPQ in plants. Major questions concern the interacting partners of PsbS and the localisation of the quenching site(s) within the antenna of PSII (Sacharz et al., 2017; Dall'Osto et al., 2017; Townsend et al., 2018; Nicol et al., 2019; Guardini et al., 2020). Moreover, the high degree of conservation through evolution of LHC subunits and their spectroscopic peculiarities suggest specific functional roles for the different members of the system. The outer PSI antenna system displays pigment organisation similar to the other members of the LHC family, but regulation of light harvesting is different in PSI compared to PSII. Exposure to sustained overexcitation did not change the PSI core:LHCI, while the long-term decrease of LHCII content is a common response under EL. Instead, compensation for excitation unbalance between PSs is reached by binding LHCII as PSI additional antenna (Bos et al., 2017) and by the long-term modulation of PSI:PSII ratio (Ballottari et al., 2007; Bonente et al., 2012). Lhca subunits display peculiar spectroscopic features including the so-called 'red spectral forms', that is, Chls with energy levels lower than the PSI RC. It is possible that red Chls provide preferential absorption of far-red (FR) photons transmitted under a canopy. Indeed, the characterisation of the *Arabidopsis* KO mutant devoid of all PSI antenna subunits indicates that Lhca optimises the photosynthetic electron transport by maintaining high PSI activity, by harvesting FR photons not absorbed by the PSII antenna (Bressan et al., 2018). These peculiar spectral properties possibly contributed to land colonisation by plants and their adaptation to life under canopies.

2 Biological constraints in light-use efficiency

2.1 Land plants

Why might enhancing light harvesting and light-use efficiency be critical to improving the yields of the major crops? To understand this point, we could consider a formula for the so-called harvestable yield 'Y', defined as the yield that a crop can achieve under optimal conditions without suffering any biotic or abiotic stresses. The original equation proposed by Monteith (1977), and adapted by Zhu et al. (2010), can be expressed as follows:

$$Y \propto S_{\mathrm{total}} \cdot \varepsilon_{\mathrm{i}} \cdot \varepsilon_{\mathrm{c}} \cdot \varepsilon_{\mathrm{p}}$$

where

- S_{total} is the total incident sunlight during the growing season;
- ε_{i} is light interception efficiency of photosynthetically active radiation (PAR, 350–730 nm), which depends on leaf absorbance, growth rate of the culture, conformation of the canopy;
- ε_{c} is the conversion efficiency, that is, the total C assimilation, net of respiratory losses; and
- ε_{p} is the harvest index, that is, the ratio of grain yield to total biomass produced by the crop.

Multiplying S_{total} by the constant 0.487, which takes into account the fraction of PAR that is either not absorbed or transmitted by leaves, the equation gives the harvestable yield in $MJ\,m^{-2}$, which can be converted into mass yield, based on the energy content of the harvested grain (Zhu et al., 2010).

In the context of the former equation, an increase in biomass yield can result from an increase in any of the terms. A number of findings suggest that the term ε_c can be further improved. For example, in soybean, an important dicotyledonous crop, artificial elevation of CO_2 concentration increased biomass yield by 18% over the growing season (Bernacchi et al., 2002), thereby (i) showing that enhancing photosynthesis in a crop does result in an increase in yield and (ii) suggesting that genetic improvements might lead to similar results without increasing [CO_2].

What is the theoretical maximum value of ε_c in higher plants? Detailed analyses of the efficiency of energy transduction, through a series of steps from the absorption of light to the biosynthesis of carbohydrates, are available and can be summarised as follow:

1. Only photons in the waveband 350–730nm can be used to fuel photosynthesis at high efficiency; moreover, the weaker absorbance of Chl in the green region and reflection further reduces the interception of PAR to approximately 90% (Zhu et al., 2008).
2. Blue photons (350–450nm) are more energetic than red photons (650–740nm) and lead to higher excited states of Chls. However, they relax rapidly, before catalysing photochemistry in the RC, thereby resulting in a further ~10% loss of energy as heat (Tredici, 2010).
3. In the RC of each photosystems, thermodynamic limits reduce the efficiency of photochemistry and charge separation, resulting in an energy loss of ~14% (Zhu et al., 2010).
4. Further energy losses are associated with photosynthetic electron transport, carbohydrate biosynthesis, and mitochondrial respiration, with the result that the theoretical maximum conversion efficiencies of PAR into biomass are 4.6% in C3 crops and 6.0% in C4 crops, and 5.4% in microalgae mass cultures (see Zhu et al., 2010; Tredici, 2010 for further details).

In addition to the theoretical calculation, values of ε_c can be determined in the field by measuring the accumulation of biomass in crops compared to the total amount of radiation intercepted by plants. This value is dependent on a number of factors, such as temperature and irrigation, while optimal nitrogen fertilisation strongly increases ε_c. Interestingly, the maximal ε_c values were found to be about a third of the theoretical for most of the species considered, even in healthy crops with optimised spacing and good nutrition (Slattery and Ort, 2015).

Productivity in field crops is limited by a number of intrinsic features within the photosynthetic mechanism (Ort, 2015). In this chapter, we will only consider inefficiencies concerning light capture and the use of excitation energy to fuel charge separation.

Three distinct phases can be identified in a typical leaf CO_2 assimilation rate in response to sunlight intensity (Fig. 3): (1) under low light, there is a linear increase of the photosynthetic rate with irradiance, that is, light is the limiting factor; (2) under higher irradiance, the photosynthetic rate undergoes inflection and rises in a nonlinear way as a function of light; and (3) when the light intensity is increased further, the assimilation rate reaches a plateau (Ye et al., 2013).

The slope of the initial, linear rise represents the amount of CO_2 molecules that can be fixed per photon absorbed. Under low light, a large fraction of the photons captured (more than 80% of the absorbed PAR) can be used while, at higher irradiance, this value rapidly falls.

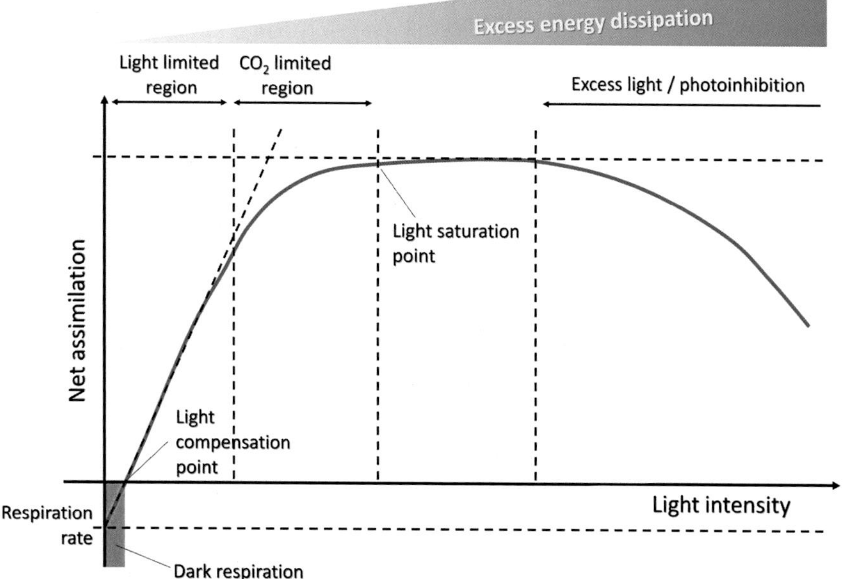

FIG. 3 Light response curves for photosynthesis. The light compensation point is the minimum irradiance at which the organism displays a gain of fixed carbon. The net assimilation rate displays a linear rise in response to increased light, within the range of light limitation. At higher light irradiances, saturation occurs as the efficiency of the photosynthetic mechanism is reduced, to which the activation of NPQ processes contributes. Under excess light, photosynthesis can decline as a result of photoxidative stress.

For example, at 50% of full sunlight, about 25% of the absorbed photons are used but, under full sunlight, this value drops to less than 15% (Oberhuber et al., 1993).

In the case of plants placed close together, the light levels inside the canopy are extremely uneven: light intensity declines towards the base of a crop and forms a gradient (Zhi et al., 2014). Consequently, in the lower parts of the crop, irradiance is too low to sustain C assimilation. Most light is absorbed by the upper leaves, leading to light saturation: in this part of the plants, the major limitation on efficient light utilisation is that leaves readily absorb more photons than they can productively use. Therefore, under EL conditions, PSII (which is usually highly efficient in PAR absorption) switches to a photoprotective state: the NPQ mechanism gets rid of the potentially harmful excess excitation energy as heat while PSII limits the production of ROS and preserves photosynthetic capacity (Niyogi, 2000). Such photoprotective energy dissipation occurs in the outer antenna system and competes with photochemistry in the de-excitation of the ^{1}Chl* states. The NPQ process involves the reversible activation of the xanthophyll cycle, which is dependent on the enzymes violaxanthin de-epoxidase (VDE) and zeaxanthin epoxidase (ZEP), and the sensing of changes in the lumen pH by the protein sensor PsbS (Li et al., 2009; Steen et al., 2020). NPQ induction occurs over a time scale of seconds to minutes and is therefore a rapid response, independent of acclimatory changes in gene expression. Photoprotection is vital for the photosynthetic apparatus: if EL cannot be dissipated safely, oxidative damage of PSII occurs, which results in reduced photosynthetic efficiency, requiring the replacement of the damaged components before quantum yield efficiency can be restored (Li et al., 2002).

On the other hand, activation of a dissipative channel decreases the quantum yield of PSII and so leads to a corresponding decrease in carbon assimilation. However, the light environment in a canopy is uneven and so plants are exposed to unpredictable extremes of high and low irradiance over the course of a day, for example, leaves that are higher in the canopy or due to passing clouds. Accordingly, plants must adapt to this intermittent shading (Kulheim et al., 2002).

It is worth noting that the recovery of C assimilation from the dissipative state to the high-efficiency state is slower than the rate of light fluctuations: activation of a dissipative reaction inevitably leads to a substantial delay in quantum yield recovery and so the reduced CO_2 uptake.

Keep in mind the metaphor of PSII as an engine: NPQ dissipates excitation pressure under high light levels but, in the shade, the pressure in the engine decreases and NPQ also declines, but not quickly enough. The system, therefore, leaks, as if a lid had been left open, and the engine cannot run as efficiently.

Algorithms were deployed to predict photosynthetic dynamics in a model canopy, and these simulations forecast that the delayed recovery of the photosynthetic efficiency on a decrease in irradiance would cause average losses in canopy carbon gain in the order of 15%, due solely to the constantly changing incident radiation on each leaf over the course of a day (Zhu et al., 2004).

2.2 Green microalgae

From the perspective of using sunlight and CO_2 to produce high-added value compounds and biomass through the mass cultivation of photoautotrophs, microalgae farming has garnered increasing interest in the last decades. Commercial production of algae on an industrial scale has been proclaimed as an environmentally sustainable strategy for energy-rich feedstock production (Yen et al., 2013). Microalgae include a large group of photosynthetic, unicellular, eukaryotic organisms; in particular, green microalgae (class Chlorophyceae) include genera that are among the most widely used for massive cultivation, such as *Chlorella vulgaris, Dunaliella salina,* and *Haematococcus pluvialis*. Microalgae are a source of genetic and metabolic diversity, since they include species adapted to diverse environmental conditions (Gimpel et al., 2015). Production in industrial-scale photobioreactors (PBRs) and biorefinery processes can yield a large variety of resources, such as bioactive compounds, recombinant proteins, wastewater treatment and biofuels, livestock feed, fertiliser, and biostimulants (Benedetti et al., 2018).

The photosynthetic apparatus of green microalgae is similar to that of plants but the light-use efficiency of algae is appreciably higher. Many reasons contribute to such a high performance level:

(i) microalgae lack heterotrophic organs, the maintenance of which is costly, and the unicellular organisation makes the whole biomass fully photosynthetically active;
(ii) algae have a short doubling time (a few hours) and are metabolically flexible, which means they undergo a metabolic shift on changes in environmental conditions, either from autotrophy to heterotrophy (using organic compounds as a source of C and energy) or to mixotrophy (carrying out photosynthesis as the main energy source and using both organic molecules and carbon dioxide as the C source) (Xie et al., 2020);
(iii) microalgae do not require fertile land and can grow on wastelands in brackish water or even sea water, so their cultivation would not compete with conventional crops (Abdel-Raouf et al., 2012); and
(iv) different species can be selected for specific growth conditions, suited to even harsh climate (Treves et al., 2013).

Although several industrial applications of microalgae have been proposed, the only successful exploitation of their potential is, so far, the production of carotenoids. Production of high-value compounds is currently undertaken in small-scale cultivations, the operating costs of which will need to be reduced if they are to be competitive with other feedstocks (Borowitzka, 2013).

The overall cost of a biorefinery depends on the cellular content of the bioproduct and the growth rate of the culture, the latter being determined by the efficiency with which light is used to drive photosynthesis.

Calculations provided both the theoretical maximum of sunlight conversion efficiency and the productivity of microalgae, which are equal to 8%–10% solar-to-biomass and 280 ton of dry biomass per ha per year, respectively (Weyer et al., 2010; Chisti, 2007). In contrast, outdoor mass cultivation of wild type strains recorded maximal productivity of around 80–100 ton of dry biomass per ha per year (Rodolfi et al., 2009). As has been shown in plants, overcoming this productivity gap is essential in order to exploit the full potential of microalgae.

Regulation of light harvesting is crucial for algal cells to balance light reactions and downstream biochemical processes. In a dilute culture of *Chlorella vulgaris*, light saturation is reached at around 1000 μmol photons $m^{-2} s^{-1}$ (Dall'Osto et al., 2019). At this irradiance, algae protect themselves from excess illumination by triggering NPQ in order to prevent the overexcitation of the RC. Moreover, algae can experience very high light, saturating photoprotective mechanisms. In this way, EL is mainly dissipated rather than contributing to biomass accumulation, or can even cause the release of ROS, which inhibits the photosynthetic machinery impairing biomass productivity (Fig. 3).

A peculiar trait of microalgae is the size of the antenna systems in both photosystems, which are larger than they are in plants and cause uneven light distribution within the culture (Formighieri et al., 2012). These large antenna systems were selected by evolution as a favourable trait since, in natural environments, light is often a limiting factor for the growth, thus implying the need of maximising light-harvesting capacity (Ort et al., 2011). Nevertheless, such a large antenna strongly contributes to the gap between theoretical and real productivity. Unlike in the natural environment, industrial cultures in PBRs need to reach a high biomass concentration per volume of installed facility (Freudenberg et al., 2021). This inevitably results in high optical density, determining a light shortage in the deeper layers of the cell suspension (Pandey et al., n.d.). On the other hand, cells in the outermost layers absorb most of the photons, leading to the saturation of photosynthesis, dissipation of excitation energy and/or photoinhibition. The innermost layers fall below the compensation point of photosynthesis, while respiration consumes energy. It transpires that dense algal cultures suffer both photodeprivation and photoinhibition, thereby reducing the overall light-to-biomass conversion efficiency to below its theoretical value. Regulated mixing of biomass may relieve the effects of light gradients, while too rapid cycles between dark and oversaturating irradiance have a deleterious effect on the photosynthetic yield (Sforza et al., 2012). Modelling of the photosynthesis light-response curve in a culture suggests that the optimal setting of optical density (OD) should minimise shading while enhancing light absorption and net photosynthesis (Formighieri et al., 2012).

In conclusion, observations reported in this section suggest that the modulation of light reactions is one of the key factors in controlling biomass yield at both saturating and subsaturating irradiance and is worthy of consideration as a domestication strategy aimed at improving productivity in both plants and algae.

3 Targets for improved light harvesting: The 'cooperative interaction' concept

The inefficiencies listed in the previous chapter are inherent to the mechanism of photosynthesis. Much of our current knowledge about these biological constraints comes from studies on model plants and algae, in which key steps of the photosynthetic mechanisms have been altered, resulting in changes to the capacity of light harvesting or biomass production. Advances in our understanding of the functional and molecular details of these processes have paved the way for a challenging and exciting prospect, namely the manipulation of specific pathways to enhance the photosynthetic productivity of either crop plants or microalgal strains for industrial applications. The 'cooperative interaction' concept underpins the 'smart' populations of photosynthetic organisms, optimised in order to interact cooperatively in accessing a widely diffuse resource such as sunlight (Ort, 2015). The final aim is to improve the light-harvesting capacity of the population as a whole, rather than of the individual organism, the biotechnological equivalent of the sentiment 'There is strength in numbers'.

Genetic engineering approaches based on detailed knowledge of potential cooperation within the population is required. Light irradiance and spectral range vary minute by minute in a plant canopy or a microalgal culture, depending on passing clouds, the wind altering the shading offered by the uppermost leaves (Kulheim et al., 2002; Zhi et al., 2014), or convective fluxes of microalgal cells within a PBR (Sforza et al., 2012). Such events lead to a significant portion of the population become increasingly shaded.

Which biophysical traits could be engineered to create a 'smart' population with improved productivity?

3.1 Lowering cell absorptivity

The antenna size, that is, the number of pigments associated with each RC, is an important trait for the efficiency of photosynthesis. Indeed, under excess irradiance, cells absorb more light than can be used by downstream metabolic processes. This means they have an antenna with excess capacity. It was therefore predicted that reducing leaf Chl content would be a promising approach in enhancing light-use efficiency (Formighieri et al., 2012) since it would improve light distribution inside the canopy (Fig. 4).

However, strategies to improve light penetration must ensure that lines with truncated antenna are not photosynthetically impaired in ways other than in reduced LHC content. Indeed, in higher plants, the depletion of a subgroup of LHCs, the so-called monomeric LHC, strongly affects the PSII light-harvesting efficiency and therefore the photoautotrophic growth, thereby cancelling out the benefits from optical density reduction (van Oort et al., 2010). Although they share some common features, LHC proteins are not all the same.

Even microalgae in mass cultures, suffer from uneven light distribution within the PBR. In the outer layers, light intensity leads to saturation (and possibly photoinhibition), while cells deeper in the culture suffer from light limitation, thus resulting in an overall decrease of productivity.

It has been proposed that a promising approach for the improvement of productivity in mass culture would be to target genes required for the synthesis of antenna pigments or their ligand LHC.

3.2 Decreasing photoprotective energy loss

In addition to considering antenna size, an increase in the quantum yield of PSII could be achieved by decreasing photoprotective energy

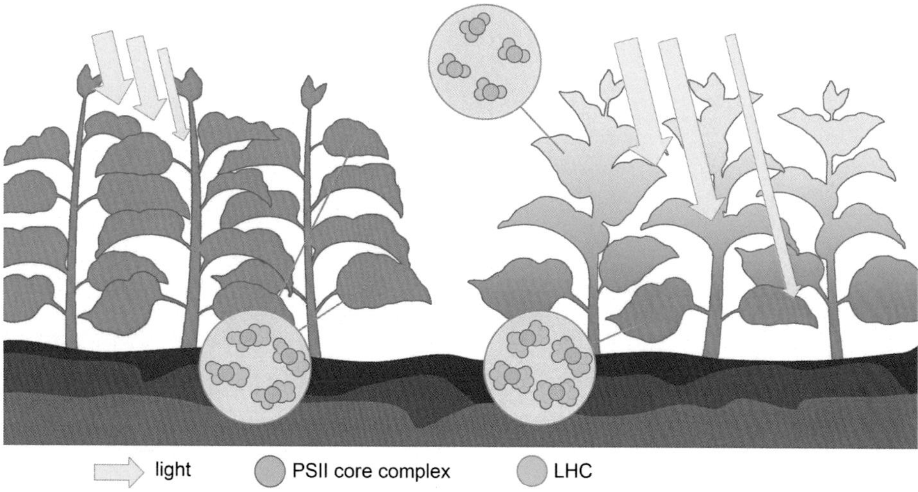

FIG. 4 Potential molecular mechanisms of plants to be targeted for improved productivity. The light environment inside a canopy is highly heterogeneous. Light declines towards the base of a crop canopy, generating gradients in the red to far-red ratio (R/FR), while cloudiness and shading by adjacent vegetation moved by the wind generate irregular fluctuations in the PAR. In a typical canopy (left), most light is absorbed by the leaves in the upper layers. A canopy with optimised light harvesting (right) would have upright, paler *green* leaves in the upper portions of the canopy and horizontal, darker leaves at the base. The light harvesting capacity would also be improved through a reduced antenna cross-section per photosystem in the upper leaf layers. The Chl *b* biosynthesis and/or the expression of LHC components could be regulated by the R/FR ratio.

loss in the photosynthetic apparatus. While excess energy must be carefully dissipated in order to avoid photooxidative stress (Fig. 2), overactivation of NPQ brings about a significant loss in efficiency. Fine-tuning of photoprotective responses would allow a greater proportion of absorbed photons to be converted to biomass. Significantly, the relative weight of other photoprotective components, such as the antioxidant carotenoids, must be considered in engineered cell lines. Indeed, the optimised accumulation of photoprotective pigments could compensate for the lack of defences and preserve PS through the effects of photoxidation (Havaux et al., 2007).

3.3 Enhancing light capture

Not all the solar energy that reaches the Earth’s surface is absorbed by the chromophores associated with photosystems. Usable PAR lies in the wavelengths ranging from 350 to 730nm, therefore comprising less than 50% of the incident solar radiation (Zhu et al., 2008). Moreover, because of the weaker absorbance of Chls and carotenoids in the green band, even the interception of usable PAR is not optimal.

However, the spectrum of absorbed PAR has been extended beyond these wavelengths in biological systems other than plants and green algae. Photosynthetic pigments such as Chl *f* (λ_{max} 706nm in methanol), Chl *d* (λ_{max} 696 nm in methanol), and bacteriochlorophyll (λ_{max} approximately 700–1000nm) absorb in the far-red wavelengths (Fig. 5) (Chen and Blankenship, 2011; Nürnberg et al., 2018).

These pigments could potentially be engineered in plants or algal species. In addition, phycobilisomes, the antenna subunits of cyanobacteria, efficiently absorb green wavelengths (550–560nm) and could also therefore be

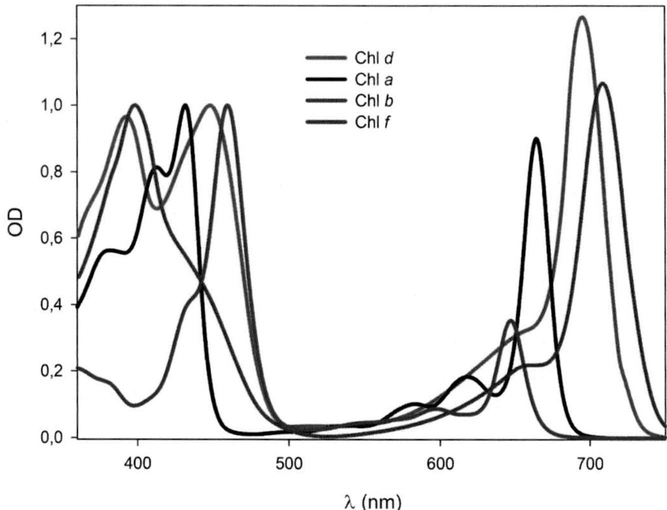

FIG. 5 Absorption spectra of Chl *a*, *b*, *d* and f in organic solvent. Each spectrum has been normalised at its Soret *(blue region)* peak.

expressed in green microalgae in order to increase the efficiency of PAR absorption (Puzorjov and McCormick, 2020). This strategy could be combined with the optimisation of LHC composition by expressing antenna isoforms that are more suited to either light-harvesting or photoprotection.

3.4 Sensing the ratio of red to far-red radiation

A peculiar feature of plants is their ability to sense the ratio of red (R) to far-red (FR) radiation. This becomes particularly important for high density cultivations. Leaves readily absorb R wavelengths but transmit most FR radiation so the light environment deeper under the canopy is enriched in FR components. The phytochrome mediates the sensing of the R/FR ratio, promoting downstream signalling, which modulates plant architecture (de Wit et al., 2016). This regulatory mechanism could be used to optimise the expression of molecular targets within the canopy, aimed at optimising photosynthesis under the prevailing light conditions. Chl *d* or Chl *f*, with greater red-shift than Chl *a*, could be expressed differently in the lower leaves of the canopy, thereby increasing their capacity to intercept FR photons. The expression of light-harvesting systems could also be modulated by the R/FR ratio. Upper leaves, which receive the most R light, would synthesise PSs with a smaller antenna, thereby enabling a higher quantum yield, while the lower leaves in limiting light, would accumulate PS with increased antenna size and improved performance in the interception of photons (Fig. 4).

Light-harvesting capacity might also be influenced by optimising the antenna composition, such as by targeting mutations in the protein sequence of the wild type LHC, in order to modulate protein–protein and/or pigment–protein interactions and shift the absorption maximum of specific chromophores to longer wavelengths (Morosinotto et al., 2005).

3.5 Optimised variation in leaf angle

In conditions of high plant density, half of the crop's photosynthesis takes place in the shade, so the reach of a more uniform distribution of sunlight inside the canopy would result in productivity boost. This could be obtained by optimising the variation of the leaf angle (Ding et al.,

2011), turning leaves in vertical in the upper portion of the canopy (under high light conditions) and those deeper in the canopy (low light conditions) turned horizontal (Fig. 4). This would optimise the distribution of sunlight by avoiding the saturation of the upper leaves and increasing the irradiance that reaches the lower leaves.

Many of the aforementioned approaches remain on the drawing board and will require extensive research work and experimental validation. Of course, there are trade-offs with all possible mutations: although a number of biophysical traits could be identified as targets for genetic engineering, there is no one mutation that surely can solve all problems. Light harvesting in both crops and algae cultures should be viewed as part of a much more complex system, including not only the light reactions but also a complex network of regulatory and acclimation processes. Thus, such manipulations should tailor to match growth conditions, locations, and seasons. Nevertheless, models were presented which demonstrate the impact of genetic manipulations in a way that produces useful outcomes (Long et al., 2006), and some promising results have already been obtained (see further sessions).

4 Optimisation of light harvesting in plants through genetic engineering

Over the last half century, the so-called 'Green Revolution' in agriculture, a domestication strategy based on breeding and phenotypic selection, has managed to boost crop productivity (Langridge, 2014). This increase in crop yield was attained mainly by increasing the amount of biomass set aside for grain, that is, the harvest index ε_P, now considered close to its upper limit. More recently, several theoretical models have been drawn up to find new ways of enhancing crop productivity. Moreover, selected targets in the photosynthetic machinery have been manipulated by genetic engineering, in order to optimise light collection and select lines that can be used under specific light conditions.

As previously discussed, the high Chl pigment content of leaves leads to uneven light distribution within the canopy with the upper leaves shading those beneath, thereby reducing the system's efficiency. In simulations of canopy photosynthesis, the reduction of Chl content per leaf was found to be a promising approach to improving canopy performance (Ort et al., 2011). Then, transgenic plants bearing small antenna systems were produced and characterised. In many cases, the results were consistent with the predictions of the modelling. For example, transgenic tobacco, the biogenesis of the peripheral antenna of which was altered, displayed a truncated light-harvesting system and resulted in plants with higher biomass yield under high-density planting conditions (Kirst et al., 2017). Similar results were obtained in rice and *A. thaliana* mutant lines with reduced Chl content (Gu et al., 2017; Jin et al., 2016). In other species, the outputs differed: photosynthetic efficiency was investigated in soybean mutants with pale-green phenotypes, and despite a significant (−50%) Chl depletion, the enhancement on biomass accumulation was very limited (Sakowska et al., 2018). While this result suggests that soybean significantly overinvests in Chl biosynthesis, the reduced pigment content failed to translate into higher yield from canopy-level processes. As previously shown, LHCs participate in the photoprotection of the photosynthetic apparatus so their reduction must be carefully calibrated. To this end, *Camelina sativa* plants were engineered to incorporate a range of reduction in Chl *b* levels, and so in LHC content, in order to pinpoint the efficiency threshold. Interestingly, plants with slightly reduced antenna size (Chl *a/b* ratio ~5.0) displayed enhanced biomass yield compared to wild type plants (Chl *a/b* ~3.0) but further reductions in this optimal antenna size impaired both light-stress tolerance and plant productivity (Friedland et al., 2019).

Recently, Kromdijk et al. (2016) produced tobacco plants that had been genetically modified to use light more efficiently, by targeting three of the gene products that plants use to catalyse NPQ. In particular, two enzymes, violaxanthin de-epoxidase and zeaxanthin epoxidase, were overexpressed to boost the xanthophyll cycle needed to maximise NPQ, together with the regulatory subunit PsbS, which enhances the rate of NPQ formation. Engineered plants were able to bounce back from a 'dissipative' into a 'productive light-harvesting' state faster than the wild type, resulting in a 14%–20% increase in the productivity measured in open field experiments. These results first displayed a genetic-engineering approach by targeting the regulatory mechanisms of light harvesting and succeeded in boosting the efficiency of photosynthesis, thereby potentially offering the opportunity for similar gains across major food crops. The finding is important because it confirmed that regulatory pathways of photon capture are not optimised for maximum yield: plants prefer to produce less in exchange for a lower risk of being damaged by too much light. This implies that there may be significant room for further improvements. For example, it will be possible to dynamically downregulate the dissipative reactions in order to make plants grow faster under lower light conditions, such as in greenhouses (Fig. 6), or to reinforce the photoprotective responses to enable plants to grow in hostile environments (Davison et al., 2002), where EL is accompanied by other stress factors, such as cold or drought.

Light is not only the essential energy source absorbed by the Chl-binding complexes that drive photosynthesis. It also participates in the coordination of a number of cellular processes that enhance photosynthetic performance. Phototropins (PHOT) are blue-light receptors that mediate a large number of responses that optimise light absorbance, such as phototropism, stomatal opening, chloroplast movement, and leaf flattening (de Wit et al., 2016). Light-induced chloroplast movements are a regulatory mechanism for light harvesting. Chloroplasts are induced to move towards a weakly illuminated area of cells to promote efficient photon capture, the so-called 'accumulation response'; in contrast, they move away from high light levels, that is, the 'avoidance response' (Fig. 7) (Suetsugu and Wada, 2007).

Under low light conditions, the accumulation response induces the alignment of the chloroplasts on the upper surfaces of the mesophyll cells while, under EL conditions, chloroplasts align along the sides of cells, producing an 'escape' response (Wada et al., 2003). Chloroplast movements affect leaf absorbance: as they accumulate, light harvesting capacity is maximised. Under EL, on the other hand, leaf absorbance of the upper cell layers is minimised, thereby providing protection against photodamage, and leaf photosynthesis is maximised due to greater light penetration within the mesophyll cell layers (Fig. 7). Under natural conditions, the chloroplast distribution constantly changes and the balance between accumulation and avoidance responses adjusts the trade-off between light harvesting and photoprotection.

In *A. thaliana*, the accumulation response is regulated by both phototropins PHOT1 and PHOT2, the avoidance response by PHOT2. The importance of these physiological responses has been investigated by using *Arabidopsis* KO mutants devoid of phototropin-mediated responses. The *phot1* plants displayed a reduced growth rate because of attenuated chloroplast accumulation (Goh et al., 2009). Several research groups demonstrated that chloroplast avoidance response limits photoxidative stress of plants under strong or fluctuating irradiances (Kasahara et al., 2002; Shang et al., 2019; Gotoh et al., 2018; Cazzaniga et al., 2013; Davis and Hangarter, 2012; Sztatelman et al., 2010), thus suggesting this mechanism is critical for photoprotection in long-term (hours) EL conditions. In contrast, a recent report showed chloroplast

FIG. 6 Genetic engineering of the light reactions of photosynthesis can potentially enhance crop productivity. Recent experimental findings give weight to the idea that photosynthetic efficiency is a trade-off between energy used in growth and energy lost (Kromdijk et al., 2016). Crops of the future might be genetically tailored for high yield in specific environments. For example, plants intended for highly controlled environments such as a greenhouse could be engineered to eliminate photoprotective responses in favour of maximum productivity.

movements did not play role in preventing PSII RC inactivation in shorter-term high light (Wilson and Ruban, 2020).

However, it is widely believed that the accumulation response is required for efficient light capture under weak light conditions. The light threshold for triggering the avoidance response is low (<100 μmol photons $m^{-2}\,s^{-1}$) so it might become activated even in leaves lying deeper under the canopy (Dutta et al., 2015). This may happen because the process is not optimised for maximum yield: in other words, plants may 'prefer' to limit productivity in return for greater safety when there is too much light.

In a recent report, Gotoh et al. (2018) demonstrated that the chloroplast accumulation response is essential for efficient photosynthetic performance in *Arabidopsis*: *phot2* mutation enhanced leaf photosynthesis and increased plant biomass production under a wide range of light conditions. Therefore, the chloroplast accumulation response was seen as a potential physiological target that could be engineered to enhance light harvesting and biomass yield. It should be noted, however, that *phot2* plants are susceptible to very high light levels (1500–1800 μmol photons $m^{-2}\,s^{-1}$), that is, the level of irradiance that the upper leaves of a canopy can easily experience in field conditions in central/south Europe (https://ec.europa.eu/jrc/en/PVGIS/tools/daily-radiation). Therefore, caution should be applied in the suppression of the avoidance response since it might increase leaf photosensitivity and cancel out the benefits for biomass production.

In a recent work, Hart et al. (2019) deployed protein engineering to improve the performance of phototropins and thereby light-use efficiency.

They suggested that the modulation of phototropin activity might offer a new opportunity to enhance the efficiency of light capture. Site-directed mutation on LOV domains of PHOT1 and PHOT2 succeeded in modulating the

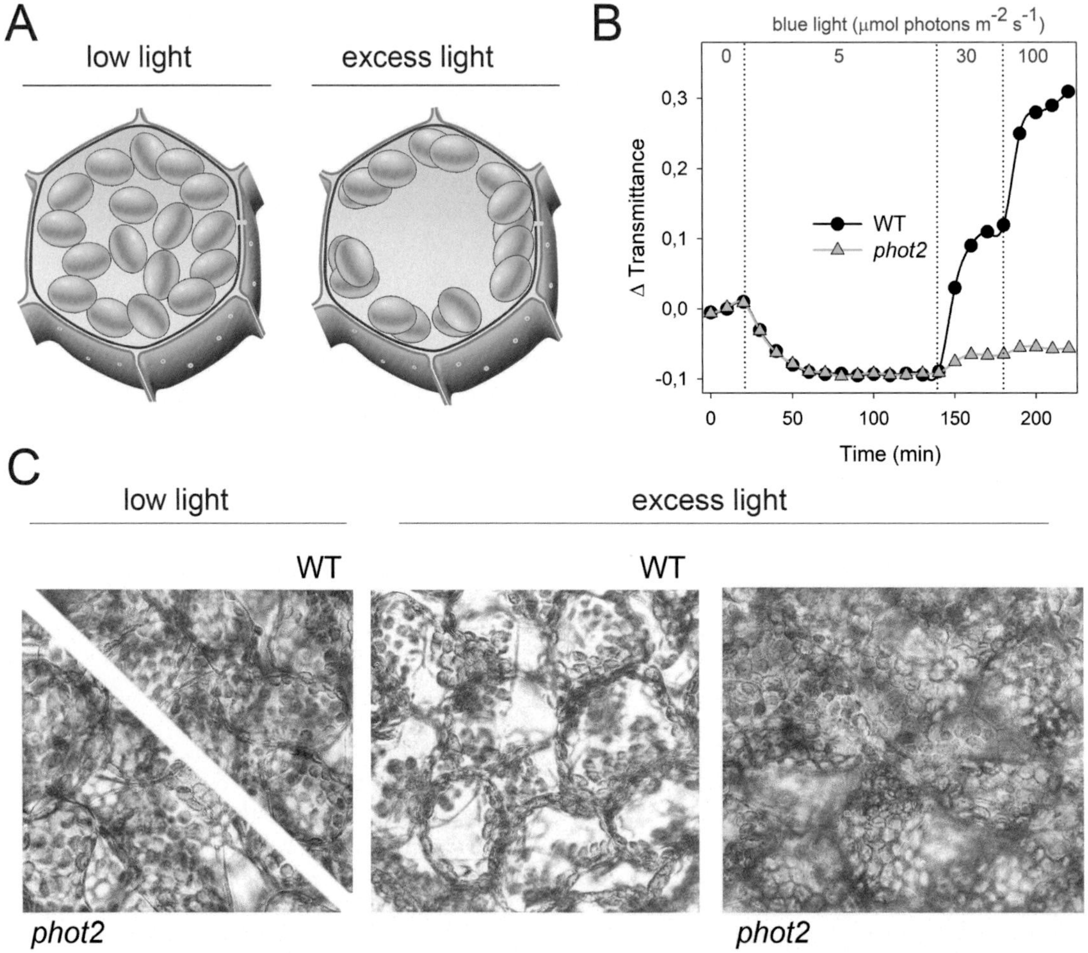

FIG. 7 Chloroplast movements in leaves. (A) Schematic drawing of chloroplasts photorelocation. Chloroplasts can change their distribution within the cell in response to the intensity of the incident light, moving from positions yielding maximum light absorption to others in which light absorption is minimised. (B) Time course of changes in the leaf transmittance induced in WT and *phot2* leaves by blue light at different intensities. Blue light at 5 μmol photons $m^{-2}\,s^{-1}$ promotes accumulation response in both genotypes, while avoidance response can be observed at higher irradiance only in WT leaves. (C) Distribution of chloroplasts in the mesophyll cells of wild type and *phot2 Arabidopsis*, as determined by light microscopy. Leaves were dark-adapted for 1 h (left panel) and then irradiated with white light at 400 μmol photons $m^{-2}\,s^{-1}$ for 30 min (right panels).

photocycle lifetime of *Arabidopsis* phototropins. These variants were used either to speed up or slow down photoreceptor activation in vivo. In this way, slowing the phototropin photocycle enhanced light harvesting responses, while accelerating it reduced phototropin's sensitivity for chloroplast accumulation movement. As a consequence, plants engineered to have increased responsiveness of PHOT1 or PHOT2 displayed higher biomass productivity under low light conditions. The explanation for this is that plants carrying engineered receptors exhibited more rapid chloroplast movement responses, together with enhanced leaf positioning and expansion (mechanisms regulated by PHOTs), resulting in improved biomass accumulation under light-limiting conditions. This finding shows that PHOTs proteins constitutes a promising new target for manipulating plant growth and optimising photosynthetic light harvesting, particularly under suboptimal light regimes.

5 Engineering of the light-harvesting system in green algae

As stated in the previous section, the high OD of cells is a condition that limits PAR penetration in the culture and this uneven light distribution accounts for the gap between the theoretical and real biomass productivity of industrial PBRs. Biomass production using wild type algal strains, endowed with features hampering high cell density, such as elevated optical density of cells, is therefore as unviable as farming with ancestral crop varieties, which in most cases showed reduced yields (Ritchie and Roser, 2013).

Microalgae biotechnology may take advantage of a domestication approach, similar to that adopted for modern crops, through the selection of strains carrying the desired traits by implementing new alleles through random mutagenesis or genetic engineering, which is expected to improve light harvesting capacity and therefore the biomass yield in industrial production systems.

Random mutagenesis has been widely employed to study the molecular basis of cellular processes and for strain improvement, thus is recognised as a powerful tool in mutation breeding. This is the 'forward genetic' strategy, which is of particular relevance for algal biotechnology since it generates mutant lines, thereby avoiding restrictions on GMO for outdoor PBRs (Snow and Smith, 2012).

Attempts to genetically improve microalgae, aimed at enhancing light-use efficiency, relied on random mutagenesis and the screening of favourable traits. The latter requires efficient selection strategies for strains with a higher biomass yield. A number of approaches succeeded in enhancing light-harvesting capacity and are illustrated below.

Due to the adverse effect of high OD on large-scale culture, mutants carrying smaller antenna size were regarding as performing better in light transmittance than wild type strains (Fig. 8).

Mutagenesis and the selection of the model green alga *C. reinhardtii* were used to isolate strains with a truncated light-harvesting system (Perrine et al., 2012; Kirst et al., 2012). All these strains displayed higher biomass productivity than the wild type in laboratory-scale PBRs. A random mutagenesis approach was applied to a thermotolerant, fast-growing strain of *Chlorella sorokiniana*, then pale-green mutants were selected by imaging Chl fluorescence since a reduced Chl content per unit cell decreases the emission of colonies. Mutants were able to perform photosynthesis more efficiently than wild types, and displayed reduced photoinhibition under EL conditions. The enhancement of photosynthetic productivity was confirmed in both bench-scale and outdoor PBRs (Cazzaniga et al., 2014). Similar results were obtained with oleaginous microalga *Nannochloropsis gaditana* strains with lower cellular pigment content (Perin et al., 2015). Finally, simultaneous down-regulation of three LHC subunits (LHCMB1, 2, and 3) in *C. reinhardtii* resulted in improved (+165%) light-to-biomass conversion yield (Oey et al., 2013).

In addition to random mutagenesis, genetic manipulation approaches include the transfer of genes identified in other species in order to generate strains with desirable productive features (Fig. 9).

The implementation of a biosynthetic pathway of additional chromophores, such as phycobilins or chlorophylls with peculiar

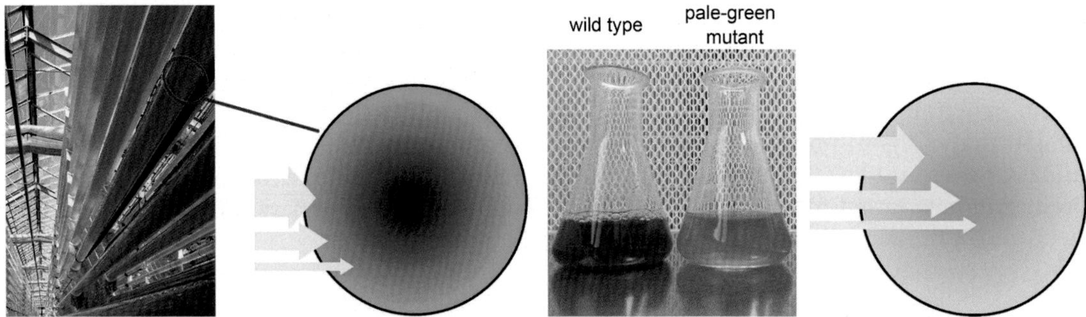

FIG. 8 Light distribution and photosynthetic activity in algal cultures. Light penetration and distribution inside a PBR *(yellow arrows)* is limited by the high optical density of the cells. Mutations that decrease Chl concentrations of cells without disrupting photoprotective responses enable increased productivity of the shaded layers and lower energy losses in the cells exposed to full light, thereby increasing overall light-use efficiency. Note that the two cultures in the flasks have the same cell concentration.

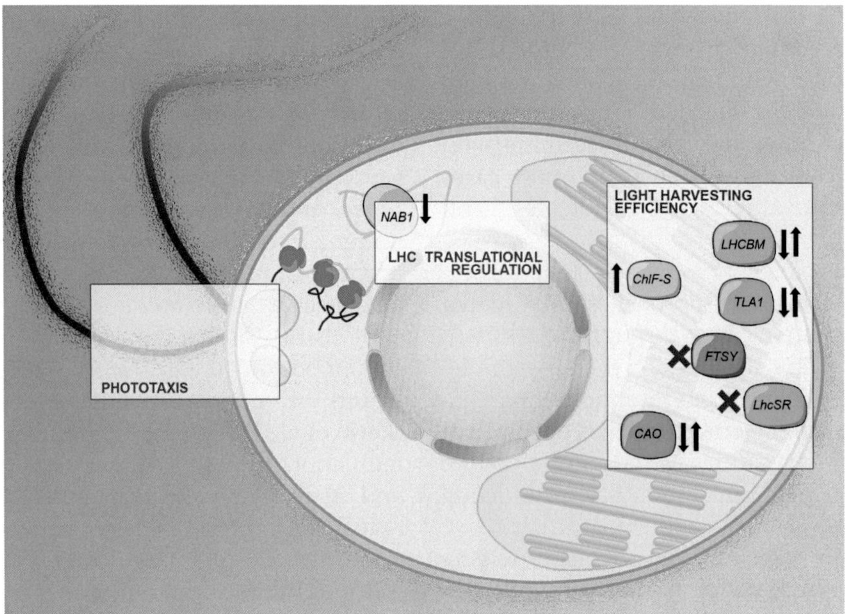

FIG. 9 Antenna-related traits to be introduced into genetically modified *C. reinhardtii* cells. The diagram displays a number of genetic strategies, aimed at enhancing productivity in the mass culture of microalgae. Traits that may result in higher productivity include increased light-use efficiency, enhanced light absorption and improved phototaxis. *Up and down arrows* mean up- and downregulation, respectively, and refer to the expression level of the corresponding endogenous enzyme. A red cross means in vivo loss of function. *Chl-f S*, chlorophyll *f* synthase; *CAO*, chlorophyll *a* oxygenase; *FTSY*, chloroplast signal recognition particle; *LHCBM/LhcSR*, light harvesting complexes; *NAB1*, RNA-binding protein; *TLA1*, truncated light-harvesting antenna 1.

spectral properties, has been proposed to enhance light harvesting over the entire PAR spectrum in genera of industrial interest. For example, Chl *f* expression in microalgae may expand the spectral range used for photosynthesis (Zamzam et al., 2020) and confer advantages for mass culture, which suffers from unfavourable sieve-effects at high cell densities. Recently, the enzyme catalysing the synthesis of Chl *f* has been identified in the cyanobacterium *C. fritschii*, and the heterologous expression of Chl *f* synthase in *Synechococcus* sp. succeeded in accumulating this chromophore (Shen et al., 2019a). However, the feasibility of this approach in improving light harvesting capacity, which assumes a correct binding of the new pigment to the existing pigment scaffolds, still awaits experimental confirmation.

As previously discussed, a high photosynthetic efficiency can be attained only in low irradiance and a stable light environment, which enable most absorbed photons to be efficiently converted into chemical energy. Outdoors, however, the efficiency drops due to rapidly fluctuating light, which easily exceeds the photosynthetic capacity. As shown in plants, microalgae also evolved mechanisms for regulating the efficiency of light harvesting, which could be the targets of domestication programmes. A number of microalgal species of industrial interest, for example, *C. reinhardtii* and *H. pluvialis*, trigger rapid phototactic responses. In order to optimise its position in the water column, algae use flagella to move towards or away from a light source. Physiological characterisation confirmed that phototaxis confers a fitness advantage and is affected by photosynthetic electron transport (Wakabayashi et al., 2011). In the attempt to select mutants with improved light-use efficiency, Kim and coworkers analysed a *C. reinhardtii* mutant population for rapid phototaxis response and identified strains with enhanced photoautotrophic growth (8.1-fold increases) compared to the wild type (Kim et al., 2016).

Photosynthetic organisms dynamically regulate the amplitude of dissipative reactions of PSII (see Section 4). By balancing the excitation energy partition between photochemistry and heat dissipation, they optimise fitness in a rapidly changing light environment. The slow relaxation rate of the NPQ response on high- to low-light transition was shown to reduce the biomass yield in microalgae, consistent with recent findings in plants. Indeed, the deletion of the OCP protein, responsible for NPQ activation in cyanobacteria, resulted in 30% more biomass productivity in mass cultures of *Synechocystis* compared to wild type strains (Peers, 2012). The random insertional mutagenesis and TILLING (targeting induced local lesions in genomes) approach, coupled with fluorescence imaging selection, produced *C. reinhardtii* mutants specifically devoid of *lhcsr* genes (Peers et al., 2009). An initial report argued that biomass yield is modulated by LhcSR protein accumulation and, indeed, *C. reinhardtii* strains lacking the two *lhcSR3* genes were found to be more productive than the wild type, therefore supporting downregulation of NPQ as a strategy for also improving light-use efficiency in microalgae. However, more recent results reported no significant differences in biomass yield between *C. reinhardtii* wild type and a mutant devoid of all LhcSR isoforms, over a wide range of light regimes (Cantrell and Peers, 2017; Barera et al., 2021).

Recently, a novel approach based on CRISPR-CAS9 genome editing technology has been successfully developed in *C. reinhardtii*, enabling the selective deletion of gene functions (Baek et al., 2016). This was achieved using a DNA-free CRISPR-Cas9 method and the outcome was the *FTSY* and *ZEP* double KO, which led to pale-green strains displaying enhanced light-use efficiency.

As previously described, a truncated antenna size led to increased productivity in green microalgae so proteins involved in the biogenesis of the light-harvesting systems are potential candidates for increasing biomass yield. Truncated light-harvesting antenna 1 (*TLA1*) is a nuclear gene involved in the regulation of both PSI and PSII outer antenna in *C. reinhardtii*. TLA1 overexpressing strains displayed a larger antenna size for both photosystems and a lower Chl *a*/*b* ratio respect to the wild type, while its downregulation resulted in the opposite phenotype (Mitra et al., 2012). As in plants, LHCII is the major light-harvesting protein associated with PSII in *C. reinhardtii*. It is nuclear-encoded, and its regulated expression finely tunes the antenna cross-section to the light environment. LHCII translation efficiency is regulated by NAB1, a cytosolic RNA-binding subunit which works as translational repressor. Under limiting light, a specific nitrosylation makes the regulator less active and thereby promotes the accumulation of LHCII and larger antenna (Berger et al., 2016). The transformation of algae with a permanently active variant of NAB1 succeeded in reducing antenna size and improved both photosynthetic efficiency at saturating light intensity and growth rate compared to wild type strains. Recently, the expression of CAO (Chl *b* biosynthetic enzyme) coupled with a NAB1 translational repressor binding site created a translational control system for dynamically tuning the antenna size in *C. reinhardtii*. In these engineered strains, Chl *b* synthesis (and therefore LHC biogenesis) was continuously adjusted depending on the light conditions. Strains expressing such light-regulated antenna systems displayed higher photosynthetic rates and twice the biomass yield than the parental strains (Negi et al., 2020).

6 Concluding remarks

(i) Population growth, by increasing demands for global food production and a sustainable bioeconomy, makes it increasingly important to generate photosynthetic organisms that will meet these needs.

(ii) Theoretical estimation and findings in the field lend support to the engineering of light-harvesting systems as a viable strategy to significantly increase the productivity of both crops and algae. One of the most striking results in this field was recently obtained with transgenic tobacco. By increasing the expression of genes involved in the regulation of light-harvesting efficiency, scientists saw increases of 14%–20% in the productivity of modified tobacco plants in field experiments. These results are strong evidence that genetic manipulation of the mechanisms involved in photon capture can bring about a significant (and possibly unprecedented) increase in crop productivity. Interestingly, engineering the regulatory switch of the antenna also affected the intracellular signalling and stomatal opening in tobacco lines, resulting in improved efficiency of water use under field conditions (Glowacka et al., 2018).

(iii) The future development of more productive cultivars will require the implementation of different approaches, including complex modelling, breeding, gene transfer and measurements under real conditions. Continuous research efforts provided a wide array of molecular tools that can be suited for the prospective redesigning of light-use efficiency in plants. Moreover, several structural and functional properties related to light perception have been identified as potential targets for optimisation. Further research is required to furnish direct proof, through genetic manipulation, of the effectiveness of these productivity targets.

(iv) In green algae, investigating the mechanisms underpinning light-use efficiency lags behind work on terrestrial plants, and a number of challenges have yet to be overcome. Currently, many of the proposed targets have not been tested in algal species of industrial interest. While a large array of molecular tools for genetic engineering are available for a model species such as *Chlamydomonas*, difficulties in efficiently transforming other industrially relevant species, such as *Chlorella* or *Dunaliella*, still limit the development of algal biotechnology. However, our knowledge is rapidly improving through the application of next-generation molecular techniques. In a future 'green revolution' of microalgal biorefineries, boosting light-use efficiency by addressing the constraints of the wild type antenna system, will be a crucial step.

References

Abdel-Raouf, N., Al-Homaidan, A.A., Ibraheem, I.B.M., 2012. Microalgae and wastewater treatment. Saudi J. Biol. Sci. 19, 257–275.

Baek, K., Kim, D.H., Jeong, J., Sim, S.J., Melis, A., Kim, J.-S., Jin, E., Bae, S., 2016. DNA-free two-gene knockout in Chlamydomonas reinhardtii via CRISPR-Cas9 ribonucleoproteins. Sci. Rep. 6, 30620.

Ballottari, M., Osto, L.D., Morosinotto, T., Bassi, R., 2007. Contrasting behavior of higher plant photosystem I and II antenna systems during acclimation. J. Biol. Chem. 282, 8947–8958.

Barera, S., Dall'Osto, L., Bassi, R., 2021. Effect of lhcsr gene dosage on oxidative stress and light use efficiency by Chlamydomonas reinhardtii cultures. J. Biotechnol. 328, 12–22.

Benedetti, M., Vecchi, V., Barera, S., Dall'Osto, L., 2018. Biomass from microalgae: the potential of domestication towards sustainable biofactories. Microb. Cell Factories 17, 173.

Berger, H., de Mia, M., Morisse, S., Marchand, C., Lemaire, S.-D., Wobbe, L., Kruse, O., 2016. A light switch based on protein S-nitrosylation fine-tunes photosynthetic light-harvesting in the microalga Chlamydomonas reinhardtii. Plant Physiol. 171, 821–832.

Bernacchi, C.J., Portis, A.R., Nakano, H., Von Caemmerer, S., Long, S.P., 2002. Temperature response of mesophyll conductance. Implications for the determination of Rubisco enzyme kinetics and for limitations to photosynthesis in vivo. Plant Physiol. 130, 1992–1998.

Bonente, G., Pippa, S., Castellano, S., Bassi, R., Ballottari, M., 2012. Acclimation of Chlamydomonas reinhardtii to different growth irradiances. J. Biol. Chem. 287, 5833–5847.

Borowitzka, M.A., 2013. High-value products from microalgae—their development and commercialisation. J. Appl. Phycol. 25, 743–756.

Bos, I., Bland, K.M., Tian, L., Croce, R., Frankel, L.K., Van Amerongen, H., Bricker, T.M., Wientjes, E., 2017. Multiple LHCII antennae can transfer energy efficiently to a single photosystem I. Biochim. Biophys. Acta 1858, 371–378.

Bressan, M., Bassi, R., Dall'Osto, L., 2018. Light harvesting complex I is essential for photosystem II photoprotection under variable light conditions in Arabidopsis thaliana. Environ. Exp. Bot. 154, 89–98.

Cantrell, M., Peers, G., 2017. A mutant of Chlamydomonas without LHCSR maintains high rates of photosynthesis, but has reduced cell division rates in sinusoidal light conditions. PLoS One 12, e0179395.

Cazzaniga, S., Dall'Osto, L., Kong, S.G., Wada, M., Bassi, R., 2013. Interaction between avoidance of photon absorption, excess energy dissipation and zeaxanthin synthesis against photooxidative stress in Arabidopsis. Plant J. 76, 568–579.

Cazzaniga, S., Dall'Osto, L., Szaub, J., Scibilia, L., Ballottari, M., Purton, S., Bassi, R., 2014. Domestication of the green alga Chlorella sorokiniana: reduction of antenna size improves light-use efficiency in a photobioreactor. Biotechnol. Biofuels 7, 157.

Chen, M., Blankenship, R.E., 2011. Expanding the solar spectrum used by photosynthesis. Trends Plant Sci. 16, 427–431.

Chisti, Y., 2007. Biodiesel from microalgae. Biotechnol. Adv. 25, 294–306.

Clark, D.A., Brown, S., Kicklighter, D.W., Chambers, J.Q., Thomlinson, J.R., Ni, J., 2001. Measuring net primary production in forests: concepts and field methods. Ecol. Appl. 11, 356–370.

Croce, R., van Amerongen, H., 2020. Light harvesting in oxygenic photosynthesis: structural biology meets spectroscopy. Science 369, 1–9.

Dall'Osto, L., Bressan, M., Bassi, R., 2015. Biogenesis of light harvesting proteins. Biochim. Biophys. Acta Bioenerg. 1847, 861–871.

Dall'Osto, L., Cazzaniga, S., Bressan, M., Paleeèk, D., Židek, K., Niyogi, K.K., Fleming, G.R., Zigmantas, D., Bassi, R., 2017. Two mechanisms for dissipation of excess light in monomeric and trimeric light-harvesting complexes. Nat. Plants 3, 1–9.

Dall'Osto, L., Cazzaniga, S., Guardini, Z., Barera, S., Benedetti, M., Mannino, G., Maffei, M.E., Bassi, R., 2019. Combined resistance to oxidative stress and reduced antenna size enhance light-to-biomass conversion efficiency in Chlorella vulgaris cultures. Biotechnol. Biofuels 12, 221.

Davis, P.A., Hangarter, R.P., 2012. Chloroplast movement provides photoprotection to plants by redistributing PSII damage within leaves. Photosynth. Res. 112, 153–161.

Davison, P.A., Hunter, C.N., Horton, P., 2002. Overexpression of beta-carotene hydroxylase enhances stress tolerance in Arabidopsis. Nature 418, 203–206.

de Wit, M., Keuskamp, D.H., Bongers, F.J., Hornitschek, P., Gommers, C.M.M., Reinen, E., Martínez-Cerón, C., Fankhauser, C., Pierik, R., 2016. Integration of phytochrome and cryptochrome signals determines plant growth during competition for light. Curr. Biol. 26, 3320–3326.

Ding, Z., Galván-Ampudia, C.S., Demarsy, E., Łangowski, Ł., Kleine-Vehn, J., Fan, Y., Morita, M.T., Tasaka, M., Fankhauser, C., Offringa, R., Friml, J., 2011. Light-mediated polarization of the PIN3 auxin transporter for the phototropic response in Arabidopsis. Nat. Cell Biol. 13, 447–453.

Dogra, V., Kim, C., 2020. Singlet oxygen metabolism: from genesis to signaling. Front. Plant Sci. 10, 1–9.

Drop, B., Yadav, K.N.S., Boekema, E.J., Croce, R., 2014. Consequences of state transitions on the structural and functional organization of photosystem i in the green alga Chlamydomonas reinhardtii. Plant J. 78, 181–191.

Dutta, S., Cruz, J.A., Jiao, Y., Chen, J., Kramer, D.M., Osteryoung, K.W., 2015. Non-invasive, whole-plant imaging of chloroplast movement and chlorophyll fluorescence reveals photosynthetic phenotypes independent of chloroplast photorelocation defects in chloroplast division mutants. Plant J. 84, 428–442.

Flannery, S.E., Hepworth, C., Wood, W.H.J., Pastorelli, F., Hunter, C.N., Dickman, M.J., Jackson, P.J., Johnson, M.P., 2021. Developmental acclimation of the thylakoid proteome to light intensity in *Arabidopsis*. Plant J. 105, 223–244.

Formighieri, C., Franck, F., Bassi, R., 2012. Regulation of the pigment optical density of an algal cell: filling the gap between photosynthetic productivity in the laboratory and in mass culture. J. Biotechnol. 162, 115–123.

Freudenberg, R.A., Baier, T., Einhaus, A., Wobbe, L., Kruse, O., 2021. High cell density cultivation enables efficient and sustainable recombinant polyamine production in the microalga Chlamydomonas reinhardtii. Bioresour. Technol. 323, 1–11.

Friedland, N., Negi, S., Vinogradova-Shah, T., Wu, G., Ma, L., Flynn, S., Kumssa, T., Lee, C.H., Sayre, R.T., 2019. Fine-tuning the photosynthetic light harvesting apparatus for improved photosynthetic efficiency and biomass yield. Sci. Rep. 9, 1–12.

Gimpel, J.A., Henríquez, V., Mayfield, S.P., 2015. In metabolic engineering of eukaryotic microalgae: potential and challenges come with great diversity. Front. Microbiol. 6, 1376.

Glowacka, K., Kromdijk, J., Kucera, K., Xie, J., Cavanagh, A.-P., Leonelli, L., Leakey, A.D.B., Ort, D.R., Niyogi, K.K., Long, S.P., 2018. Photosystem II subunit S overexpression increases the efficiency of water use in a field-grown crop. Nat. Commun. 9, 868.

Goh, C.-H., Jang, S., Jung, S., Kim, H.-S., Kang, H.-G., Park, Y.-I., Bae, H.-J., Lee, C.-H., An, G., 2009. Rice phot1a mutation reduces plant growth by affecting photosynthetic responses to light during early seedling growth. Plant Mol. Biol. 69, 605–619.

Gotoh, E., Suetsugu, N., Yamori, W., Ishishita, K., Kiyabu, R., Fukuda, M., Higa, T., Shirouchi, B., Wada, M., 2018. Chloroplast accumulation response enhances leaf photosynthesis and plant biomass production. Plant Physiol. 178, 1358–1369.

Gu, J., Zhou, Z., Li, Z., Chen, Y., Wang, Z., Zhang, H., 2017. Rice (Oryza sativa L.) with reduced chlorophyll content exhibit higher photosynthetic rate and efficiency, improved canopy light distribution, and greater yields than normally pigmented plants. Field Crop. Res. 200, 58–70.

Guardini, Z., Bressan, M., Caferri, R., Bassi, R., Dall'Osto, L., 2020. Identification of a pigment cluster catalysing fast photoprotective quenching response in CP29. Nat. Plants 6, 303–313.

Hart, J.E., Sullivan, S., Hermanowicz, P., Petersen, J., Aranzazú Diaz-Ramos, L., Hoey, D.J., Łabuz, J., Christie, J. M., 2019. Engineering the phototropin photocycle improves photoreceptor performance and plant biomass production. Proc. Natl. Acad. Sci. U. S. A. 116, 12550–12557.

Havaux, M., Dall'Osto, L., Cuine, S., Giuliano, G., Bassi, R., 2004. The effect of zeaxanthin as the only xanthophyll on the structure and function of the photosynthetic apparatus in Arabidopsis thaliana. J. Biol. Chem. 279, 13878–13888.

Havaux, M., Dall'Osto, L., Bassi, R., 2007. Zeaxanthin has enhanced antioxidant capacity with respect to all other xanthophylls in Arabidopsis leaves and functions independent of binding to PSII antennae. Plant Physiol. 145, 1506–1520.

Jin, H., Li, M., Duan, S., Fu, M., Dong, X., Liu, B., Feng, D., Wang, J., Wang, H.B., 2016. Optimization of light-harvesting pigment improves photosynthetic efficiency. Plant Physiol. 172, 1720–1731.

Kasahara, M., Kagawa, T., Oikawa, K., Suetsugu, N., Miyao, M., Wada, M., 2002. Chloroplast avoidance movement reduces photodamage in plants. Nature 420, 829–832.

Kim, J.Y.H., Kwak, H.S., Sung, Y.J., Choi, H.I., Hong, M.E., Lim, H.S., Lee, J.-H., Lee, S.Y., Sim, S.J., 2016. Microfluidic high-throughput selection of microalgal strains with superior photosynthetic productivity using competitive phototaxis. Sci. Rep. 6, 21155.

Kirst, H., Garcia-Cerdan, J.G., Zurbriggen, A., Ruehle, T., Melis, A., 2012. Truncated photosystem chlorophyll antenna size in the green microalga Chlamydomonas reinhardtii upon deletion of the TLA3-CpSRP43 gene. Plant Physiol. 160, 2251–2260.

Kirst, H., Gabilly, S.T., Niyogi, K.K., Lemaux, P.G., Melis, A., 2017. Photosynthetic antenna engineering to improve crop yields. Planta 245, 1009–1020.

Krieger-Liszkay, A., Fufezan, C., Trebst, A., 2008. Singlet oxygen production in photosystem II and related protection mechanism. Photosynth. Res. 98, 551–564.

Kromdijk, J., Głowacka, K., Leonelli, L., Gabilly, S.T., Iwai, M., Niyogi, K.K., Long, S.P., 2016. Improving photosynthesis and crop productivity by accelerating recovery from photoprotection. Science (80-) 354, 857–861.

Kulheim, C., Agren, J., Jansson, S., 2002. Rapid regulation of light harvesting and plant fitness in the field. Science (80-) 297, 91–93.

Langridge, P., 2014. Reinventing the green revolution by harnessing crop mutant resources. Plant Physiol. 166, 1682–1683.

Li, X.P., Muller-Moule, P., Gilmore, A.M., Niyogi, K.K., 2002. PsbS-dependent enhancement of feedback de-excitation protects photosystem II from photoinhibition. Proc. Natl. Acad. Sci. U. S. A. 99, 15222–15227.

Li, Z., Wakao, S., Fischer, B.B., Niyogi, K.K., 2009. Sensing and responding to excess light excess light (EL): a relative term that describes the absorption of light that exceeds photosynthetic capacity. Annu. Rev. Plant Biol. 60, 239–260.

Li, Y., Liu, B., Zhang, J., Kong, F., Zhang, L., Meng, H., Li, W., Rochaix, J.D., Li, D., Peng, L., 2019. OHP1, OHP2, and HCF244 form a transient functional complex with the photosystem II reaction center. Plant Physiol. 179, 195–208.

Liu, Z., Yan, H., Wang, K., Kuang, T., Zhang, J., Gui, L., An, X., Chang, W., 2004. Crystal structure of spinach major light-harvesting complex at 2.72 Å resolution. Nature 428, 287.

Long, S.P., Zhu, X.G., Naidu, S.L., Ort, D.R., 2006. Can improvement in photosynthesis increase crop yields? Plant Cell Environ. 29, 315–330.

Mitra, M., Kirst, H., Dewez, D., Melis, A., 2012. Modulation of the light-harvesting chlorophyll antenna size in Chlamydomonas reinhardtii by TLA1 gene over-expression and RNA interference. Philos. Trans. R. Soc. Lond. B Biol. Sci. 367, 3430–3443.

Monteith, J.L., 1977. Climate and the efficiency of crop production in Britain. Philos. Trans. R. Soc. Lond. Ser. B 281, 277–294.

Morosinotto, T., Mozzo, M., Bassi, R., Croce, R., 2005. Pigment-pigment interactions in Lhca4 antenna complex of higher plants photosystem I. J. Biol. Chem. 280, 20612–20619.

Negi, S., Perrine, Z., Friedland, N., Kumar, A., Tokutsu, R., Minagawa, J., Berg, H., Barry, A.N., Govindjee, G., Sayre, R., 2020. Light regulation of light-harvesting antenna size substantially enhances photosynthetic efficiency and biomass yield in green algae†. Plant J. 103, 584–603.

Neilson, J.A.D., Durnford, D.G., 2010. Structural and functional diversification of the light-harvesting complexes in photosynthetic eukaryotes. Photosynth. Res. 106, 57–71.

Nelson, N., Ben-Shem, A., 2004. The complex architecture of oxygenic photosynthesis. Nat. Rev. Mol. Cell Biol. 5, 239–250.

Nicol, L., Nawrocki, W.J., Croce, R., 2019. Disentangling the sites of non-photochemical quenching in vascular plants. Nat. Plants 5, 1177–1183.

Niyogi, K.K., 2000. Safety valves for photosynthesis. Curr. Opin. Plant Biol. 3, 455–460.

Nürnberg, D.J., Morton, J., Santabarbara, S., Telfer, A., Joliot, P., Antonaru, L.A., Ruban, A.V., Cardona, T., Krausz, E., Boussac, A., Fantuzzi, A., William Rutherford, A., 2018. Photochemistry beyond the red limit in chlorophyll f–containing photosystems. Science (80-) 360, 1210–1213.

Oberhuber, W., Dai, Z.-Y., Edwards, G.E., 1993. Light Dependence of Quantum Yields of Photosystem II and CO_2 Fixation in C_3 and C_4 Plants*. Kluwer Academic Publishers.

Oey, M., Ross, I.L., Stephens, E., Steinbeck, J., Wolf, J., Radzun, K.A., Kügler, J., Ringsmuth, A.K., Kruse, O., Hankamer, B., 2013. RNAi knock-down of LHCBM1, 2 and 3 increases photosynthetic H2 production efficiency of the green alga Chlamydomonas reinhardtii. PLoS One 8, e61375.

Ort, D.R., et al., 2015. Redesigning photosynthesis to sustainably meet global food and bioenergy demand. Proc. Natl. Acad. Sci. 112 (28), 8529–8536.

Ort, D.R., Zhu, X., Melis, A., 2011. Optimizing antenna size to maximize photosynthetic efficiency. Plant Physiol. 155, 79–85.

Pan, X., Li, M., Wan, T., Wang, L., Jia, C., Hou, Z., Zhao, X., Zhang, J., Chang, W., 2011. Structural insights into energy regulation of light-harvesting complex CP29 from spinach. Nat. Struct. Mol. Biol. 18, 309–315.

Pan, X., Ma, J., Su, X., Cao, P., Chang, W., Liu, Z., Zhang, X., Li, M., 2018. Structure of the maize photosystem I supercomplex with light-harvesting complexes I and II. Science (80-) 360, 1109–1113.

Pandey, R., Sahu, A., Vasumathi, K.K., Premalatha, M., 2015. Studies on light intensity distribution inside an open pond photo-bioreactor. Bioprocess Biosyst. Eng. 38, 1547–1557.

Peers, G., 2012. Enhancement of biomass production by disruption of light energy dissipation pathways. Patent Application Publication USA, Pub. No. US 2012/0178134 A1, July 12.

Peers, G., Truong, T.B., Ostendorf, E., Busch, A., Elrad, D., Grossman, A.R., Hippler, M., Niyogi, K.K., 2009. An ancient light-harvesting protein is critical for the regulation of algal photosynthesis. Nature 462, 518–521.

Perin, G., Bellan, A., Segalla, A., Meneghesso, A., Alboresi, A., Morosinotto, T., 2015. Generation of random mutants to improve light-use efficiency of Nannochloropsis gaditana cultures for biofuel production. Biotechnol. Biofuels 8, 161.

Perrine, Z., Negi, S., Sayre, R.T., 2012. Optimization of photosynthetic light energy utilization by microalgae. Algal Res. 1, 134–142.

Pesaresi, P., Pribil, M., Wunder, T., Leister, D., 2011. Dynamics of reversible protein phosphorylation in thylakoids of flowering plants: the roles of STN7, STN8 and TAP38. Biochim. Biophys. Acta Bioenerg. 1807, 887–896.

Pietrzykowska, M., Suorsa, M., Semchonok, D.A., Tikkanen, M., Boekema, E.J., Aro, E.M., Jansson, S., 2014. The light-harvesting chlorophyll a/b binding proteins Lhcb1 and Lhcb2 play complementary roles during state transitions in Arabidopsis. Plant Cell 26, 3646–3660.

Puzorjov, A., McCormick, A.J., 2020. Phycobiliproteins from extreme environments and their potential applications. J. Exp. Bot. 71, 3827–3842.

Redekop, P., Rothhausen, N., Rothhausen, N., Melzer, M., Mosebach, L., Dülger, E., Bovdilova, A., Caffarri, S., Hippler, M., Jahns, P., 2020. PsbS contributes to photoprotection in Chlamydomonas reinhardtii independently of energy dissipation. Biochim. Biophys. Acta Bioenerg. 1861, 1–12.

Ritchie, H., Roser, M., 2013. Crop Yields. OurWorldIn Data.org.

Roach, T., Na, C.S., Stöggl, W., Krieger-Liszkay, A., 2020. The non-photochemical quenching protein LHCSR3 prevents oxygen-dependent photoinhibition in Chlamydomonas reinhardtii. J. Exp. Bot. 71, 2650–2660.

Rochaix, J.-D., Bassi, R., 2019. LHC-like proteins involved in stress responses and biogenesis/repair of the photosynthetic apparatus. Biochem. J. 476, 581–593.

Rodolfi, L., Zittelli, G.C., Bassi, N., Padovani, G., Biondi, N., Bonini, G., Tredici, M.R., 2009. Microalgae for oil: strain selection, induction of lipid synthesis and outdoor mass cultivation in a low-cost photobioreactor. Biotechnol. Bioeng. 102, 100–112.

Ruban, A.V., 2016. Nonphotochemical chlorophyll fluorescence quenching: mechanism and effectiveness in protecting plants from photodamage. Plant Physiol. 170, 1903–1916.

Sacharz, J., Giovagnetti, V., Ungerer, P., Mastroianni, G., Ruban, A.V., 2017. The xanthophyll cycle affects reversible interactions between PsbS and light-harvesting complex II to control non-photochemical quenching. Nat. plants 3, 16225.

Sakowska, K., et al., 2018. Leaf and canopy photosynthesis of a chlorophyll deficient soybean mutant. Plant Cell Environ. 41, 1427–1437.

Sforza, E., Simionato, D., Giacometti, G.M., Bertucco, A., Morosinotto, T., 2012. Adjusted light and dark cycles can optimize photosynthetic efficiency in algae growing in photobioreactors. PLoS One 7 (6), e38975.

Shang, B., Zang, Y., Zhao, X., Zhu, J., Fan, C., Guo, X., Zhang, X., 2019. Functional characterization of GhPHOT2 in chloroplast avoidance of Gossypium hirsutum. Plant Physiol. Biochem. 135, 51–60.

Shen, G., Canniffe, D.P., Ho, M.Y., Kurashov, V., van der Est, A., Golbeck, J.H., Bryant, D.A., 2019a. Characterization of chlorophyll f synthase heterologously produced in Synechococcus sp. PCC 7002. Photosynth. Res. 140, 77–92.

Shen, L., Huang, Z., Chang, S., Wang, W., Wang, J., Kuang, T., Han, G., Shen, J.R., Zhang, X., 2019c. Structure of a C2S2M2N2-type PSII–LHCII supercomplex from the green alga Chlamydomonas reinhardtii. Proc. Natl. Acad. Sci. U. S. A. 116, 21246–21255.

Slattery, R.A., Ort, D.R., 2015. Photosynthetic energy conversion efficiency: setting a baseline for gauging future improvements in important food and biofuel crops. Plant Physiol. 168, 383–392.

Snow, A.A., Smith, V.H., 2012. Genetically engineered algae for biofuels: a key role for ecologists. Bioscience 62, 765–768.

Steen, C.J., Morris, J.M., Short, A.H., Niyogi, K.K., Fleming, G.R., 2020. Complex roles of PsbS and Xanthophylls in the regulation of nonphotochemical quenching in Arabidopsis thaliana under fluctuating light. J. Phys. Chem. B 124, 10311–10325.

Su, X., Ma, J., Wei, X., Cao, P., Zhu, D., Chang, W., Liu, Z., Zhang, X., Li, M., 2017. Structure and assembly mechanism of plant $C_2S_2M_2$-type PSII-LHCII supercomplex. Science (80-) 357, 815–820.

Suetsugu, N., Wada, M., 2007. Chloroplast photorelocation movement mediated by phototropin family proteins in green plants. Biol. Chem. 388, 927–935.

Suga, M., Ozawa, S.I., Yoshida-Motomura, K., Akita, F., Miyazaki, N., Takahashi, Y., 2019. Structure of the green algal photosystem I supercomplex with a decameric light-harvesting complex I. Nat. Plants 5, 626–636.

Sztatelman, O., Waloszek, A., Banas, A.K., Gabrys, H., 2010. Photoprotective function of chloroplast avoidance movement: in vivo chlorophyll fluorescence study. J. Plant Physiol. 167, 709–716.

Tibiletti, T., Auroy, P., Peltier, G., Caffarri, S., 2016. Chlamydomonas reinhardtii PsbS protein is functional and accumulates rapidly and transiently under high light. Plant Physiol. 171, 2717–2730.

Tokutsu, R., Iwai, M., Minagawa, J., 2009. CP29, a monomeric light-harvesting complex II protein, is essential for state transitions in Chlamydomonas reinhardtii. J. Biol. Chem. 284, 7777–7782.

Tokutsu, R., Fujimura-Kamada, K., Yamasaki, T., Matsuo, T., Minagawa, J., 2019. Isolation of photoprotective signal transduction mutants by systematic bioluminescence screening in Chlamydomonas reinhardtii. Sci. Rep. 9, 2028.

Townsend, A.J., Saccon, F., Giovagnetti, V., Wilson, S., Ungerer, P., Ruban, A.V., 2018. The causes of altered chlorophyll fluorescence quenching induction in the Arabidopsis mutant lacking all minor antenna complexes. Biochim. Biophys. Acta Bioenerg. 1859, 666–675.

Tredici, M.R., 2010. Photobiology of microalgae mass cultures: understanding the tools for the next green revolution. Biofuels 1, 143–162.

Treves, H., Raanan, H., Finkel, O.M., Berkowicz, S.M., Keren, N., Shotland, Y., Kaplan, A., 2013. A newly isolated Chlorella sp. from desert sand crusts exhibits a unique resistance to excess light intensity. FEMS Microbiol. Ecol. 86, 373–380.

Triantaphylides, C., Krischke, M., Hoeberichts, F.A., Ksas, B., Gresser, G., Havaux, M., Van Breusegem, F., Mueller, M.-J., 2008. Singlet oxygen is the major reactive oxygen species involved in photooxidative damage to plants. Plant Physiol. 148, 960–968.

van Oort, B., Alberts, M., de Bianchi, S., Dall'Osto, L., Bassi, R., Trinkunas, G., Croce, R., Van Amerongen, H., 2010. Effect of antenna-depletion in photosystem II on excitation energy transfer in Arabidopsis thaliana. Biophys. J. 98, 922–931.

Wada, M., Kagawa, T., Sato, Y., 2003. Chloroplast movement. Annu. Rev. Plant Biol. 54, 455–468.

Wakabayashi, K., Misawa, Y., Mochiji, S., Kamiya, R., 2011. Reduction-oxidation poise regulates the sign of phototaxis in Chlamydomonas reinhardtii. Proc. Natl. Acad. Sci. U. S. A. 108, 11280–11284.

Weyer, K.M., Bush, D.R., Darzins, A., Willson, B.D., 2010. Theoretical maximum algal oil production. Bioenergy Res. 3, 204–213.

Wilson, S., Ruban, A.V., 2020. Rethinking the influence of chloroplast movements on non-photochemical quenching and photoprotection. Plant Physiol. 183, 1213–1223.

Xie, Y., Li, J., Ho, S.H., Ma, R., Shi, X., Liu, L., Chen, J., 2020. Pilot-scale cultivation of Chlorella sorokiniana FZU60 with a mixotrophy/photoautotrophy two-stage strategy for efficient lutein production. Bioresour. Technol. 314, 123767.

Ye, Z.-P., Suggett, D.J., Robakowski, P., Kang, H.-J., 2013. A mechanistic model for the photosynthesis-light response based on the photosynthetic electron transport of photosystem II in C_3 and C_4 species. New Phytol. 199, 110–120.

Yen, H.-W., Hu, I.-C., Chen, C.-Y., Ho, S.-H., Lee, D.-J., 2013. Microalgae-based biorefinery – from biofuels to natural products. Bioresour. Technol. 135, 166–174.

Zamzam, N., Rakowski, R., Kaucikas, M., Dorlhiac, G., Viola, S., Nürnberg, D.J., Fantuzzi, A., Rutherford, A.W., van Thor, J.J., 2020. Femtosecond visible transient absorption spectroscopy of chlorophyll-f-containing photosystem II. Proc. Natl. Acad. Sci. U. S. A. 117, 23158–23164.

Zhi, X., Han, Y., Mao, S., Wang, G., Feng, L., Yang, B., Fan, Z., Du, W., Lu, J., Li, Y., 2014. Light spatial distribution in the canopy and crop development in cotton. PLoS One 9, 113409.

Zhu, X.G., Ort, D.R., Whitmarsh, J., Long, S.P., 2004. The slow reversibility of photosystem II thermal energy dissipation on transfer from high to low light may cause large losses in carbon gain by crop canopies: a theoretical analysis. J. Exp. Bot. 55, 1167–1175.

Zhu, X.G., Long, S.P., Ort, D.R., 2008. What is the maximum efficiency with which photosynthesis can convert solar energy into biomass? Curr. Opin. Biotechnol. 19, 153–159.

Zhu, X.-G., Long, S.P., Ort, D.R., 2010. Improving photosynthetic efficiency for greater yield. Annu. Rev. Plant Biol. 61, 235–261.

CHAPTER

8

Improving the transport of electrons

Conrad W. Mullineaux

School of Biological and Behavioural Sciences, Queen Mary University of London, London, United Kingdom

1 Layout of photosynthetic electron transport chains in plants and cyanobacteria

1.1 What do chloroplast and cyanobacterial electron transport chains have in common?

Chloroplasts are the descendants of free-living cyanobacteria, and it is no surprise that their photosynthetic electron transport chains retain many common features. The core membrane-integral electron transport complexes of the chloroplast photosynthetic electron transport chain, photosystem II, photosystem I, and the cytochrome b_6f complex, have diverged only subtly from their cyanobacterial counterparts, despite the fact that their evolutionary paths separated around 2 billion years ago. Among the subtle differences are some changes in the complement of the more peripheral photosystem subunits, including subunits associated with the water-oxidising complex on the lumenal side of photosystem II, and some of the smaller membrane-integral photosystem I subunits. The latter changes have the consequence that photosystem I in green plant chloroplasts is monomeric, whereas in cyanobacteria it is predominantly trimeric (although monomeric and tetrameric forms are also possible). Chloroplast and cyanobacterial thylakoid membranes also share a very similar lipid composition dominated by galactolipids, and both use plastoquinone as a lipid-soluble electron carrier. Chloroplasts and cyanobacteria use similar electron carriers in the main electron transport pathway downstream of photosystem I: ferredoxin and FNR. Both electron transport chains use the lumenal electron carrier plastocyanin, although cyanobacteria have retained the use of cytochrome c_6 as an alternative to plastocyanin, expressed under low-copper conditions. This is one of several instances where the cyanobacterial electron transport chain shows flexible features and alternative pathways that are absent in chloroplasts, as illustrated in Fig. 1. Alternative options may have been lost in chloroplasts as they became unnecessary due to the more stable and protected environment provided by the eukaryotic host. Although cyanobacterial and chloroplast thylakoid membranes retain many common features, cyanobacterial electron chains are much more complex, with many more competing options for electron transport (Fig. 1). In addition, there are striking differences between green plant chloroplasts and cyanobacteria in the complement of light-harvesting complexes and in the larger-scale

Photosynthesis in Action
https://doi.org/10.1016/B978-0-12-823781-6.00006-X

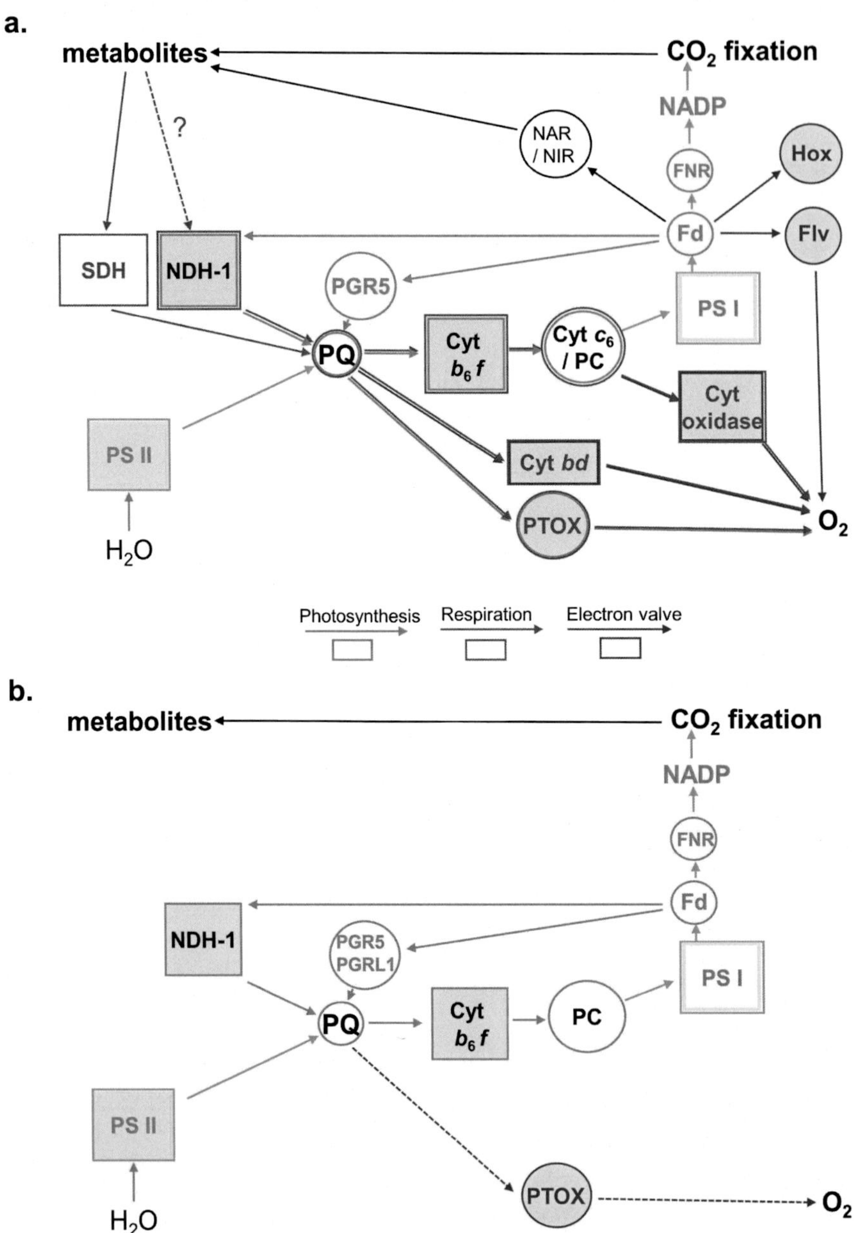

FIG. 1 Major electron transport routes in and around the thylakoid membranes of (A) a typical cyanobacterium and (B) a typical higher plant (angiosperm) chloroplast. Large membrane-integral complexes are shown as rectangles, smaller and more mobile components as circles. *Blue shading* indicates involvement in generating the proton gradient across the membrane (either by proton pumping or uptake/release of protons on one side of the membrane or the other). A *yellow box* indicates a site of light-energy input. *Cyt bd*, cytochrome *bd*; *complex Cyt* b_6f, cytochrome b_6f complex; *Cyt* c_6, cytochrome c_6; *Cyt oxidase*, cytochrome oxidase (aa_3 type); *Fd*, ferredoxin; *FNR*, ferredoxin-NADP oxidoreductase; *Flv*, flavodiiron protein (Flv1/3 or Flv2/4); *Hox*, bidirectional NiFe hydrogenase; *NAR*, nitrate reductase; *NIR*, nitrite reductase; *NDH-1*, 'NAD(*P*)H dehydrogenase', complex I; *PC*, plastocyanin; *PGR5*, proton gradient regulation protein 5; *PGRL1*, PGR5-like protein 1; *PQ*, plastoquinone; *PSI*, photosystem I; *PSII*, photosystem II; *PTOX*, plastoquinol terminal oxidase; *SDH*, succinate dehydrogenase. *Adapted and updated from Mullineaux, C.W., 2014. Co-existence of photosynthetic and respiratory activities in cyanobacterial thylakoid membranes. Biochim. Biophys. Acta Bioenerg. 1837, 503–511. https://doi.org/10.1016/j.bbabio.2013.11.017.*

architecture and layout of the thylakoid membrane. These differences are summarised in the next section.

1.2 What is different about cyanobacterial electron transport chains?

In cyanobacteria, the thylakoid membrane is the major site of respiratory electron transport as well as photosynthetic electron transport. This means that cyanobacterial thylakoid membranes incorporate a full set of respiratory complexes, including succinate dehydrogenase, NDH-1, and multiple terminal oxidases. The electron transport activities of these complexes are not rigorously segregated from the photosynthetic complexes, and components including plastoquinone, cytochrome b_6f, plastocyanin/cytochrome c_6, and NDH-1 take part in major pathways of both photosynthetic and respiratory electron transport (Fig. 1A). In the light, when both photosynthetic and respiratory electron transport can be active, there is a possibility of 'hybrid' electron transport pathways where electrons switch between photosynthetic and respiratory routes. What determines the relative prevalence of different electron transport pathways in this complex situation? Later sections in this chapter will discuss some factors that must be influential, but we certainly do not yet have a complete answer to the question. The possibilities in chloroplasts are more limited (Fig. 1B). In addition to linear electron transport, there is cyclic electron transport around photosystem I, which can be promoted by the formation of a supercomplex including photosystem I and cytochrome b_6f (Iwai et al., 2010). There is a chloroplast NDH that appears to be involved mainly in cyclic electron transport around photosystem I and a plastoquinol terminal oxidase (PTOX), which serves as an electron sink during plastid development and under stress conditions (McDonald et al., 2011).

A major difference between the cyanobacterial and the chloroplast photosynthetic apparatus is in the light-harvesting complexes. 'Typical' cyanobacteria have phycobilisomes, large membrane-extrinsic pigment-protein complexes associated with the cytoplasmic side of the thylakoid membrane, although these are absent in the very abundant marine cyanobacteria known as prochlorophytes. Phycobilisomes have been retained in the chloroplasts of red algae (rhodophytes) but are absent in other chloroplasts, including those of green plants. Green plant chloroplasts have functionally replaced the phycobilisomes with membrane-integral light-harvesting proteins binding chlorophylls *a* and *b*: these appear to have been repurposed from cyanobacterial proteins that transiently bind chlorophylls during the assembly and turnover of photosystems. The switch in light-harvesting strategies during the evolution of green plant chloroplasts has had major consequences for the overall architecture and organisation of the thylakoid membrane.

Green plant thylakoid membranes have a very distinctive organisation, with a complex three-dimensional architecture and lateral segregation of the membrane into grana and stroma lamellae. Photosystem II is concentrated in the stacked grana membranes, while photosystem I is excluded from the grana and therefore found exclusively in the stroma lamellae. A consequence of this organisation is that linear photosynthetic electron transfer (which requires photosystem I and photosystem II to act in series) must occur over extended length-scales, as will be discussed later. Cyanobacterial thylakoid membranes lack grana stacks. Thylakoids is some species form rather smooth, regular cylinders, while in other species the membranes may be more curved and irregular. However, structures resembling chloroplast grana have not been observed in any cyanobacterium. This may be partly a consequence of the presence of phycobilisomes (in most cyanobacteria) on the cytoplasmic face of the thylakoid membrane, since clearly the bulky phycobilisomes will prevent the tight appression of the membrane surfaces that is seen in grana. However, the phycobilisomes do not provide a complete answer, since they are absent in cyanobacteria such as *Prochlorococcus*, which also lack grana membranes. Interactions between LHCII molecules and photosystem II stabilise the grana in green plant chloroplasts. Other factors that help to shape the thylakoid membrane are incompletely understood, but the membrane-integral protein CURT1 (or CurT) has been shown to have an important in promoting membrane curvature in both chloroplasts (Armbruster et al., 2013) and cyanobacteria (Heinz et al., 2016).

Although cyanobacterial thylakoid membranes lack the sharp and defined lateral heterogeneity of green plant thylakoids, multiple experimental approaches have shown that electron transport complexes and other membrane components are not evenly and randomly mixed in the membrane. Hyperspectral fluorescence imaging showed distinct radial distributions of photosystem II and photosystem I in *Synechocystis* sp. PCC6803, with photosystem II more concentrated towards the cell periphery (Vermaas et al., 2008). Fluorescent protein tagging showed patchy distribution of multiple thylakoid membrane complexes in *Synechococcus* sp. PCC7942 (Casella et al., 2017) and atomic force and electron microscopy on isolated thylakoid membranes have shown the presence of domains that may be dominated by either photosystem II (Folea et al., 2008) or photosystem I (MacGregor-Chatwin et al., 2017). Cryo-electron tomography has revealed the presence of biogenic regions where the membrane is decorated with ribosomes rather than phycobilisomes (Rast et al., 2019), and fluorescence microscopy also provides evidence for specialised regions dedicated to biogenesis and repair (Mahbub et al., 2020). Patchy and heterogeneous distribution of complexes must certainly influence electron transport function, a point that will be explored further below.

2 Time-scales, length-scales, and constraints in electron transport

2.1 How far do electrons travel?

The lateral segregation of photosystem II and photosystem I in green plant chloroplasts imposes a requirement for lateral diffusion of an electron carrier in order to complete the pathway of linear photosynthetic electron transport. The typical diameter of a granum is about 300–600 nm, which suggests that an electron extracted from a water molecule by a photosystem II complex at the centre of a granal disc may have to travel at least 150–300 nm before completing its journey in the thylakoid membrane. Since the cytochrome b_6f complex is found in both the grana and the stroma lamellae, the electron carrier mediating longer-range diffusion might be plastoquinone diffusing within the lipid bilayer from photosystem II to cytochrome b_6f, or plastocyanin diffusing within the lumen from cytochrome b_6f to photosystem I, or some combination of both. Recent results suggest that plastoquinone predominantly mediates short-range electron transfer within photosystem II-cytochrome b_6f microdomains in the grana (Johnson et al., 2014), while the longer-range electron carrier is plastocyanin diffusing from cytochrome b_6f in the grana to photosystem I in the stroma lamellae (Höhner et al., 2020). The lumenal space undergoes dynamic swelling and shrinkage depending on illumination, and is significantly swollen in the light: this effect may be crucial to provide plastocyanin with sufficient diffusion space to allow the rapid long-distance migration required for linear electron transport (Höhner et al., 2020). The other major mode of electron transport in plant thylakoids is cyclic electron transport around photosystem I, in which there is no obvious requirement for long-range electron transport. Indeed, this mode of electron transport sometimes appears to take place within tightly associated protein supercomplexes (Iwai et al., 2010). It is therefore likely that the long-range diffusion of plastocyanin is a key constraint on the efficiency of linear electron transport vs cyclic electron transport, with consequences for the ATP/NADPH ratio and the supply of reducing power from the photosynthetic electron transport chain.

The less-defined organisation of cyanobacterial thylakoid membranes means that there are fewer clues to the length-scales of electron transport in cyanobacteria. There is evidence for some supercomplexes containing electron transport components: a complex of photosystem I and NDH-1 (Gao et al., 2016) and a complex of phycobilisomes with photosystem II and photosystem I (Liu et al., 2013). Furthermore, one form of FNR is tethered to the phycobilisomes in cyanobacteria, potentially bringing it into a supercomplex with one or both reaction centres (Liu et al., 2019). However, none of the cyanobacterial supercomplexes characterised thus far includes cytochrome b_6f, which is central to all the major electron transport pathway in the thylakoid membrane (Fig. 1A). Therefore, electrons would have to leave any of the known supercomplexes at some point in order to complete the pathway. This leaves the length of the electron's journey in the membrane as an open question, although the various supercomplexes could play major roles in steering electrons down specific pathways. The larger-scale organisation of cyanobacterial thylakoid membranes has been probed by fluorescence microscopy in combination with fluorescent protein tagging of specific complexes (Casella et al., 2017; Liu et al., 2012). This approach reveals that multiple thylakoid membrane protein complexes have a patchy distribution on length scales of the order of 100–300 nm (Casella et al., 2017; Liu et al., 2012). Redistribution of NDH-1 and SDH can be induced by a redox trigger, with the complexes being concentrated in localised spots under conditions where the plastoquinone pool is oxidised, and becoming more evenly dispersed when the plastoquinone pool is reduced (Liu et al., 2012). This change in distribution correlates with a significant enhancement of the cyclic electron transport pathway (Liu et al., 2012), which gives a clue that length scales of about 100–300 nm are significant for electron transport pathways, suggesting that a typical electron's journey in the membrane may be on this scale.

2.2 How fast do electrons travel?

A comprehensive estimate for the time taken for an electron to travel from the water-oxidising complex of photosystem II to P_{700} of photosystem I in green plant chloroplasts suggests an average total time of about 8.5 ms (Höhner et al., 2020). This includes two steps involving electron carriers: the plastoquinone-mediated electron transfer from photosystem II to the cytochrome b_6f complex is estimated to take about 3.2 ms, while plastocyanin-mediated electron transfer from cytochrome b_6f to photosystem I is estimated to take about 220–300 μs (Höhner et al., 2020). It is these diffusion-mediated electron transfer steps that must be dependent on the architecture of the thylakoid membranes and the supramolecular organisation of the electron transport complexes. Indeed, Höhner et al. show that the time required for the plastocyanin-mediated electron transfer step is drastically increased to 6 ms or more in an *Arabidopsis* mutant lacking the CURT1 protein (Höhner et al., 2020). CURT1 helps to shape the architecture of thylakoid membranes (Armbruster et al., 2013), and the mutant has granal discs enlarged in diameter from around 500 to 1600 nm, leading to the greatly increased diffusion time for plastocyanin (Höhner et al., 2020).

The time taken for the electron's journey will impact on the competition between different electron transport pathways, and also on the competition between productive electron transport and wasteful and possibly dangerous side reactions such as the production of reaction oxygen species by reaction of O_2 with reduced electron carriers. The factors that control electron transfer times will be further discussed in sections below.

3 Factors that determine where electrons go

3.1 Light-harvesting

Linear electron transport requires the activity of both photosystems. Classical cyclic electron transport pathways require photosystem I only, and modes of electron transport involving photosystem II only are possible. For example, electron transport from photosystem II to oxygen via a plastoquinol terminal oxidase is the major pathway in the marine green alga *Ostreococcus* (Cardol et al., 2008). It follows that the relative excitation of the two photosystems must influence the prevalence of different electron transfer pathways, with balanced excitation of photosystem I and photosystem II favouring linear electron transport while higher excitation of photosystem I favours cyclic electron transport and higher excitation of photosystem II could favour a water–water cycle as in *Ostreococcus* (Cardol et al., 2008). Cyclic electron transport and the water–water cycle both generate a proton gradient (and therefore ATP) without net production of reducing power, while linear electron transport generates both ATP and NADPH. Therefore, the balance of excitation of the two photosystems must influence the relative generation of ATP vs reducing power. The relative excitation of the photosystems could be regulated on short timescales by mechanisms that moves light-harvesting complexes from one photosystem to the other, or which selectively quench energy destined for one photosystem or the other. On longer timescales, adjustment of the populations of light-harvesting complexes associated with each photosystem (and the relative abundance of the two photosystems) will have similar effects.

State 1–state 2 transitions are a posttranslational mechanism that shuttles the LHCII light-harvesting complexes between photosystem II and photosystem I in green plants in response to changes in the redox state of the plastoquinone pool. This could serve as a mechanism to regulate linear vs cyclic electron transport, and such effects are well-documented in the green alga *Chlamydomonas*, for example (Minagawa, 2011). State transitions in cyanobacteria similarly adjust the relative excitation of the two photosystems, although debate continues as to the relative importance of regulation of the location of the phycobilisome light-harvesting complexes vs selective energy quenching at photosystem II (Bhatti et al., 1861; Joshua and Mullineaux, 2004). State transitions provide an effective method to adjust the relative excitation of the photosystems in both plants and cyanobacteria, with concomitant effects on linear vs cyclic electron transport. By contrast, photoprotective energy quenching mechanisms appear less effective in controlling the relative activity of the photosystems. The major energy quenching mechanism in cyanobacteria is via the orange carotenoid-binding protein, which quenches energy at the phycobilisomes (Wilson et al., 2006). However, this affects energy transfer to both photosystems, and hence may not be an effective way to rebalance electron flow. In green plants, energy quenching in the LHCII antenna could potentially provide a more selective way to reduce photosystem II activity, but in practice nonphotochemical quenching mainly affects the energy supply to photosystem II reaction centres that are already inactive in electron transfer because they are photochemically closed (Belgio et al., 2014).

3.2 Concentrations of electron transport components

The effective concentration of the electron acceptor must be one of the key factors controlling the rate of an electron transfer step and hence the probability of an electron following a particular path. For example, in cyanobacteria a major fork in the road concerns the fate of electrons carried by ferredoxin, which might go on the linear path to reduce NADP via FNR, or which might go to several other destinations, including a return to the photosynthetic electron transport chain via ferredoxin's interaction with NDH-1, resulting in cyclic electron flow (Fig. 1A). This crucial choice must surely be strongly influenced by the concentrations of FNR (in the cytoplasm, or anchored to the phycobilisomes) and NDH-1 in the membrane. Cyanobacteria have a multitude of alternative electron acceptors that can divert electrons away from the linear pathway leading to $NADP^+$ reduction to NADPH. These include oxidases that take electrons from cytochrome *c*/plastocyanin, cytochrome *bd* that takes electrons from plastoquinol and flavodiiron proteins and the bidirectional hydrogenase that take electrons from the acceptor side of photosystem I (Fig. 1A). Physiologically, these complexes are important as electron valves that prevent overreduction of electron carriers under fluctuating conditions. The active concentrations of these alternative electron acceptors must influence the yield of NADPH and the fate of electrons from the photosynthetic electron transport chain, and indeed the removal of terminal oxidases has been shown to increase the yield of a competing pathway leading to the extracellular export of electrons (Bradley et al., 2013).

Similarly, photosystem II must compete as an electron donor to plastoquinone with the NDH-1 and SDH complexes involved in cyclic and respiratory electron transport pathways. The prevalence of linear photosynthetic electron transport over the other modes must be influenced by the concentration of active photosystem II complexes relative to the other electron transport complexes in the membrane, and especially complexes such as NDH-1 and SDH that are exclusively involved in competing pathways. Therefore, there are multiple ways in which the manipulation of the concentration of electron carriers in the membrane could be manipulated to enhance the supply of reducing power to drive a biotechnologically useful electron transport reaction, for example. This point will be discussed further below.

3.3 Location of electron transport components

The physical proximity of donor and acceptor complexes appears to be one of the most effective ways to influence the prevalence of different possible electron transport pathways. A mobile electron carrier like plastoquinone, plastocyanin, or ferredoxin will diffuse away from the electron donor complex until it is captured by an electron acceptor complex, with the fate of the electron largely depending on which acceptor complex captures the electron carrier first. In random Brownian motion, the mean displacement from the starting point is proportional to the square root of the time elapsed. If electron donor complexes and electron acceptor complexes were evenly dispersed in a membrane, the mean time taken for a mobile electron carrier to shuttle an electron between them would be proportional to the square of the distance between the donor and the acceptor. So, where there are competing electron acceptors at similar concentrations, the one nearest the donor complex will have a strong advantage.

One way in which nature exploits the proximity principle to favour specific electron transport pathways is simply by building supercomplexes in which multiple electron transport components are tightly associated. These supercomplexes need not contain all the components of an electron transport pathway to be effective in controlling pathways of electron transport: it could be enough to put together two complexes either side of one of the crucial switch points in electron transport. An example is the cyanobacterial supercomplex of photosystem I and NDH-1, which clearly promotes cyclic electron transport at the expense of linear (Gao et al., 2016). This supercomplex does not contain all the ingredients for cyclic electron transport since cytochrome b_6f is absent, and therefore both plastoquinone and plastocyanin/cytochrome *c* must diffuse out of the supercomplex to complete the cyclic electron transport pathway. However, the proximity of photosystem I and NDH-1 is presumably enough to ensure that ferredoxin reduced at the acceptor side of photosystem I is likely to interact next with NDH-1 rather than with FNR. Thus, electrons are returned to the cyclic pathway in the thylakoid membrane rather than going on to reduce $NADP^+$ in the linear pathway. A cyanobacterial megacomplex of phycobilisomes with photosystem II and photosystem I (Liu et al., 2013) could play a more subtle role in promoting linear electron transport. Although mobile electron carriers will clearly have to leave the supercomplex to complete the electron transport pathway, the proximity principle means that a cytochrome b_6f complex that passes electrons to photosystem I is likely to have taken its electrons from photosystem II rather than from NDH-1, thus favouring linear rather than cyclic electron transport.

The organisation of electron carriers in cyanobacteria appears flexible and diverse. As well as a supercomplex containing photosystem II and photosystem I, there is strong evidence for segregated domains containing only one of the photosystems (Folea et al., 2008; MacGregor-Chatwin et al., 2017) and patchy distributions of multiple electron transport complexes (Casella et al., 2017; Liu et al., 2012). The larger-scale distribution of NDH-1 complexes in cyanobacterial thylakoids has been shown to change on a timescale of around 30 min in response to redox signals. Reduction of the plastoquinone pool appears to trigger a redistribution of NDH-1 complexes from a very patchy distribution in the membrane to being more evenly dispersed, and this change correlates with an increase in the extent of cyclic vs linear electron transport (Liu et al., 2012). Larger-scale redistribution of the bidirectional hydrogenase has also been observed, with the hydrogenase becoming concentrated in localised spots close to the distal thylakoid membranes under the anoxic conditions in which hydrogenase activity us normally observed (Burroughs et al., 2014). This raises the possibility that such a distribution of the hydrogenase favours electron supply from photosystem I (Burroughs et al., 2014). Another intriguing control on the localisation of a cyanobacterial electron transport component concerns FNR, which can be produced in two isoforms. The larger isoform contains an additional domain that links it to the phycobilisomes, which places it in proximity to the thylakoid surface and potentially adjacent to photosystem I (Liu et al., 2019). This could serve as a way to steer electrons into the linear pathway rather than the cyclic one by favouring electron donation from ferredoxin to FNR rather than to NDH-1. However, any regulation and its physiological consequences are not yet well characterised. A recent study on plant chloroplasts indicates that FNR is always thylakoid membrane associated and relocates to the grana margins during light adaptation, through a switch in interaction partners. This may serve as a mechanism to promote linear electron transport at the expense of cyclic (Kramer et al., 2021).

4 Ways to manipulate electron transport

4.1 Removal of competing pathways

The most straightforward way to promote a particular electron transport pathway is to remove components that are essential for competing pathways. In cyanobacteria, the plethora of alternative electron acceptors make an obvious target for this approach. A nice example is given by the deletion of three terminal oxidases in *Synechocystis* sp. PCC6803, which resulted in a 24-fold increase in the rate of extracellular electron transport to ferricyanide (Bradley et al., 2013).

However, this large increase is only seen in the dark, reflecting the fate of respiratory electrons originating from stored metabolites: in the light, the major pathway for electron transfer remains through photosystem I to generate reductant for CO_2 fixation (Bradley et al., 2013). The bidirectional hydrogenase Hox in *Synechocystis* takes electrons from ferredoxin, and therefore, the removal of competing ferredoxin electron acceptors is an obvious approach to boost photohydrogen production. The major pathway through FNR is essential for cell viability, but a merodiploid mutant with lower FNR levels showed higher photohydrogen production (Gutekunst et al., 2014). Other sinks for electrons from ferredoxin include nitrate reductase and nitrite reductase, both involved in the nitrate assimilation pathway. A double mutant lacking both these enzymes shows greatly increased hydrogen production (Baebprasert et al., 2011). This appears to result from improved electron supply rather than from any other effects on the abundance or activity of Hox (Baebprasert et al., 2011). The examples discussed above indicate that a reliable strategy to steer electrons down a preferred route is to eliminate nonessential competing pathways, combined with downregulation of the essential route to CO_2 fixation. There are further competing pathways that can be eliminated in cyanobacteria, since the flavodiiron proteins act as additional alternative electron sinks. In fact, elimination of flavodiiron proteins has been shown to boost the production of polyhydroxybutyrate, glycogen, and sucrose in cyanobacteria (Thiel et al., 2019; Santos-Merino et al., 2021). This indicates that CO_2 fixation via the Calvin cycle can be significantly enhanced by eliminating competing pathways whose physiological role is to act as emergency electron sinks to prevent overreduction of the electron transport chain. Promoting the production and export of sucrose partially compensates for the loss of these protective mechanisms (Santos-Merino et al., 2021).

In effect, cyclic electron transport acts as another competing sink for electrons. Diversion of electrons from the photosystem I acceptor side back to photosynthetic electron transport chain will keep the photosynthetic electron carriers more reduced, effectively inhibiting the supply of 'new' electrons extracted from water by photosystem II, and decreasing the supply of electrons available to be supplied to a preferred sink at the photosystem I acceptor side. Several factors are known that promote cyclic electron transport in plants and cyanobacteria. For example, the thylakoid proteins PGR5 and PGRL1 promote cyclic electron transport in plants (Fig. 1B). Downregulation of any of these factors would provide another route to boosting the reduction of a preferred electron acceptor.

4.2 Controlling the location of electron acceptors

As discussed in previous sections, the location of electron carriers is critical for efficient electron supply and competition between alternative electron transport pathways. Therefore, manipulation of the supramolecular organisation of electron transport components is a very plausible route to promoting a specific, favoured electron transport pathway. A simple way to achieve this would be to produce a mutant in which two normally separate complexes are tethered together. The acceptor side of photosystem I is an obvious location for tethering of an acceptor complex, since it is the site of maximum reducing power, with several subunits conveniently exposed on the cytoplasmic/stromal side of the thylakoid membrane. A successful example is given by the tethering of the native bidirectional hydrogenase to a truncated version of the PsaD subunit of photosystem I in the cyanobacterium *Synechocystis* (Appel et al., 2020). The mutant shows greatly enhanced photohydrogen production under anaerobic conditions, likely due to direct electron transfer from an iron–sulphur cluster at the PSI acceptor side to the hydrogenase. The mutant is also capable of photoautotrophic growth, albeit slower than the wild type, which suggests that the hydrogenase competes effectively with ferredoxin for electrons at the PSI acceptor side, but does not completely block the pathway to ferredoxin and FNR, which is essential for cell viability (Appel et al., 2020). A practical problem is the oxygen sensitivity of the hydrogenase, which means that the culture must be kept anaerobic to allow sustained photoproduction of hydrogen. This obviously requires continuous removal of oxygen from a culture that is growing by oxygenic photosynthesis. Nevertheless, the concept has obvious potential for boosting the photosynthetic production of any product that requires photosynthetic electron input, while still allowing for photoautotrophic growth and proliferation of the culture.

In some filamentous cyanobacteria, the oxygen problem could be solved by engineering the heterologous electron transport pathway to take place specifically in heterocysts, which are cells specialised for nitrogen fixation that maintain a microaerobic environment in their cytoplasm. Heterocysts retain photosystem I activity but lose active photosystem II: reducing power is derived from sugars imported from the neighbouring vegetative cells. It appears that electrons derived from the imported carbohydrate are fed into the photosynthetic electron transport chain, which means that photosystem I in heterocysts remains a potent source of electrons and is probably the main source of reducing power required by the nitrogenase for nitrogen fixation. A drawback of the use of heterocysts as anaerobic cell factories is that they do not normally constitute more than 10% of the cells in the filament, and electron supply may be limited by provision of sugars from the vegetative cells, which occurs by diffusion through 'septal junction' channels linking the cells (Nürnberg et al., 2015).

4.3 Addition of heterologous electron acceptors

The example discussed above in *Synechocystis* used a version of the native bidirectional hydrogenase, but, in principle, the approach could be extended to heterologous electron acceptors. For example, the hydrogenase from the β-proteobacterium *Ralstonia eutropha* is unusually oxygen-tolerant and could potentially circumvent the need to keep a cyanobacterial culture anaerobic for hydrogen photoproduction. In fact, fusion of the *Ralstonia* hydrogenase to the *Synechocystis* photosystem I acceptor side was shown to result in efficient photohydrogen production in vitro, using a complex assembled from *Synechocystis* photosystem I and a fusion of the hydrogenase and *Synechocystis* PsaE isolated from a *Ralstonia* transformant (Ihara et al., 2006). The practical difficulty with achieving this in a cyanobacterium in vivo is that a functional hydrogenase requires a nickel and iron containing metal centre and has a complex assembly pathway requiring multiple maturation factors. Therefore, engineering the efficient heterologous production of the *Ralstonia* hydrogenase in a cyanobacterium is likely to be challenging. This is particularly the case given that cyanobacteria keep very tight control of their internal metal content. Similar considerations are likely to apply to the introduction of other heterologous electron carriers and electron acceptors into photosynthetic organisms.

5 How much can electron transport be 'improved'?

The photosynthetic electron transport pathways of cyanobacteria and chloroplasts have been honed by evolution over billions of years, and therefore, the prospect of engineering any fundamental improvements might seem remote. The basic electron transport pathway involving photosystem II, cytochrome b_6f, and photosystem I has been highly conserved over evolutionary time, and it does a remarkably efficient job of using light energy to extract reducing power from water, coupled with generation of a proton gradient that powers ATP synthesis. There looks to be little scope for engineering improvements in this core process. However, there may well be scope for using add-ons to tailor electron transport to function better in specific environments. An example is given by the flavodiiron proteins, which act as emergency electron valves and confer tolerance to fluctuating light conditions in many photosynthetic organisms. These proteins appear to have been lost at some point during the evolutionary trajectory leading to angiosperm plants, and restoring them can promote tolerance to fluctuating light conditions (Yamamoto et al., 2016). It has also been shown that engineering enhanced electron sinks can improve the robustness of photosynthesis in cyanobacteria (Santos-Merino et al., 2021).

The second way in electron transport can be 'improved' is by tailoring electron pathways away from the natural priorities of the organism to favour the production of a desired product for biotechnology. Here, the idea is to divert reducing power away from the core pathway leading to CO_2 fixation for direct synthesis of the desired product. Examples of work in this direction have been discussed in sections above. A complete strategy for engineering the most efficient electron supply for photoproduction of a product might involve the following steps, which are illustrated in Fig. 2 for the specific

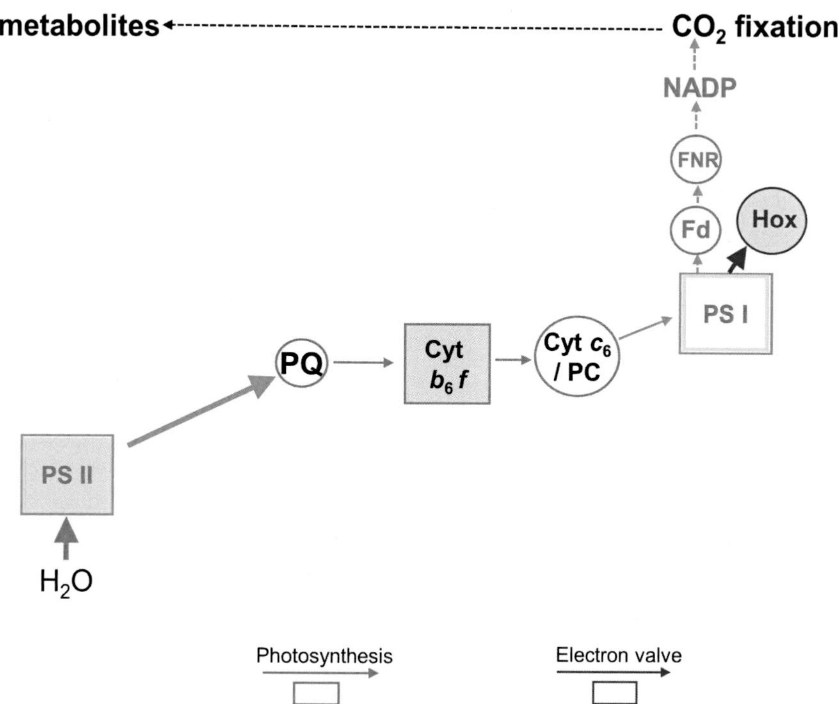

FIG. 2 Strategy for optimising photohydrogen production in a cyanobacterium. Starting from the native situation (Fig. 1A), competing electron transfer pathways (cyclic electron transport, respiration and electron valves) have been eliminated. Photosystem II turnover has been enhanced (by increasing its concentration and/or enhancing its light-harvesting capacity) to generate an excess electron supply. The Hox hydrogenase has been anchored to the photosystem I acceptor side (as in Appel et al., 2020) where it competes with ferredoxin for electrons. The engineered cyanobacterium will grow slowly due to the decreased input into the Calvin cycle for CO_2 fixation.

case of a photohydrogen production in a cyanobacterium.

1. Maximising linear electron transport at the expense of cyclic electron transport to generate an excess of reductant in the cell or chloroplast. This could involve modifying the balance of light-harvesting and the ratio of photosystem II to photosystem I to ensure sufficient extraction of electrons from water by photosystem II, and disabling or downregulating factors required specifically for cyclic electron transport.
2. Disabling or downregulating alternative electron sinks such as oxidases and flavodiiron proteins. Where the desired electron transport route is in direct competition with CO_2 fixation via the Calvin cycle (as with the hydrogenase discussed above), the Calvin cycle or its electron supply could be downregulated. This could be achieved, for example, by reducing the level of FNR as in Gutekunst et al. (2014).
3. Locating a specific electron acceptor in the best place to receive photosynthetic electrons, for example, at the acceptor side of photosystem I as in Appel et al. (2020).

The idea would be to generate a viable but slow-growing organism where a high proportion of photosynthetic electrons are channelled to production of the desired product rather than to the growth of the organism. Excessive growth of the culture is undesirable anyway when the aim is to generate a specific exported product. However, a significant issue in this scenario is that evolutionary pressures will always favour mutations that divert electrons back to CO_2 fixation for faster growth. It will be particularly a problem with cyanobacteria and other phototrophic microbes due to their short generation time and rapid evolution. One way to minimise the problem would be to set up conditions where rapid growth is prevented by shortage of an essential mineral like iron, phosphate, or nitrate. Mutants that acquire enhanced CO_2 fixation capabilities will then be prevented from proliferating rapidly and taking over the bioreactor. A complementary strategy involves making the organism dependent on its re-engineered electron sink to compensate for the loss of native electron valves such as flavodiiron proteins, which normally confer robustness and resistance to fluctuating light conditions. In fact, it has been shown that engineering sucrose export and cytochrome P450 as an extra electron sink in a cyanobacterium can increase the robustness of photosynthesis and partially compensate for the loss of flavodiiron proteins (Santos-Merino et al., 2021).

References

Appel, J., Hueren, V., Boehm, M., Gutekunst, K., 2020. Cyanobacterial in vivo solar hydrogen production using a photosystem I–hydrogenase (PsaD-HoxYH) fusion complex. Nat. Energy 5, 458–467. https://doi.org/10.1038/s41560-020-0609-6.

Armbruster, U., Labs, M., Pribil, M., Viola, S., Xu, W., Scharfenberg, M., Hertle, A.P., Rojahn, U., Jensen, P.E., Rappaport, F., Joliot, P., Dörmann, P., Wanner, G., Leister, D., 2013. Arabidopsis CURVATURE THYLAKOID1 proteins modify thylakoid architecture by inducing membrane curvature. Plant Cell 25, 2661–2678. https://doi.org/10.1105/tpc.113.113118.

Baebprasert, W., Jantaro, S., Khetkorn, W., Lindblad, P., Incharoensakdi, A., 2011. Increased H_2 production in the cyanobacterium *Synechocystis* sp. strain PCC 6803 by redirecting the electron supply via genetic engineering of the nitrate assimilation pathway. Metab. Eng. 13, 610–616. https://doi.org/10.1016/j.ymben.2011.07.004.

Belgio, E., Kapitonova, E., Chmeliov, J., Duffy, C.D.P., Ungerer, P., Valkunas, L., Ruban, A.V., 2014. Economic photoprotection in photosystem II that retains a complete light-harvesting system with slow energy traps. Nat. Commun. 5, 4433. https://doi.org/10.1038/ncomms5433.

Bhatti, A.F., Choubeh, R.R., Kirilovsky, D., Wientjes, E., van Amerongen, H., 1861. State transitions in cyanobacteria studied with picosecond fluorescence at room temperature. Biochim. Biophys. Acta Bioenerg. 2020, 148255. https://doi.org/10.1016/j.bbabio.2020.148255.

Bradley, R.W., Bombelli, P., Lea-Smith, D.J., Howe, C.J., 2013. Terminal oxidase mutants of the cyanobacterium Synechocystis sp. PCC 6803 show increased electrogenic activity in biological photo-voltaic systems. Phys. Chem. Chem. Phys. 15, 13611–13618. https://doi.org/10.1039/C3CP52438H.

Burroughs, N.J., Boehm, M., Eckert, C., Mastroianni, G., Spence, E.M., Yu, J., Nixon, P.J., Appel, J., Mullineaux, C.-W., Bryan, S.J., 2014. Solar powered biohydrogen production requires specific localization of the hydrogenase. Energy Environ. Sci. 7, 3791–3800. https://doi.org/10.1039/c4ee02502d.

Cardol, P., Bailleul, B., Rappaport, F., Derelle, E., Béal, D., Breyton, C., Bailey, S., Wollman, F.A., Grossman, A., Moreau, H., Finazzi, G., 2008. An original adaptation of photosynthesis in the marine green alga *Ostreococcus*. Proc. Natl. Acad. Sci. 105, 7881–7886. https://doi.org/10.1073/pnas.0802762105.

Casella, S., Huang, F., Mason, D., Zhao, G.-Y., Johnson, G.N., Mullineaux, C.W., Liu, L.-N., 2017. Dissecting the native architecture and dynamics of cyanobacterial photosynthetic machinery. Mol. Plant 10, 1434–1448. https://doi.org/10.1016/j.molp.2017.09.019.

Folea, I.M., Zhang, P., Aro, E.-M., Boekema, E.J., 2008. Domain organization of photosystem II in membranes of the cyanobacterium *Synechocystis* PCC6803 investigated by electron microscopy. FEBS Lett. 582, 1749–1754. https://doi.org/10.1016/j.febslet.2008.04.044.

Gao, F., Zhao, J., Chen, L., Battchikova, N., Ran, Z., Aro, E.-M., Ogawa, T., Ma, W., 2016. The NDH-1L-PSI supercomplex is important for efficient cyclic electron transport in cyanobacteria. Plant Physiol. 172, 1451–1464. https://doi.org/10.1104/pp.16.00585.

Gutekunst, K., Chen, X., Schreiber, K., Kaspar, U., Makam, S., Appel, J., 2014. The bidirectional NiFe-hydrogenase in *Synechocystis* sp. PCC 6803 is reduced by flavodoxin and ferredoxin and is essential under mixotrophic, nitrate-limiting conditions. J. Biol. Chem. 289, 1930–1937. https://doi.org/10.1074/jbc.M113.526376.

Heinz, S., Rast, A., Shao, L., Gutu, A., Gügel, I.L., Heyno, E., Labs, M., Rengstl, B., Viola, S., Nowaczyk, M.M., Leister, D., Nickelsen, J., 2016. Thylakoid membrane architecture in *Synechocystis* depends on CurT, a homolog of the granal CURVATURE THYLAKOID1 proteins. Plant Cell 28, 2238–2260. https://doi.org/10.1105/tpc.16.00491.

Höhner, R., Pribil, M., Herbstová, M., Lopez, L.S., Kunz, H.-H., Li, M., Wood, M., Svoboda, V., Puthiyaveetil, S., Leister, D., Kirchhoff, H., 2020. Plastocyanin is the long-range electron carrier between photosystem II and photosystem I in plants. Proc. Natl. Acad. Sci. 117, 15354–15362. https://doi.org/10.1073/pnas.2005832117.

Ihara, M., Nishihara, H., Yoon, K.-S., Lenz, O., Friedrich, B., Nakamoto, H., Kojima, K., Honma, D., Kamachi, T., Okura, I., 2006. Light-driven hydrogen production by a hybrid complex of a [NiFe]-hydrogenase and the cyanobacterial photosystem I. Photochem. Photobiol. 82, 676–682. https://doi.org/10.1562/2006-01-16-RA-778.

Iwai, M., Takizawa, K., Tokutsu, R., Okamuro, A., Takahashi, Y., Minagawa, J., 2010. Isolation of the elusive supercomplex that drives cyclic electron flow in photosynthesis. Nature 464, 1210–1213. https://doi.org/10.1038/nature08885.

Johnson, M.P., Vasilev, C., Olsen, J.D., Hunter, C.N., 2014. Nanodomains of cytochrome b_6f and photosystem II complexes in spinach grana thylakoid membranes. Plant Cell 26, 3051–3061. https://doi.org/10.1105/tpc.114.127233.

Joshua, S., Mullineaux, C.W., 2004. Phycobilisome diffusion is required for light-state transitions in cyanobacteria. Plant Physiol. 135, 2112–2119. https://doi.org/10.1104/pp.104.046110.

Kramer, M., Rodriguez-Heredia, M., Saccon, F., Mosebach, L., Twachtmann, M., Krieger-Liszkay, A., Duffy, C., Knell, R.-J., Finazzi, G., Hanke, G.T., 2021. Regulation of photosynthetic electron flow on dark to light transition by ferredoxin:NADP(H) oxidoreductase interactions. Elife 10. https://doi.org/10.7554/eLife.56088, e56088.

Liu, L.-N., Bryan, S.J., Huang, F., Yu, J., Nixon, P.J., Rich, P.R., Mullineaux, C.W., 2012. Control of electron transport routes through redox-regulated redistribution of respiratory complexes. Proc. Natl. Acad. Sci. U. S. A. 109, 11431–11436. https://doi.org/10.1073/pnas.1120960109.

Liu, H., Zhang, H., Niedzwiedzki, D.M., Prado, M., He, G., Gross, M.L., Blankenship, R.E., 2013. Phycobilisomes supply excitations to both photosystems in a megacomplex in cyanobacteria. Science (80-) 342, 1104–1107. https://doi.org/10.1126/science.1242321.

Liu, H., Weisz, D.A., Zhang, M.M., Cheng, M., Zhang, B., Zhang, H., Gerstenecker, G.S., Pakrasi, H.B., Gross, M.L., Blankenship, R.E., 2019. Phycobilisomes harbor FNR_L in cyanobacteria. MBio 10. https://doi.org/10.1128/mBio.00669-19, e00669-19.

MacGregor-Chatwin, C., Sener, M., Barnett, S.F.H., Hitchcock, A., Barnhart-Dailey, M.C., Maghlaoui, K., Barber, J., Timlin, J.A., Schulten, K., Hunter, C.N., 2017. Lateral segregation of photosystem I in cyanobacterial thylakoids. Plant Cell 29, 1119–1136. https://doi.org/10.1105/tpc.17.00071.

Mahbub, M., Hemm, L., Yang, Y., Kaur, R., Carmen, H., Engl, C., Huokko, T., Riediger, M., Watanabe, S., Liu, L.-N., Wilde, A., Hess, W.R., Mullineaux, C.W., 2020. mRNA localization, reaction centre biogenesis and thylakoid membrane targeting in cyanobacteria. Nat. Plants 6, 1179–1191. https://doi.org/10.1038/s41477-020-00764-2.

McDonald, A.E., Ivanov, A.G., Bode, R., Maxwell, D.P., Rodermel, S.R., Hüner, N.P.A., 2011. Flexibility in photosynthetic electron transport: the physiological role of plastoquinol terminal oxidase (PTOX). Biochim. Biophys. Acta Bioenerg. 1807, 954–967. https://doi.org/10.1016/j.bbabio.2010.10.024.

Minagawa, J., 2011. State transitions—the molecular remodeling of photosynthetic supercomplexes that controls energy flow in the chloroplast. Biochim. Biophys. Acta Bioenerg. 1807, 897–905. https://doi.org/10.1016/j.bbabio.2010.11.005.

Nürnberg, D.J., Mariscal, V., Bornikoel, J., Nieves-Morión, M., Krauß, N., Herrero, A., Maldener, I., Flores, E., Mullineaux, C.W., 2015. Intercellular diffusion of a fluorescent sucrose analog via the septal junctions in a filamentous cyanobacterium. MBio 6. https://doi.org/10.1128/mBio.02109-14, e02109-14.

Rast, A., Schaffer, M., Albert, S., Wan, W., Pfeffer, S., Beck, F., Plitzko, J.M., Nickelsen, J., Engel, B.D., 2019. Biogenic regions of cyanobacterial thylakoids form contact sites with the plasma membrane. Nat. Plants 5, 436–446. https://doi.org/10.1038/s41477-019-0399-7.

Santos-Merino, M., Torrado, A., Davis, G.A., Röttig, A., Bibby, T.S., Kramer, D.M., Ducat, D.C., 2021. Improved photosynthetic capacity and photosystem I oxidation via heterologous metabolism engineering in cyanobacteria. Proc. Natl. Acad. Sci. 118. https://doi.org/10.1073/pnas.2021523118, e2021523118.

Thiel, K., Patrikainen, P., Nagy, C., Fitzpatrick, D., Pope, N., Aro, E.-M., Kallio, P., 2019. Redirecting photosynthetic electron flux in the cyanobacterium *Synechocystis* sp. PCC 6803 by the deletion of flavodiiron protein Flv3. Microb. Cell Fact. 18, 189. https://doi.org/10.1186/s12934-019-1238-2.

Vermaas, W.F.J., Timlin, J.A., Jones, H.D.T., Sinclair, M.B., Nieman, L.T., Hamad, S.W., Melgaard, D.K., Haaland, D.-M., 2008. *In vivo* hyperspectral confocal fluorescence imaging to determine pigment localization and distribution in cyanobacterial cells. Proc. Natl. Acad. Sci. 105, 4050–4055. https://doi.org/10.1073/pnas.0708090105.

Wilson, A., Ajlani, G., Verbavatz, J.-M., Vass, I., Kerfeld, C.A., Kirilovsky, D., 2006. A soluble carotenoid protein involved in phycobilisome-related energy dissipation in cyanobacteria. Plant Cell 18, 992–1007. https://doi.org/10.1105/tpc.105.040121.

Yamamoto, H., Takahashi, S., Badger, M.R., Shikanai, T., 2016. Artificial remodelling of alternative electron flow by flavodiiron proteins in *Arabidopsis*. Nat. Plants 2, 16012. https://doi.org/10.1038/nplants.2016.12.

CHAPTER

9

Improving carbon fixation

Christine A. Raines[a], Amanda P. Cavanagh[a], and Andrew J. Simkin[b]

[a]School of Life Sciences, University of Essex, Colchester, Essex, United Kingdom [b]School of Biosciences, University of Kent, Canterbury, Kent, United Kingdom

1 Introduction

The vast percentage of plant species (85%) on the planet fix CO_2 from the atmosphere using the Calvin–Benson (CB or C3) cycle, which then converts this inorganic form of C into organic C compounds, which are used as the starting point for the synthesis of sugars, starch, nucleotides, thiamine, isoprenoids, and phenylpropanoids (Fig. 4, Lawson et al., Chapter 3). The CB cycle assimilates approximately 100 billion tons of carbon a year, representing 15% of the carbon in the atmosphere, and is the single largest flux of organic carbon in the biosphere. The theoretical maximum efficiency of total available light into biomass after losses due is 4.6% for C3 species (Zhu et al., 2010). After decreases due to limits imposed by light absorption, the major uses of the energy are for carbohydrate production, photorespiration, and respiration. In the field, the maximum conversion efficiency of 4.6% is not attained and often efficiencies of less than 50% of this are observed. What is it that poses this limitation on CO_2 assimilation reaching its maximum potential? Although the CB cycle is complex having 11 different enzymes, it has been known for some time that a large proportion of the limitation to CO_2 fixation in C3 plants is due to the catalytic properties of the enzyme rubisco (Portis and Parry, 2007).

Ribulose-1,5-bisphosphate carboxylase oxygenase (rubisco) catalyses the first step of the CB cycle, which fixes a CO_2 molecule to the five-carbon sugar phosphate molecule, ribulose-1,5-bisphosphate (RuBP), generating two molecules of 3-phosphoglycerate (3-PGA), which then enters the reductive phase of the CB cycle. Photosynthetic rate correlates with rubisco content (von Caemmerer, 2020), and the overexpression of rubisco has shown improvements in growth and photosynthesis and biomass in both the C3 crop rice (Yoon et al., 2020) and the C4 crop maize (Salesse-Smith et al., 2018). However, rubisco is a bifunctional enzyme, and it also catalyses a competing reaction in which oxygen from the atmosphere is fixed instead of CO_2 resulting in the oxygenation of RuBP and the production of one molecule of 2-phosphoglycolate (2-PG) in addition to one molecule of 3-PGA. The 2PG produced by this oxygenation reaction is not used in the CB cycle and is instead recycled through another process called photorespiration. The photorespiratory pathway requires energy, and although it is able to recover a portion of the previously fixed

Photosynthesis in Action
https://doi.org/10.1016/B978-0-12-823781-6.00009-5

carbon and return it to the CB cycle, there is a high energy cost, which lowers the efficiency of CO_2 assimilation by as much as 48% (Bowes et al., 1971; Ogren, 2003). When photorespiration is modelled at the canopy and regional scale, the impact of rubisco oxygenation on agricultural productivity becomes apparent, as the process decreases US wheat and soybean yield by 20% and 36%, respectively (Walker et al., 2016). Consequently, lowering the cost of photorespiration by modifying the amount or activity of rubisco has been an important biotechnology target.

2 Improving carbon fixation by manipulation carboxylation (rubisco)

Engineering a faster version of rubisco that is more specific for CO_2 than O_2 has been a longstanding goal to improve carbon fixation, but in all plant species surveyed, rubisco exhibits a trade-off between catalytic turnover rate (speed) and substrate specificity. This apparent trade-off, where improvements in one parameter come at the expense of others, is typical of almost all plant rubiscos characterised to date, hindering attempts to engineer an improved enzyme (von Caemmerer, 2020). Modern plant rubisco appears to have evolved to represent a compromise between catalytic rate and substrate specificity that is optimised for previous atmospheric CO_2 concentrations of ~220 ppm, rather than our current >400 ppm (Zhu et al., 2004). Despite this, surveys of both plant species and other photosynthetic organisms indicate that there is exploitable variation in rubisco performance that could be an important target for crop improvement (Orr et al., 2016). However, directly manipulating rubisco in plants is complicated by both the genetic and structural features of this enzyme, which requires the coordinated assembly of eight chloroplast encoded large subunits (LSUs) and eight nuclear encoded small subunits (SSUs). The catalytic site of rubisco is located on the LSU, which is encoded in the chloroplast making attempts to manipulate enzyme activity through mutagenesis more technically challenging. Modifications to the chloroplastic genome require the use of chloroplast transformation and this is only currently an option for a small number of species. A more accessible approach for directed engineering of rubisco activity may be through the nuclear-encoded SSU genes, as specific amino acid residues on SSU isoforms are known to impact substrate specificity and enzyme turnover rate (Khumsupan et al., 2020; Lin et al., 2020; Martin-Avila et al., 2020; Yoon et al., 2020).

An alternate approach to manipulating native rubisco is to facilitate direct replacement of the endogenous plant rubisco with an enzyme from another organism. For example, it has been shown that red algae and cyanobacteria have rubiscos that are more specific for CO_2 than higher plants and introducing these nonnative forms of rubisco into crop plants could enhance leaf carboxylation (McGrath and Long, 2014). Experiments have been done substituting tobacco rubisco with faster but less CO_2 specific versions of rubisco from the proteobacterium *Rhodospirillum rubrum* (Whitney and Andrews, 2001) the cyanobacterium *Synechococcus elongatus* (Lin et al., 2014; Occhialini et al., 2016) as well as plant rubiscos from closely related species (Martin-Avila et al., 2020). Promising lines from these manipulations show successful rubisco assembly in plant chloroplasts and higher rates of carbon fixation per unit of enzyme, but display impaired growth at ambient [CO_2] due to reduced rubisco substrate specificity or content relative to untransformed controls. Overcoming this barrier will likely require careful co-expression of accessory and chaperone proteins, or combinations with other bioengineering strategies discussed below, such as installing chloroplastic carbon concentrating mechanisms (Section 3). Improvement of rubisco as outlined above remains high on the list of targets to improve the CB cycle, but there are still challenges to this approach being widely applicable. As mentioned earlier, the requirement for chloroplast transformation, as well as unique folding and chaperone requirements, imposes species limits on this technique.

Rubisco carboxylation can also be improved by means other than direct engineering of its catalytic parameters. The activation of rubisco *in vivo* is mediated by the regulatory protein rubisco activase (Rca) which removes the sugar phosphate inhibitors from both inactive uncarbamylated and inhibited carbamylated rubisco (Chapter 4). The role of Rca is of particular importance during transitions from dark to full light, and heat stress. Transgenic overexpression of Rca alone has been shown to increase the rate of photosynthetic light induction, but this often occurs at the expense of steady state carbon fixation due to concomitant declines in rubisco content (Yamori et al., 2011). Overexpressing both rubisco and Rca together increases both rubisco content and overall activation relative to control plants, leading to enhanced photosynthesis at steady state (Suganami et al., 2021). Another manipulation strategy seeks to exploit differences in redox regulation between different Rca isoforms, overexpressing only a nonredox regulated form of the enzyme to promote rapid induction of rubisco carboxylation when plants are transferred from low to high light (Carmo-Silva et al., 2015). Proof of concept studies in Arabidopsis have achieved this by knocking down only the redox-regulated isoform of Rca, leading to rapid photosynthetic induction responses, and increased carbon gain in fluctuating light environments (Carmo-Silva and Salvucci, 2013). The thermal sensitivity of Rca leads to deactivation of rubisco at moderately high temperatures, leading to declines in carbon assimilation. Expressing a more thermostable version of Rca improved photosynthesis and growth rate under a moderate heat stress in *A. thaliana* (Kumar et al., 2009; Kurek et al., 2007), and expressing a more thermostable version of Rca from a wild relative improves seed yield in rice (Scafaro et al., 2018). In vitro mutation of wheat Rca has shown that a conserved amino acid residue identified in warm-adapted species can impart thermostability to the enzyme (Degen et al., 2020), presenting a compelling target for manipulation.

3 Improving carbon fixation by concentrating CO_2 in leaf chloroplasts

Alongside strategies to enhance rubisco performance directly, other approaches aim to lower the cost of photorespiration by decreasing the rubisco oxygenation reaction. Two main approaches are actively being pursued: (i) introduction of biophysical CCMs from cyanobacteria and green algae (Fig. 1) and (ii) engineering of C4 anatomy and metabolism (Fig. 2).

Algae and cyanobacteria have evolved highly efficient biophysical CCMs that can enhance CO_2 levels around rubisco by up to 1000-fold. Biophysical carbon concentration mechanisms (CCMs) share key functional features: (1) transporters for the uptake of CO_2 in the form of bicarbonate; (2) carbonic anhydrase with catalyses the release of CO_2 form bicarbonate close to the site of rubisco; (3) a subcellular compartment in cyanobacteria—carboxysomes and in algae—pyrenoids; and (4) cellular features preventing the loss of CO_2 from the site of carboxylation by rubisco (Rae et al., 2017). Evidence from modelling indicates that in order to get the maximum benefit from introduction of a cyanobacterial CCM, it will be necessary to introduce all four of these characteristics. However, it may also be possible to obtain some improvement of the CB cycle by installing only the bicarbonate transporters. It has been shown that it is possible to introduce inorganic carbon transporters into the chloroplastic inner membrane using either the algal and cyanobacterial genes. However, fully function transporters have not been demonstrated in the plants, and more work is needed to resolve the activation and energy requirements of these transporters (Fig. 1).

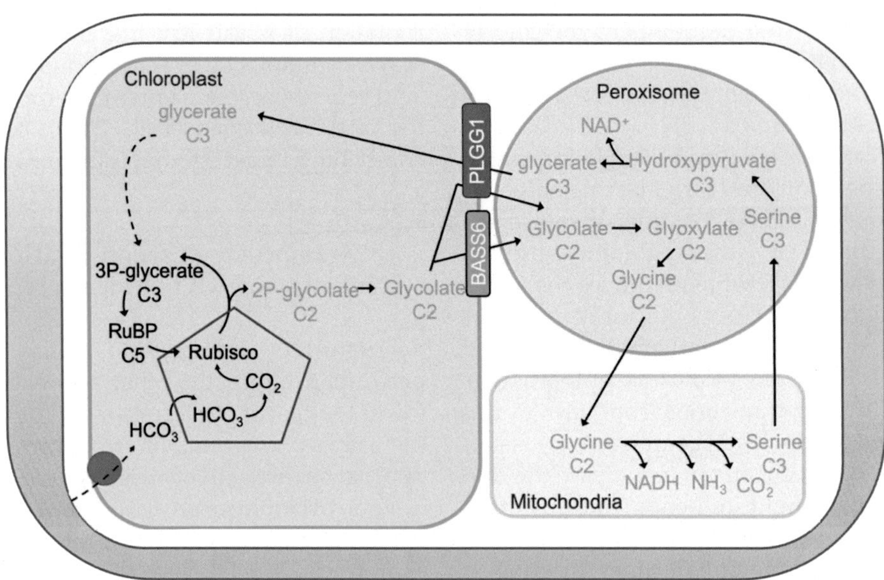

FIG. 1 Biophysical carbon concentrating mechanisms (CCMs). Algae and cyanobacteria have evolved highly efficient biophysical CCMs that can enhance CO_2 levels around rubisco by up to 1000-fold. The installation of the CCMs (pyrenoid or carboxysome structures represented here as a *blue* pentagram) and the expression of bicarbonate transporters can prevent rubisco oxygenation and enrich CO_2 at the rubisco active site. *From South, P.F., Cavanagh, A.P., Lopez-Calcagno, P.E., Raines, C.A., Ort, D.R., 2018. Optimizing photorespiration for improved crop productivity. J. Integr. Plant Biol. 60 (12), 1217–1230.*

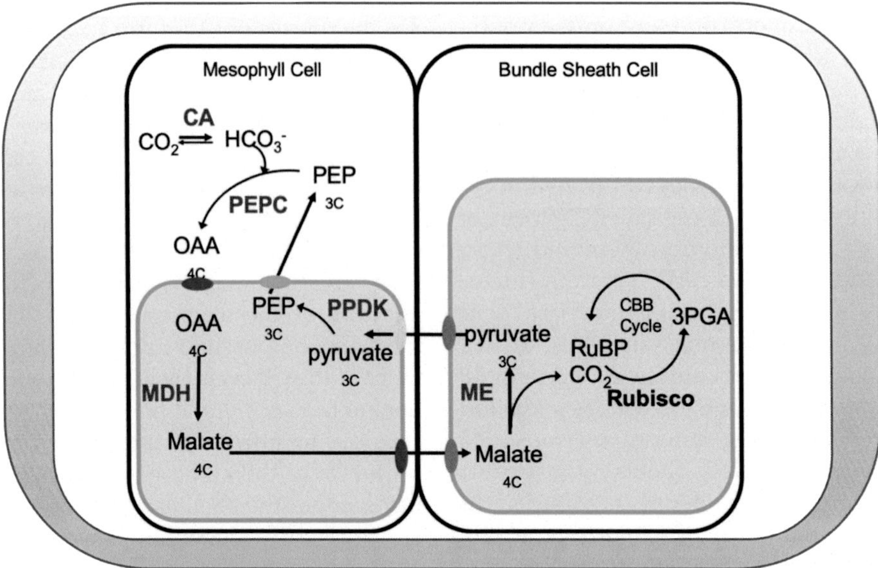

FIG. 2 Introducing C4 metabolism in C3 plants. The C_4 pathway is characterised by biochemical and anatomical specialisation that ensures high CO_2 partial pressure at the site of rubisco. This requires the expression of enzymes represented in *red*, including: carbonic anhydrase (CA), PEP carboxylase (PEPC), malate dehydrogenase (MDH), NADP-malic enzyme (ME), and pyruvate Pi dikinase (PPDK).

To date, fully assembled carboxysomes and pyrenoid assembly in plant chloroplasts has remained elusive, but significant progress has been made in establishing both a functional carboxysome (Long et al., 2018) and a proto-pyrenoid in plant chloroplasts (Atkinson et al., 2020). Further engineering efforts will need to identify the minimum number of genes responsible for assembly, transport proteins, and proper targeting of rubisco into the CCM microcompartments.

An alternative method to increase CO_2 concentration at rubisco is to introduce C4 photosynthesis into C3 crops (Fig. 2). In C4 plants, the photosynthesis processes are spatially separated between two adjacent leaf cell types (mesophyll and bundle sheath cells) (see Chapter 4). In mesophyll cells, CO_2 from the atmosphere is fixed by the enzyme phospho*enol*-pyruvate (PEP) carboxylase and converted into a four-carbon organic, oxalo-acetic acid (OAA). PEP carboxylase does not catalyse an oxygenase reaction and is therefore more efficient than rubisco at capturing atmospheric CO_2. The four-carbon dicarboxylic acid, OAA, is actively transported into the bundle sheath where it is decarboxylated forming and releasing CO_2, increasing the CO_2 concentration near rubisco. Engineering C4 photosynthesis in C3 plants has been outlined as a stepwise process (Schuler et al., 2016) that includes alteration of plant tissue anatomy, establishment of bundle sheath morphology, as well as ensuring a cell type-specific enzyme expression. Engineering efforts to introduce C4 photosynthesis into rice through the C4 Rice project (https://c4rice.com), have recently established a two-cell prototype for an NADP-malic enzyme type C4 rice through transformation of a single construct containing the coding regions of carbonic anhydrase, phospho*enol*pyruvate (PEP) carboxylase, NADP-malate dehydrogenase, pyruvate orthophosphate dikinase, and NADP-malic enzyme from *Zea mays*, driven by cell-preferential promoters. Gene expression, protein accumulation, and enzyme activity were confirmed for all five transgenes, although protein expression levels require optimization to support significant metabolic flux through the pathway (Ermakova et al., 2020). Current models predict that introducing the C4 biochemical pathway into the cells around existing leaf veins in C3 plants will provide modest benefit to carbon gain, although it is likely that anatomical changes including increasing vein density will ultimately be necessary to realise high efficiency C4 rice.

4 Re-engineer the photorespiratory pathway using nonnative genes and alternative metabolic pathways

The reactions of the photorespiratory pathway that result in the recycling of 2PG, produced in the oxygenase reaction of rubisco, into 3PGA which can re-enter the CB cycle take place in three separate organelles (chloroplast, peroxisome, and mitochondria) and the cytosol (Fig. 3). This pathway is able to recover 75% of the carbon lost by the rubisco oxygenation reaction, but releases the remaining 25% as CO_2 in the mitochondria and each rubisco oxygenation consumes 4.75 ATP and three reducing equivalents (Fig. 3; Bauwe and Kolukisaoglu, 2003; Peterhansel et al., 2010; Walker et al., 2016). Early attempts to reduce the losses due to the photorespiratory process focused on downregulating the pathway, but most mutations in photorespiration enzyme encoding genes are lethal, or display poor growth under ambient air conditions (Levey et al., 2018; Messant et al., 2018; Somerville and Ogren, 1982). Attempts to increase photorespiratory activity are discussed in Section 8. In recent years, attempts to reengineer photorespiration have emerged to introduce novel metabolic pathways that use less energy and release photorespired CO_2 in the chloroplast, where it can be re-fixed more efficiently by rubisco (Fig. 4).

To date, several alternative photorespiratory bypasses have been designed and tested in

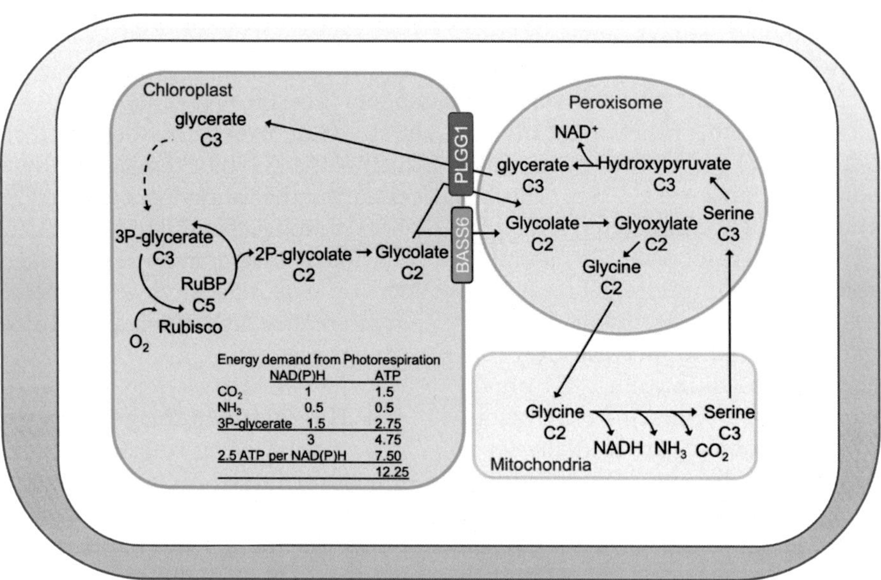

FIG. 3 Schematic representation of photorespiration. For every two oxygenation reactions catalysed by rubisco in the chloroplast, one molecule of glycerate is generated in the peroxisome and transported to the chloroplast for reintroduction into the C3 cycle and one carbon is released as CO_2 in the mitochondria. Number of carbons per molecule are indicated. Energy demand of photorespiration depicted in reducing equivalents (NAD(P)H) and ATP. *From South, P.F., Cavanagh, A.P., Lopez-Calcagno, P.E., Raines, C.A., Ort, D.R., 2018. Optimizing photorespiration for improved crop productivity. J. Integr. Plant Biol. 60 (12), 1217–1230.*

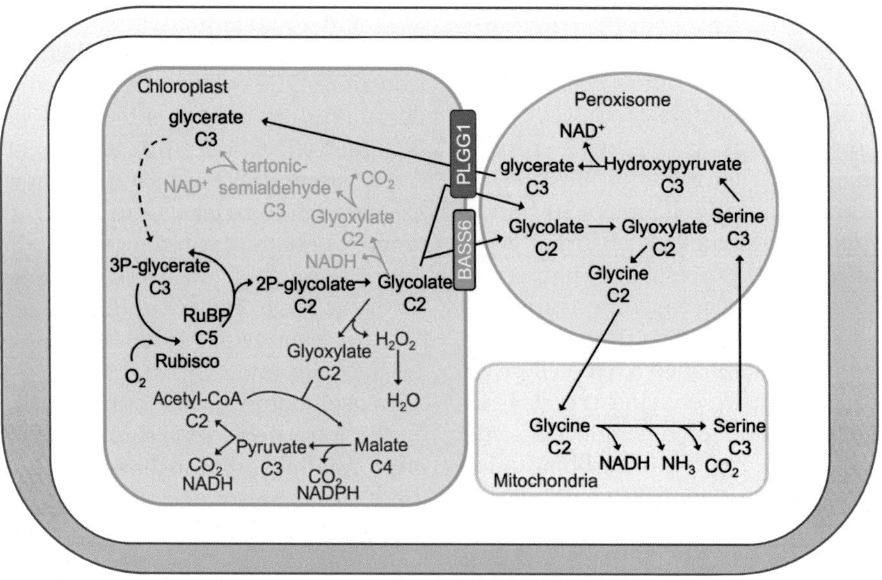

FIG. 4 Strategies for bypassing photorespiration using nonnative genes. These nonnative pathways (shown in *blue* and *purple*) are used to more efficiently process glycolate either back to glycerate similar to native photorespiration or by fully decarboxylating glycolate to CO_2 to be re-fixed by rubisco (Dalal et al., 2015; Kebeish et al., 2007; Nolke et al., 2014). *From South, P.F., Cavanagh, A.P., Lopez-Calcagno, P.E., Raines, C.A., Ort, D.R., 2018. Optimizing photorespiration for improved crop productivity. J. Integr. Plant Biol. 60 (12), 1217–1230.*

planta using transgenic technology. The first pathway was inspired by the cyanobacterial glyoxylate oxidation pathway, where glyoxylate is reduced to glycerate directly in the chloroplast. In plants, this was achieved by the simultaneous introduction of chloroplast-targeted glycolate dehydrogenase, glyoxylate carboxyligase, and tartronic semialdehyde reductase from *Escherichia coli* (Fig. 4). The main benefit of this pathway is that it uses less ATP than the native photorespiratory pathway and shifts the release of CO_2 to the chloroplast, where it can be re-fixed by rubisco. Of the published alternative photorespiratory pathways, this has been tested most extensively, and increases in photosynthesis and biomass have been observed in several species, including Arabidopsis, potato, and camelina (Dalal et al., 2015; Kebeish et al., 2007; Nolke et al., 2014). However, when tested in field grown tobacco, plants expressing this pathway showed no difference in photosynthesis or harvested biomass (South et al., 2019).

Another nonnative photorespiratory pathway tested in plants uses the glycolate oxidase pathway intended to fully decarboxylate glycolate within the chloroplast. This glycolate oxidase pathway requires chloroplast-targeted expression of the glycolate oxidase normally expressed in the peroxisome, malate synthase to convert glyoxylate to malate, and a catalase enzyme because the conversion of glycolate to glyoxylate, by glycolate oxidase, generates hydrogen peroxide as a byproduct (Fig. 3). In addition to confining all steps of glycolate metabolism to the chloroplast, this alternative pathway also increases the CO_2 concentration around rubisco, thereby decreasing oxygenation reactions. Expression of this pathway led to increased growth in Arabidopsis, and when the pathway was modified to remove the need for catalase showed increased photosynthesis and biomass in field-grown tobacco (Maier et al., 2012; South et al., 2019). However, this alternative pathway is expected to expend more energy compared to the native photorespiratory pathway (Xin et al., 2015) and fails to return any P-glycerate to the photosynthetic carbon reduction cycle, suggesting some alternative metabolism not yet understood is at play (Maier et al., 2012; Peterhansel et al., 2013).

5 Identifying enzymes, other than rubisco that limit photosynthetic carbon fixation

Although it is clear that rubisco remains an important target for manipulation, with the potential to increase CO_2 assimilation, it is only one of 11 enzymes in the CB cycle, and this raises the question—Is there an opportunity to improve photosynthesis through manipulation of one of the other enzymes in the cycle? Traditionally, analysis of metabolic pathways, such as the CB cycle, focused on the study of the kinetic properties of individual enzymes. Early attempts to identify 'key' limiting steps in the CB cycle were based on modelling approaches that focused on the catalytic and regulatory properties of single enzymes based on in vitro data. Using this approach, a number of 'key' enzymes were identified, rubisco, SBPase, FBPase, and PRKase, as those most likely to have the greatest influence on CB cycle CO_2 assimilation (Laisk et al., 1989; Pettersson and Ryde, 1988; Poolman et al., 2000). This classification was based primarily on the fact that the activity of these enzymes was regulated by a number of factors, including light, stromal pH [Mg^{2+}], and also that the reactions they catalysed were irreversible (Portis et al., 1977; Woodrow and Berry, 1988). The conclusion from this biochemical analysis was that these 'key' enzymes were likely to have the greatest importance in controlling the rate of CO_2 fixation in the cycle. However, these modelling studies based on in vitro data provided no information on the relative importance of the enzymes compared to each other or any quantitative information about the extent to which any single CB cycle enzyme controlled the rate of CO_2 fixation in vivo.

The maximum rate of CO_2 fixation in the CB cycle and the subsequent flow of fixed C through the cycle is determined by the slowest enzyme reaction, similar to movement of traffic on a single track road, which is limited by the slowest car. To explore the question on steps limiting the fixation and flow of carbon in the CB cycle an alternative methodology was used taking a whole system approach, termed metabolic control analysis. This can be used to determine the relative importance of an individual enzyme in controlling the flux through a pathway (Fell, 1997). To undertake metabolic control analysis of a pathway, it is necessary to be able to reduce specifically the amount of individual enzymes in that pathway, while all the others remain as normal. The effect of this reduction in enzyme level on the fixation of CO_2 (flux into the pathway) can then be compared to the control with the normal level of enzyme activity (for fuller description of this approach, see Fell, 1997). This analysis provides a quantitative measure of the control exerted by a single enzyme over the flux through a pathway and can be defined mathematically:

$$C = \delta J / J / E / \delta E$$

where C is the flux control co-efficient; J the original flux through the pathway; δJ change in flux; E original enzyme activity; and δE change in enzyme activity.

The flux control coefficient value can vary from 0, for an enzyme that makes no contribution to control, to 1, for an enzyme that exerts total control. The flux control value for any single enzyme is not a constant, and the numerical value can change depending on the conditions under which the analysis is carried out. However, the sum of the flux control co-efficients for all of the enzymes in a single pathway should equal 1, the implication of which is, that if the flux control value for one enzyme in the system increases, then the flux control value(s) for one or more of the other enzymes in the system must decrease. One fundamental difference between this approach and that of earlier studies based on the kinetics of individual enzymes is that metabolic control analysis allows for all enzymes in a pathway to share control of flux in that pathway. The control each enzyme exerts is not equal and the contribution that an individual enzyme makes to the control of carbon flux can vary dependent on developmental stage of the plants and the growth conditions.

The control of CO_2 flux through the CB cycle has been investigated using transgenic plants that have been manipulated to reduce the levels of single enzymes in the cycle. The first application of metabolic control analysis to identify enzymes limiting the fixation of CO_2 into the CB cycle was initiated by Stitt and co-workers using transgenic plants with reductions in the levels of the rubisco protein (Rodermel et al., 1988). Transgenic plants with reductions in rubisco protein levels were produced using a gene construct targeted specifically to interfere with the RNA expressing the rubisco SSU protein. A number of independent transgenic plants were produced that displayed differing levels of rubisco protein, and these plants were used to assess the relative contribution that rubisco imposed on carbon flux. This approach has demonstrated clearly that the limitation imposed by rubisco on C3 carbon fixation is greatest in high light and temperature conditions (Stitt and Schulze, 1994). In contrast, as expected, this is reduced in plants grown in elevated CO_2 (Masle et al., 1993). Importantly, these transgenic studies also showed that rubisco does not limit the CB cycle under all conditions and that enzymes of the regeneration phase of the CB cycle also play a role in determining the rate of photosynthesis (Raines, 2003, 2006; Stitt et al., 2010). This approach was extended to include 7 of the 11 enzymes in the Calvin cycle, providing new information on the relative control they exert on CO_2 fixation and identified those that exert the greatest control.

The reduction and regeneration steps of the CB cycle involve 10 enzymes, and following on from this early study, antisense plants with reduced levels of seven individual enzymes were created and analysed (Raines, 2003; Stitt and Schulze, 1994). Under many conditions, reductions in the activities of enzymes catalysing highly regulated, effectively irreversible reactions, glyceraldehyde-3-phosphate dehydrogenase (GAPDH), fructose 1,6-bisphosphatase (FBPase), and ribulose 5-phosphate kinase (phosphoribulokinase, PRK) had little impact on carbon assimilation (for review, see Raines, 2003; Stitt et al., 2010). In contrast, small reductions in the enzyme sedoheptulose 1,7-bisphosphatase (SBPase) resulted in a decrease in CO_2 fixation and growth, identifying this enzyme as a major control point in the CB cycle. The relative importance of SBPase in determining flux through this cycle varied with development and environmental conditions, confirming that no one enzyme limits the C3 cycle under all conditions (Harrison et al., 1998, 2001; Olcer et al., 2001) This fits with the result from the model of Poolman et al. (2000) provided evidence that the control of flux in the CB cycle is shared mainly between SBPase and rubisco, dependent on the environmental conditions in which the plants are grown.

A somewhat unexpected result was obtained from the analysis of plants with reductions in the activity of either transketolase (TK) or aldolase, enzymes that are not highly regulated and operate close to equilibrium (Haake et al., 1998, 1999; Henkes et al., 2001). It was shown that relatively small reductions in the levels of either of these enzymes resulted in a reduction in the capacity of the plants to fix CO_2 into the CB cycle. This was unexpected, as it was previously thought that enzymes that were subject to regulation would be most important in determining the rate of carbon flow through the cycle. In addition, the analyses of the transgenic plants with reduced levels of CB cycle plants identified enzymes other than rubisco that can limit CO_2 fixation. The interesting aspect of these results is that it raised the question that if the reduction in the level of ability of the plant to take up CO_2 could increase the levels of these enzymes improve the CB cycle?

6 Evidence that transgenic manipulation of RuBP regeneration can increase CO_2 fixation

The results presented below demonstrate that increasing SBPase and FBPA enzyme activities can increase in CO_2 assimilation and also lead to increases in growth and vegetative biomass under controlled conditions.

The overexpression of SBPase in tobacco in 2005 showed that increasing SBPase enzyme activity resulted in an increase the rates of CO_2 assimilation in young expanding leaves (Fig. 5). As a direct consequence of this manipulation, both sucrose and starch accumulated resulting in a 30% increase in biomass yield (Lefebvre et al., 2005). These early results, again 10 years later, showed that these increases were maintained across several generations and when grown under either high or low light conditions (Simkin et al., 2015). Importantly, when these plants were grown under field conditions, high SBPase activity was shown to increase biomass yield under open-air elevation of CO_2 (Rosenthal et al., 2011).

More recently, it has been shown in a number of species that increasing the level of SBPase enzyme activity increases carbon assimilation, biomass, and seed/grain yield. In tomato (Ding et al., 2016), *Arabidopsis thaliana* (Arabidopsis) (Simkin et al., 2017) and wheat (Driever et al., 2017), for example, in Arabidopsis a 42% increase in biomass, and a 53% increase in seed yield (Simkin et al., 2017). More importantly, increasing SBPase in wheat, one of the major world crops, resulted in a 30%–40% increase in grain yield in addition to an increase in overall biomass (Driever et al., 2017). As a consequence of the results obtained in wheat, and the importance of this crop to the human

FIG. 5 Control and transgenic tomato and wheat plants overexpressing sedoheptulose-1,7-bisphosphatase (SB) (Simkin, 2019). *From Simkin, A.J., Lopez-Calcagno, P.E., Raines, C.A., 2019. Feeding the world: improving photosynthetic efficiency for sustainable crop production. J. Exp. Bot. 70(4), 1119–1140.*

food chain, Driever et al. (2017) cultivated these plants under two different growth regimes to determine if the growth phenotype is robustly conserved. In the first experiment, plants were grown at high density where the production of side shoots (tillering) is limited. Under these higher growing densities, the wheat plants with increased SBPase activity developed fewer tillers with increased numbers of seed per ear. In contrast, under lower density conditions the wheat, plants were shown to produce more tillers, but no significant increase in seed number per ear was observed (Fig. 5). In both experiments, total seed weight, seed number, and whole plant biomass were increased demonstrating that increasing SBPase enzyme activity can have a positive impact independent of growing density of the plants (Driever et al., 2017).

Furthermore, overexpression of fructose 1,6-bisphosphate aldolase (Ald) in tobacco in 2012 resulted in an up to twofold increase in aldolase activity, resulting in an increase in photosynthetic CO_2 fixation and a 10%–30% increase in biomass under ambient [~400 ppm] CO_2 (Uematsu et al., 2012), and when these plants were grown in elevated [700 ppm] CO_2, a 70%–120% increase in biomass was observed (Fig. 6). More recent experiments with Arabidopsis with a 46%–80% increase in Ald activity resulted in a 31% increase in CO_2 assimilation. This increased ability to assimilate carbon resulted in a 32% increase in dry weight and more importantly a 35% increase in seed yield (Simkin et al., 2017).

In addition to the use of plant enzymes, some authors have expressed cyanobacterial derived photosynthetic enzymes in planta. For example, expression of the cyFBPase in plants has also been shown to significant effect plant development. In tobacco, plants expressing cyFBPase were also shown to accumulate +1.3-fold more biomass than control plants. This increase in biomass was attributed to a 15% increase in CO_2 fixation rates (Tamoi et al., 2006). Furthermore,

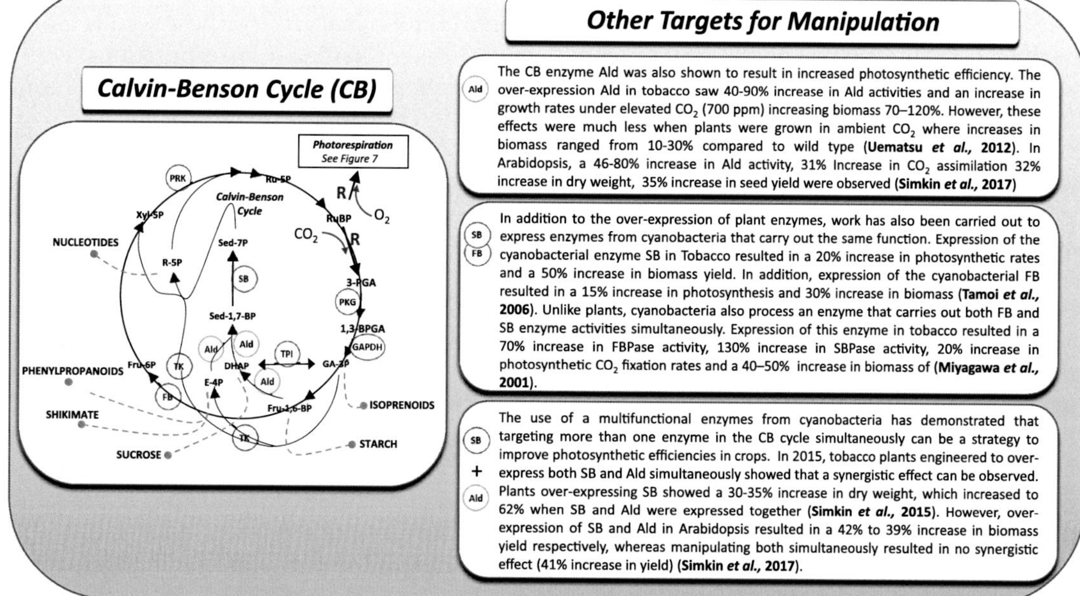

FIG. 6 Schematic representation of photosynthetic Calvin–Benson (CB) cycle. Sedoheptulose-1,7-bisphosphatase (SB), fructose-1,6-bisphosphate aldolase (Ald), fructose-1,6-bisphosphatases (FB), triosephosphate isomerase (TPI), transketolase (TK), phosphoribulokinase (PRK), phosphoglycerate kinase (PGK), and ribulose-bisphosphate carboxylase large chain (R) (Simkin, 2019). *From Simkin, A.J., Lopez-Calcagno, P.E., Raines, C.A., 2019. Feeding the world: improving photosynthetic efficiency for sustainable crop production. J. Exp. Bot. 70(4), 1119–1140.*

expression of the cyanobacterial FBP/SBPase enzyme in tobacco, a bifunctional enzyme with both SBPase and FBPase activity was shown to increase in the rate of photosynthetic carbon fixation by 20% and result in a 1.5-fold increase in final biomass (Tamoi et al., 2006) (Fig. 6). These results demonstrate that nonplant enzymes with the same biological activity can be used to improve photosynthetic efficiency in planta.

7 Evidence that transgenic multiple target manipulation of photosynthesis could result in a cumulative increase in yield

Initial attempts to improve CO_2 fixation in the CB cycle focussed on manipulation of single enzymes, and the results discussed above provide evidence that manipulating single enzymes in the CB cycle can improve efficiency of the CB cycle and result in increased yields. However, there may be additional gains in the CB cycle that could be achieved by manipulating multiple steps simultaneously (Fig. 6). Evidence suggesting that this might be a useful approach came from the output from a dynamic model of carbon metabolism (Zhu et al., 2007). The work of Zhu et al. (2007) used an evolutionary algorithm together with a model using existing kinetic data. Based on this, it was proposed that increasing SBPase, FBPA, and ADPglucose pyrophosphorylase (AGPase) in the same plant, together with a modest reduction in photorespiration, could lead to an increase in the efficiency of photosynthetic carbon assimilation. The importance of this model is that it highlighted the fact that more than one target is likely to be needed and that modelling has the potential to allow the most promising combination of targets to be identified.

For example, the overexpression of both SBPase and FBP aldolase (Ald) in transgenic tobacco resulted in a cumulative increase in biomass yield (Fig. 6). Plants expressing SBPase showed a +34% increase in biomass compared to a +62% compared in plants expressing SBPase and Ald together (Simkin et al., 2015). This result provided the first experiment evidence that manipulation of two or more enzymes the CB cycle has the potential to lead to additional increases in biomass. However, when SBPase and Ald together were overexpressed in transgenic Arabidopsis, no significant differences were observed between plants expressing a single gene and those expressing both indicating that no cumulative effect is observed in these plants (Simkin et al., 2017). These results indicate that different targeted manipulations will be needed, depending on the plant species under evaluation (Fig. 6).

8 Increasing photorespiratory activity increases biomass yield

An alternative strategy explored in the quest to increase the efficiency of the CB cycle has been the overexpression of components of the glycine decarboxylase system (GCS). The rationale for this was the aim to increase photorespiration flow, release CO_2 more rapidly for uptake by rubisco and to decrease accumulation of potentially toxic photorespiratory intermediates (Fig. 7). In plants, the GCS releases CO_2 and, together with

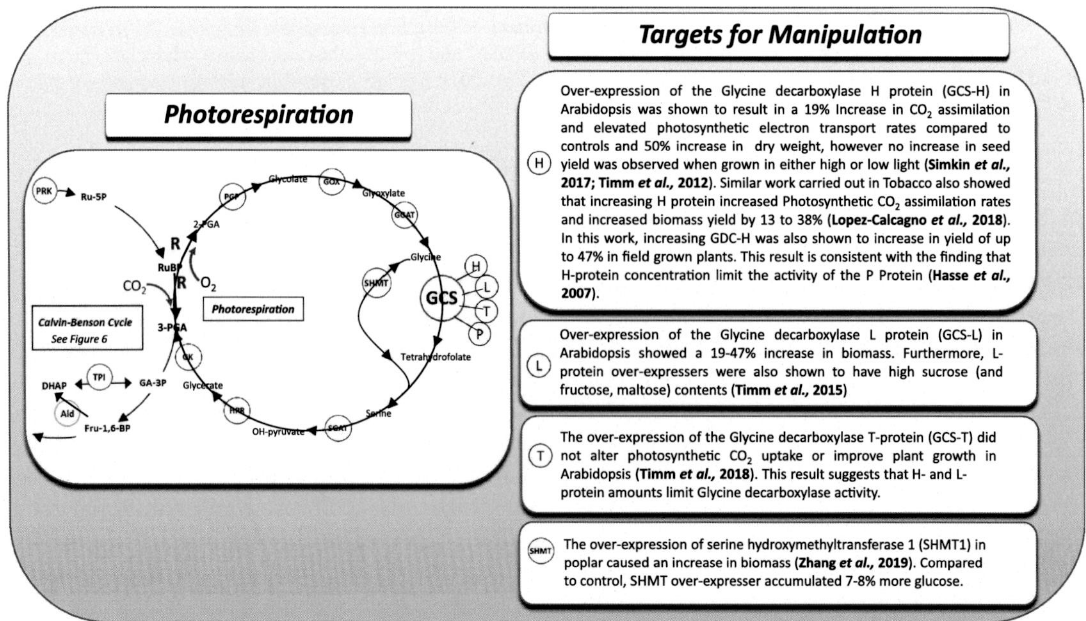

FIG. 7 Schematic representation of photorespiration. Ribulose-bisphosphate carboxylase large chain (R0, 2-phosphoglycerate phosphatase (PGP), glycolate oxidase (GOX), glycine glycolate transaminase (GGAT), serine hydroxymethyltransferase (SHMT), serine glycolate transaminase (SGAT), hydroxypyruvate reductase (HPR), glycine kinase (GK), and glycine decarboxylase system (GDS), which is composed of four proteins (P, T, L, and H) (Simkin, 2019). *From Simkin, A.J., 2019. Genetic engineering for global food security: photosynthesis and biofortification. Plants 8, 586.*

serine hydroxymethyltransferase (SHMT), is responsible for the interconversion of glycine and serine (Bauwe and Kolukisaoglu, 2003). The GCS comprises four proteins, three enzymes (P-protein, T-protein, and L-protein) and a small lipoylated protein known as H-protein. The H protein has no catalytic activity, it acts as a substrate for the P-, T-, and L-proteins and has been demonstrated, in vitro, to be capable of enhancing the activity of the GCS (Hasse et al., 2009).

Work in Arabidopsis has shown that increasing the expression of the H-protein or L-protein results in an increase in photosynthesis and higher vegetative biomass compared to controls (Simkin et al., 2017; Timm et al., 2012, 2015, 2016). L-Protein overexpressers were equally shown to have elevated sugar content (sucrose, fructose, and maltose) suggesting that the overexpression of the L-protein alters carbon flow through the TCA cycle (Timm et al., 2015). In contrast, it has been shown that increasing the expression of the T-protein does not lead to an increase in biomass yield (Timm et al., 2017). Furthermore, overexpression of one or other of the GCS proteins does not lead to an increase in the expression or protein accumulation of the other GCS proteins indicating that they are independently regulated, consistent with the belief that they are not functioning as a static protein complex with a fixed stoichiometry. More recent work in tobacco has also shown that the overexpression of the H-protein increases biomass yield by as much as 47% in both the greenhouse and importantly under field conditions (López-Calcagno et al., 2018). From existing data, it is not clear what the mechanism is that enables the H protein to have this positive effect, but it has been suggested that photorespiratory intermediates such as 2PG inhibit the activities of CBC enzyme TPI, PRK and SBPase and glycolate has been shown to inhibit rubisco activation (see Simkin, 2019; Simkin et al., 2019, and references therein). It has been proposed that the increased flux through the photorespiratory pathway by the overexpression of GCS proteins depletes these intermediates, thereby preventing them from accumulating to levels capable of inhibiting CBC activity.

In addition to targeting the glycine decarboxylase system, other authors have looked to overexpress SHMT (Fig. 7). SHMT, together with GCS, is responsible for the interconversion of glycine and serine (Bauwe and Kolukisaoglu, 2003). In potato, plants with a reduction in SHMT were shown to have lower photosynthetic activity. Zhang et al. (2019) demonstrated that the overexpression of SMHT in poplar resulted in an increase in photosynthesis, an increase in biomass and 7%–8% increase in glucose.

9 Combining overexpression of the glycine decarboxylase H subunit and CB cycle enzymes

It has previously been shown that increasing the activity of SBPase and Ald in transgenic Arabidopsis resulted an increase in biomass yield (Fig. 6). Furthermore, overexpression of GDC-H in Arabidopsis led to a similar increase in biomass yield (Fig. 7). Simkin et al. (2017) simultaneously increased GDC-H and increased the activity of two CB cycle enzymes (SBPase and Ald). Plants expressing SBPase, Ald, and GDC H-protein were generated. The results demonstrated a synergistic effect with a positive impact on biomass yield where the increase in biomass was greater in plants overexpressing all three proteins compared to plants overexpressing a single protein or both CBC enzymes. Surprisingly, this work also showed that whilst increasing photorespiration alone resulted in an increase in biomass yield, it did not result in an increase in seed yield in Arabidopsis. In contrast, overexpression of CBC genes results in both an increase in biomass and an increase in seed yield. More interestingly, the simultaneous increase in both CBC and GDC-H activity resulted in a synergistic increase in seed yield compared with plants overexpressing CBC enzymes alone (Simkin et al., 2017). The reason behind these differences in seed yield remains elusive; however, changes in starch and sucrose levels observed in GCS overexpressing lines suggest that changes in carbon sink/source allocation may play a part; however, further study is required.

10 Unexpected outcomes

Recent work in tobacco showed that the overexpression of some CBC enzymes is not beneficial (Fig. 3). The overexpression of transketolase, for example, led to a negative effect on plant development and plants regularly showed a chlorotic phenotype and retarded growth (Khozaei et al., 2015).

11 Future prospects and conclusion

Previous sections have outlined current progress on research aimed at improving carbon fixation, but even the most promising of these targets are unlikely to feature in growers fields at scale for at least 25–30 years due to requirements of field testing and regulatory procedures. At the same time, global anthropogenic emissions are currently tracking the worst-case 'business as usual' scenario (RCP 8.5), which will likely result in a midcentury [CO_2] of 550 ppm and unprecedented warming from preindustrial (1850–1990) levels of 3–5°C by the end of the century (IPCC, 2014). The CB cycle of current crops appears optimised for a much lower [CO_2] than our current atmosphere, and even less optimal for future atmospheric [CO_2] (Zhu et al., 2004).

As atmospheric [CO_2] increases, a direct stimulation of carbon assimilation arises as the rubisco oxygenation reaction and photorespiration is suppressed (Long et al., 2004). However, it has been observed that growth under elevated [CO_2] conditions causes some C3 plant species to decrease rubisco content in leaves through an acclimatory response. In C3 plants, rubisco content decreased by ~20% under growth at elevated [CO_2], which partially offsets this gain (Ainsworth and Long, 2005). Strategies discussed in previous sections to increase rubisco content in crops could mitigate against this loss and maximise the response to increased atmospheric [CO_2] (Yoon et al., 2020). Most engineering strategies have thus far aimed to increase rubisco specificity without negatively impacting rubisco catalytic rate, but improving enzyme catalytic rate at the expense of specificity may increase daily carbon gain by as much as 10% for the same amount of rubisco in plants grown under future [CO_2] conditions (Zhu et al., 2004).

As both [CO_2] and temperature rise, theoretical models predict that the limitation of carbon assimilation shifts from rubisco to RuBP regeneration (Long et al., 2004). Therefore, modifications that improve RuBP regeneration are predicted to stimulate photosynthesis and yield under elevated atmospheric [CO_2]. This is supported by experimental evidence using plants grown in free-air CO_2 enrichment (FACE) facilities, which permit both open-air elevation of CO_2 as well as canopy warming using heating arrays in an agricultural setting. When grown at a [CO_2] of 585 ppm, transgenic tobacco plants overexpressing SBPase exhibit a greater stimulation in instantaneous and daily carbon gain than do unmodified plants, leading to greater yield increases at elevated [CO_2] (Rosenthal et al., 2011). When both [CO_2] and temperature are manipulated to better represent future conditions, transgenic overexpression of cyanobacterial bifunctional FBP/SBPase in soybean protects against temperature-induced yield loss under elevated [CO_2] (Köhler et al., 2016). These results confirm that improving carbon fixation in food crops can mitigate the effects of climate change on yield and also demonstrate the importance of testing future manipulations under midcentury climate conditions.

The potential for improving the CB cycle to achieve higher yielding plants has been demonstrated, mainly in model species of plant but with some evidence now emerging in crop plants. Although our understanding of the enzymes and the reactions in the CB cycle are well understood and are evolutionary conserved between C3 species the variation in the response of this cycle to environmental conditions and how this is regulated remains to be fully explored. Future work on the CB cycle will need a combination of curiosity driven research, asking basic questions on the regulation of individual enzymes and of how the carbon flow from the CB cycle is regulated, together with the application of knowledge to deliver biotechnological solution for crop improvement.

References

Ainsworth, E.A., Long, S.P., 2005. What have we learned from 15 years of free-air CO_2 enrichment (FACE)? A meta-analytic review of the responses of photosynthesis, canopy properties and plant production to rising CO_2. New Phytol. 165, 351–372.

Atkinson, N., Mao, Y., Chan, K.X., McCormick, A.J., 2020. Condensation of Rubisco into a proto-pyrenoid in higher plant chloroplasts. Nat. Commun. 11, 6303.

Bauwe, H., Kolukisaoglu, Ü., 2003. Genetic manipulation of glycine decarboxylation. J. Exp. Bot. 54, 1523–1535.

Bowes, G., Ogren, W.L., Hageman, R.H., 1971. Phosphoglycolate production catalyzed by ribulose diphosphate carboxylase. Biochem. Biophys. Res. Commun. 45, 716–722.

Carmo-Silva, A.E., Salvucci, M.E., 2013. The regulatory properties of rubisco activase differ among species and affect photosynthetic induction during light transitions. Plant Physiol. 161, 1645–1655.

Carmo-Silva, E., Scales, J.C., Madgwick, P.J., Parry, M.A.J., 2015. Optimizing Rubisco and its regulation for greater resource use efficiency. Plant Cell Environ. 38, 1817–1832.

Dalal, J., Lopez, H., Vasani, N.B., Hu, Z.H., Swift, J.E., Yalamanchili, R., Dvora, M., Lin, X.L., Xie, D.Y., Qu, R.D., Sederoff, H.W., 2015. A photorespiratory bypass increases plant growth and seed yield in biofuel crop *Camelina sativa*. Biotechnol. Biofuels 8, 175.

Degen, G.E., Worrall, D., Carmo-Silva, E., 2020. An isoleucine residue acts as a thermal and regulatory switch in wheat Rubisco activase. Plant J. 103, 742–751.

Ding, F., Wang, M., Zhang, S., Ai, X., 2016. Changes in SBPase activity influence photosynthetic capacity, growth, and tolerance to chilling stress in transgenic tomato plants. Sci. Rep. 6, 32741.

Driever, S.M., Simkin, A.J., Alotaibi, S., Fisk, S.J., Madgwick, P.J., Sparks, C.A., Jones, H.D., Lawson, T., Parry, M.A.J., Raines, C.A., 2017. Increased SBPase activity improves photosynthesis and grain yield in wheat grown in greenhouse conditions. Philos. Trans. R. Soc. B 372, 1730.

Ermakova, M., Arrivault, S., Giuliani, R., Danila, F., Alonso-Cantabrana, H., Vlad, D., Ishihara, H., Feil, R., Guenther, M., Borghi, G.L., Covshoff, S., Ludwig, M., Cousins, A.B., Langdale, J.A., Kelly, S., Lunn, J.E., Stitt, M., von Caemmerer, S., Furbank, R.T., 2020. Installation of C4 photosynthetic pathway enzymes in rice using a single construct. Plant Biotechnol. J. 19, 575–588.

Fell, D., 1997. Understanding the Control of Metabolism. Portland Press Ltd., London, UK.

Haake, V., Zrenner, R., Sonnewald, U., Stitt, M., 1998. A moderate decrease of plastid aldolase activity inhibits photosynthesis, alters the levels of sugars and starch, and inhibits growth of potato plants. Plant J. 14, 147–157.

Haake, V., Geiger, M., Walch-Liu, P., Engels, C., Zrenner, R., Stitt, M., 1999. Changes in aldolase activity in wild-type potato plants are important for acclimation to growth irradiance and carbon dioxide concentration, because plastid aldolase exerts control over the ambient rate of photosynthesis across a range of growth conditions. Plant J. 17, 479–489.

Harrison, E.P., Willingham, N.M., Lloyd, J.C., Raines, C.A., 1998. Reduced sedoheptulose-1,7-bisphosphatase levels in transgenic tobacco lead to decreased photosynthetic capacity and altered carbohydrate accumulation. Planta 204, 27–36.

Harrison, E.P., Olcer, H., Lloyd, J.C., Long, S.P., Raines, C.A., 2001. Small decreases in SBPase cause a linear decline in the apparent RuBP regeneration rate, but do not affect Rubisco carboxylation capacity. J. Exp. Bot. 52, 1779–1784.

Hasse, D., Mikkat, S., Hagemann, M., Bauwe, H., 2009. Alternative splicing produces an H-protein with better substrate properties for the P-protein of glycine decarboxylase. FEBS J. 276, 6985–6991.

Henkes, S., Sonnewald, U., Badur, R., Flachmann, R., Stitt, M., 2001. A small decrease of plastid transketolase activity in antisense tobacco transformants has dramatic effects on photosynthesis and phenylpropanoid metabolism. Plant Cell 13, 535–551.

IPCC, 2014. In: Field, C.B., Barros, V.R., Dokken, D.J., Mach, K.J., Mastrandrea, M.D., Bilir, T.E., et al. (Eds.), Climate Change 2014 Impacts, Adaptation and Vulnerability: Part A: Global and Sectoral Aspects: Working Group II Contribution to the Fifth Assessment Report of the Intergovernmental Panel on Climate Change. Cambridge University Press, Cambridge, UK and New York, NY, doi:10.1017/CBO9781107415379.

Kebeish, R., Niessen, M., Thiruveedhi, K., Bari, R., Hirsch, H.-J., Rosenkranz, R., Stabler, N., Schonfeld, B., Kreuzaler, F.,

Peterhansel, C., 2007. Chloroplastic photorespiratory bypass increases photosynthesis and biomass production in Arabidopsis thaliana. Nat. Biotechnol. 25, 593–599.

Khozaei, M., Fisk, S., Lawson, T., Gibon, Y., Sulpice, R., Stitt, M., Lefebvre, S.C., Raines, C.A., 2015. Overexpression of plastid transketolase in tobacco results in a thiamine auxotrophic phenotype. Plant Cell 27, 432–447.

Khumsupan, P., Kozlowska, M.A., Orr, D.J., Andreou, A.I., Nakayama, N., Patron, N., Carmo-Silva, E., McCormick, A.J., 2020. Generating and characterizing single-and multigene mutants of the Rubisco small subunit family in Arabidopsis. J. Exp. Bot. 71 (19), 5963–5975.

Köhler, I.H., Ruiz-Vera, U.M., VanLoocke, A., Thomey, M.L., Clemente, T., Long, S.P., Ort, D.R., Bernacchi, C.J., 2016. Expression of cyanobacterial FBP/SBPase in soybean prevents yield depression under future climate conditions. J. Exp. Bot. 68, 715–726.

Kumar, A., Li, C., Portis, A.R., 2009. Arabidopsis thaliana expressing a thermostable chimeric Rubisco activase exhibits enhanced growth and higher rates of photosynthesis at moderately high temperatures. Photosynth. Res. 100, 143–153.

Kurek, I., Chang, T.K., Bertain, S.M., Madrigal, A., Liu, L., Lassner, M.W., Zhu, G., 2007. Enhanced thermostability of *Arabidopsis* rubisco activase improves photosynthesis and growth rates under moderate heat stress. Plant Cell 19, 3230–3241.

Laisk, A., Oja, V., Kiirats, O., Raschke, K., Heber, U., 1989. The state of the photosynthetic aparatus in leaves as analyzed by rapid gas-exchange and optical methods – the pH of the chloroplast stroma and activation of enzymes *in-vivo*. Planta 177, 350–358.

Lefebvre, S., Lawson, T., Fryer, M., Zakhleniuk, O.V., Lloyd, J.C., Raines, C.A., 2005. Increased sedoheptulose-1,7-bisphosphatase activity in transgenic tobacco plants stimulates photosynthesis and growth from an early stage in development. Plant Physiol. 138, 451–460.

Levey, M., Timm, S., Mettler-Altmann, T., Luca Borghi, G., Koczor, M., Arrivault, S., PM Weber, A., Bauwe, H., Gowik, U., Westhoff, P., 2018. Efficient 2-phosphoglycolate degradation is required to maintain carbon assimilation and allocation in the C4 plant Flaveria bidentis. J. Exp. Bot. 70, 575–587.

Lin, M.T., Occhialini, A., Andralojc, P.J., Parry, M.A.J., Hanson, M.R., 2014. A faster Rubisco with potential to increase photosynthesis in crops. Nature 513, 547–550.

Lin, M.T., Stone, W.D., Chaudhari, V., Hanson, M.R., 2020. Small subunits can determine enzyme kinetics of tobacco Rubisco expressed in Escherichia coli. Nat. Plants 6, 1289–1299.

Long, S.P., Ainsworth, E.A., Rogers, A., Ort, D.R., 2004. Rising atmospheric carbon dioxide: plants FACE the future. Annu. Rev. Plant Biol. 55, 591–628.

Long, B.M., Hee, W.Y., Sharwood, R.E., Rae, B.D., Kaines, S., Lim, Y.-L., Nguyen, N.D., Massey, B., Bala, S., von Caemmerer, S., Badger, M.R., Price, G.D., 2018. Carboxysome encapsulation of the CO_2-fixing enzyme Rubisco in tobacco chloroplasts. Nat. Commun. 9, 3570.

López-Calcagno, P.E., Fisk, S.J., Brown, K., Bull, S.E., South, P.F., Raines, C.A., 2018. Overexpressing the H-protein of the glycine cleavage system increases biomass yield in glasshouse and field-grown transgenic tobacco plants. Plant Biotechnol. J. 17, 141–151.

Maier, A., Fahnenstich, H., von Caemmerer, S., Engqvist, M.-K.M., Weber, A.P.M., Flugge, U.I., Maurino, V.G., 2012. Transgenic introduction of a glycolate oxidative cycle into *A. thaliana* chloroplasts leads to growth improvement. Front. Plant Sci. 3, 38.

Martin-Avila, E., Lim, Y.-L., Birch, R., Dirk, L.M.A., Buck, S., Rhodes, T., Sharwood, R.E., Kapralov, M.V., Whitney, S.M., 2020. Modifying plant photosynthesis and growth via simultaneous chloroplast transformation of rubisco large and small subunits. Plant Cell 32, 2898–2916.

Masle, J., Hudson, G.S., Badger, M.R., 1993. Effects of ambient CO_2 concentration on growth and nitrogen use in Tobacco (Nicotiana tabacum) plants transformed with an antisense gene to the small subunit of ribulose-1,5-bisphosphate carboxylase/oxygenase. Plant Physiol. 103, 1075–1088.

McGrath, J.M., Long, S.P., 2014. Can the cyanobacterial carbon-concentrating mechanism increase photosynthesis in crop species? A theoretical analysis. Plant Physiol. 164, 2247–2261.

Messant, M., Timm, S., Fantuzzi, A., Weckwerth, W., Bauwe, H., Rutherford, A.W., Krieger-Liszkay, A., 2018. Glycolate induces redox tuning of photosystem II in vivo: study of a photorespiration mutant. Plant Physiol. 177, 1277–1285.

Miyagawa, Y., Tamoi, M., Shigeoka, S., 2001. Overexpression of a cyanobacterial fructose-1,6-sedoheptulose-1,7-bisphosphatase in tobacco enhances photosynthesis and growth. Nat. Biotechnol. 19, 965–969.

Nolke, G., Houdelet, M., Kreuzaler, F., Peterhansel, C., Schillberg, S., 2014. The expression of a recombinant glycolate dehydrogenase polyprotein in potato (*Solanum tuberosum*) plastids strongly enhances photosynthesis and tuber yield. Plant Biotechnol. J. 12, 734–742.

Occhialini, A., Lin, M.T., Andralojc, P.J., Hanson, M.R., Parry, M.A.J., 2016. Transgenic tobacco plants with improved cyanobacterial Rubisco expression but no extra assembly factors grow at near wild-type rates if provided with elevated CO_2. Plant J. 85, 148–160.

Ogren, W.L., 2003. Affixing the O to Rubisco: discovering the source of photorespiratory glycolate and its regulation. Photosynth. Res. 76, 53–63.

Olcer, H., Lloyd, J.C., Raines, C.A., 2001. Photosynthetic capacity is differentially affected by reductions in sedoheptulose-1,7-bisphosphatase activity during leaf development in transgenic tobacco plants. Plant Physiol. 125, 982–989.

Orr, D.J., Alcântara, A., Kapralov, M.V., Andralojc, P.J., Carmo-Silva, E., Parry, M.A.J., 2016. Surveying rubisco diversity and temperature response to improve crop photosynthetic efficiency. Plant Physiol. 172, 707–717.

Peterhansel, C., Horst, I., Niessen, M., Blume, C., Kebeish, R., Kürkcüoglu, S., Kreuzaler, F., 2010. Photorespiration. Arabidopsis Book 8, e0130.

Peterhansel, C., Krause, K., Braun, H.P., Espie, G.S., Fernie, A.R., Hanson, D.T., Keech, O., Maurino, V.G., Mielewczik, M., Sage, R.F., 2013. Engineering photorespiration: current state and future possibilities. Plant Biol. (Stuttg.) 15, 754–758.

Pettersson, G., Ryde, P.U., 1988. A mathematical-model of the calvin photosynthesis cycle. Eur. J. Biochem. 175, 661–672.

Poolman, M.G., Fell, D.A., Thomas, S., 2000. Modelling photosynthesis and its control. J. Exp. Bot. 51, 319–328.

Portis, A.R., Parry, M.A.J., 2007. Discoveries in Rubisco (Ribulose 1,5-bisphosphate carboxylase/oxygenase): a historical perspective. Photosynth. Res. 94, 121–143.

Portis, A.R., Chon, C.J., Mosbach, A., Heldt, H.W., 1977. Fructose-and sedoheptulosebisphosphatase. The sites of a possible control of CO_2 fixation by light-dependent changes of the stromal Mg^{2+} concentration. Biochim. Biophys. Acta Bioenerg. 461, 313–325.

Rae, B.D., Long, B.M., Förster, B., Nguyen, N.D., Velanis, C.-N., Atkinson, N., Hee, W.Y., Mukherjee, B., Price, G.D., McCormick, A.J., 2017. Progress and challenges of engineering a biophysical CO_2-concentrating mechanism into higher plants. J. Exp. Bot. 68, 3717–3737.

Raines, C.A., 2003. The Calvin cycle revisited. Photosynth. Res. 75, 1–10.

Raines, C.A., 2006. Transgenic approaches to manipulate the environmental responses of the C3 carbon fixation cycle. Plant Cell Environ. 29, 331–339.

Rodermel, S.R., Abbott, M.S., Bogorad, L., 1988. Nuclear-organelle interactions: nuclear antisense gene inhibits ribulose bisphosphate carboxylase enzyme levels in transformed tobacco plants. Cell 55, 673–681.

Rosenthal, D.M., Locke, A.M., Khozaei, M., Raines, C.A., Long, S.P., Ort, D.R., 2011. Over-expressing the C3 photosynthesis cycle enzyme Sedoheptulose-1-7 Bisphosphatase improves photosynthetic carbon gain and yield under fully open air CO_2 fumigation (FACE). BMC Plant Biol. 11, 123.

Salesse-Smith, C.E., Sharwood, R.E., Busch, F.A., Kromdijk, J., Bardal, V., Stern, D.B., 2018. Overexpression of Rubisco subunits with RAF1 increases Rubisco content in maize. Nat. Plants 4, 802–810.

Scafaro, A.P., Atwell, B.J., Muylaert, S., Reusel, B.V., Ruiz, G.-A., Rie, J.V., Gallé, A., 2018. A thermotolerant variant of rubisco activase from a wild relative improves growth and seed yield in rice under heat stress. Front. Plant Sci. 9, 1663.

Schuler, M.L., Mantegazza, O., Weber, A.P.M., 2016. Engineering C4 photosynthesis into C3 chassis in the synthetic biology age. Plant J. 87, 51–65.

Simkin, A.J., 2019. Genetic engineering for global food security: photosynthesis and biofortification. Plants 8, 586.

Simkin, A.J., McAusland, L., Headland, L.R., Lawson, T., Raines, C.A., 2015. Multigene manipulation of photosynthetic carbon assimilation increases CO_2 fixation and biomass yield in tobacco. J. Exp. Bot. 66, 4075–4090.

Simkin, A.J., Lopez-Calcagno, P.E., Davey, P.A., Headland, L.R., Lawson, T., Timm, S., Bauwe, H., Raines, C.A., 2017. Simultaneous stimulation of sedoheptulose 1,7-bisphosphatase, fructose 1,6-bisphophate aldolase and the photorespiratory glycine decarboxylase H-protein increases CO_2 assimilation, vegetative biomass and seed yield in Arabidopsis. Plant Biotechnol. J. 15, 805–816.

Simkin, A.J., Lopez-Calcagno, P.E., Raines, C.A., 2019. Feeding the world: improving photosynthetic efficiency for sustainable crop production. J. Exp. Bot. 70, 1119–1140.

Somerville, C.R., Ogren, W.L., 1982. Genetic modification of photorespiration. Trends Biochem. Sci. 7, 171–174.

South, P.F., Cavanagh, A.P., Liu, H.W., Ort, D.R., 2019. Synthetic glycolate metabolism pathways stimulate crop growth and productivity in the field. Science 363, eaat9077.

Stitt, M., Schulze, D., 1994. Does Rubisco control the rate of photosynthesis and plant-growth – an exercise in molecular ecophysiology. Plant Cell Environ. 17, 465–487.

Stitt, M., Lunn, J., Usadel, B., 2010. Arabidopsis and primary photosynthetic metabolism—more than the icing on the cake. Plant J. 61, 1067–1091.

Suganami, M., Suzuki, Y., Tazoe, Y., Yamori, W., Makino, A., 2021. Co-overproducing Rubisco and Rubisco activase enhances photosynthesis in the optimal temperature range in rice. Plant Physiol. 185 (1), 108–119.

Tamoi, M., Nagaoka, M., Miyagawa, Y., Shigeoka, S., 2006. Contribution of fructose-1,6-bisphosphatase and sedoheptulose-1,7-bisphosphatase to the photosynthetic rate and carbon flow in the Calvin cycle in transgenic plants. Plant Cell Physiol. 47, 380–390.

Timm, S., Florian, A., Arrivault, S., Stitt, M., Fernie, A.R., Bauwe, H., 2012. Glycine decarboxylase controls photosynthesis and plant growth. FEBS Lett. 586, 3692–3697.

Timm, S., Wittmiss, M., Gamlien, S., Ewald, R., Florian, A., Frank, M., Wirtz, M., Hell, R., Fernie, A.R., Bauwea, H., 2015. Mitochondrial dihydrolipoyl dehydrogenase activity shapes photosynthesis and photorespiration of *Arabidopsis thaliana*. Plant Cell 27, 1968–1984.

Timm, S., Florian, A., Fernie, A.R., Bauwe, H., 2016. The regulatory interplay between photorespiration and photosynthesis. J. Exp. Bot. 67, 2923–2929.

Timm, S., Giese, J., Engel, N., Wittmiss, M., Florian, A., Fernie, A.R., Bauwea, H., 2017. T-protein is present in large excess over the other proteins of the glycine cleavage system in leaves of Arabidopsis. Planta 247, 41–51.

Timm, S., Giese, J., Engel, N., Wittmiß, M., Florian, A., FernieBauwe, A.R.H., 2018. T-protein is present in large excess over the other proteins of the glycine cleavage system in leaves of *Arabidopsis*. Planta 247 (1), 41–51.

Uematsu, K., Suzuki, N., Iwamae, T., Inui, M., Yukawa, H., 2012. Increased fructose 1,6-bisphosphate aldolase in plastids enhances growth and photosynthesis of tobacco plants. J. Exp. Bot. 63, 3001–3009.

von Caemmerer, S., 2020. Rubisco carboxylase/oxygenase: from the enzyme to the globe: a gas exchange perspective. J. Plant Physiol. 252, 153240.

Walker, B.J., VanLoocke, A., Bernacchi, C.J., Ort, D.R., 2016. The costs of photorespiration to food production now and in the future. Annu. Rev. Plant Biol. 67, 107–109.

Whitney, S.M., Andrews, T.J., 2001. Plastome-encoded bacterial ribulose-1,5-bisphosphate carboxylase/oxygenase (RubisCO) supports photosynthesis and growth in tobacco. Proc. Natl. Acad. Sci. U. S. A. 98, 14738–14743.

Woodrow, I., Berry, J., 1988. Enzymatic regulation of photosynthetic CO_2, fixation in C3 plants. Annu. Rev. Plant Biol. 39, 533–594.

Xin, C.P., Tholen, D., Devloo, V., Zhu, X.G., 2015. The benefits of photorespiratory bypasses: how can they work? Plant Physiol. 167, 574–585.

Yamori, W., Nagai, T., Makino, A., 2011. The rate-limiting step for CO_2 assimilation at different temperatures is influenced by the leaf nitrogen content in several C3 crop species. Plant Cell Environ. 34, 764–777.

Yoon, D.-K., Ishiyama, K., Suganami, M., Tazoe, Y., Watanabe, M., Imaruoka, S., Ogura, M., Ishida, H., Suzuki, Y., Obara, M., Mae, T., Makino, A., 2020. Transgenic rice overproducing Rubisco exhibits increased yields with improved nitrogen-use efficiency in an experimental paddy field. Nat. Food 1, 134–139.

Zhang, J., Li, M., Bryan, A.C., Yoo, C.G., Rottmann, W., Winkeler, K.A., Collins Cassandra, M., Singan, V., Lindquist, E.A., Jawdy, S.S., Gunter, L.E., Engle, N.L., Yang, X., Barry, K., Tschaplinski, T.J., Schmutz, J., Pu, Y., Ragauskas, A.J., Tuskan, G.A., Muchero, W., Chen, J.-G., 2019. Overexpression of a serine hydroxymethyltransferase increases biomass production and reduces recalcitrance in the bioenergy crop Populus. Sustainable Energy Fuels 3, 195–207.

Zhu, X.G., de Sturler, E., Long, S.P., 2007. Optimizing the distribution of resources between enzymes of carbon metabolism can dramatically increase photosynthetic rate: a numerical simulation using an evolutionary algorithm. Plant Physiol. 145, 513–526.

Zhu, X.G., Long, S.P., Ort, D.R., 2010. Improving photosynthetic efficiency for greater yield. Annu. Rev. Plant Biol. 61, 235–261.

Zhu, X.G., Portis, A.R., Long, S.P., 2004. Would transformation of C3 crop plants with foreign Rubisco increase productivity? A computational analysis extrapolating from kinetic properties to canopy photosynthesis. Plant Cell Environ. 27, 155–165.

PART IV

Synthesis

CHAPTER

10

Integrating the stages of photosynthesis

Jeremy Harbinson[a], *Elias Kaiser*[b], *and Alejandro Sierra Morales*[c]

[a]Laboratory of Biophysics, Helix Building, Wageningen University, Wageningen, The Netherlands [b]Horticulture and Product Physiology Group, Radix Building, Wageningen University, Wageningen, The Netherlands [c]Center for Crop Systems Analysis, Radix Building, Wageningen University, Wageningen, The Netherlands

1 The integration of processes in photosynthesis under steady state conditions

1.1 Introduction

Photosynthesis is a complex process, which is comprised of several distinct subprocesses that have been described from various perspectives in other chapters in this book. Within this chapter, we will deal with the integration of these subprocesses to form the 'top-level' phenomena that are observed within photosynthetic organisms in nature, including plants grown in field agriculture and other more controlled environments, such as greenhouses. Photosynthetic organisms are diverse, and we will focus on photosynthesis in the embryophytes—the land plants—and more specifically largely on the photosynthetic properties of leafy angiosperms making use of the C3 photosynthetic pathway. This bias is driven by the much greater knowledge and understanding there is of photosynthesis in green, leafy C3 land plants. An implication of this is that much of the diversity of photosynthesis is not so well understood as is photosynthesis in green land plant group. When trying to show something of how photosynthesis arises from the interaction and integration of its component parts, it is necessary to define some boundary to photosynthesis. Parasitic plants that no longer depend on photosynthesis as a source of energy and biomass are typically much reduced in comparison to plants that photosynthesise. The parasitic orchid *Rhizanthella gardneri*, for example, has no above ground parts and even flowers underground. By implication, therefore, much of the above ground structure and architecture of plants is there to serve the needs of photosynthesis by positioning and supporting the leaves, supplying water and mineral nutrients to the leaves, and transporting photosynthate away from the leaves. The roots are similarly important for photosynthesis, supplying the mineral nutrients that are essential for the photosynthetic machine, and the water to replace lost by transpiration. The needs of photosynthesis, therefore, shape much of plant form and function. Within this chapter,

Photosynthesis in Action
https://doi.org/10.1016/B978-0-12-823781-6.00001-0

we will, however, largely confine ourselves to a more restricted and conventional view of photosynthesis and consider the interaction and integration of the following:

1. The absorption of light and therefore energy.
2. The transformation of absorbed light energy into a metabolically useful form of chemical energy (ATP and NADPH).
3. The chemical conversion of carbon dioxide to carbohydrate, a process that requires reductant (NADPH) and energy (ATP).
4. Photoprotection to reduce the formation of reactive oxygen species and deal with then once formed.

The supply of carbon dioxide—an essential raw material for assimilation—is also essential for the sustained operation of photosynthesis and is covered in Chapter 3.

The ordering of items 1–4 reflects how photosynthesis is often portrayed—a process that begins with light absorption and ends with the conversion of CO_2 to carbohydrate. In this chapter, we will order things differently as this makes the description of their integration easier—the basics of the assimilation of CO_2 and problems associated with the assimilation process have a strong influence on the operation and regulation of the light-driven processes that provide the ATP and NADPH required for assimilation.

The conversion of carbon dioxide to carbohydrate is often viewed as reversing the respiration or oxidation of glucose (Eq. 1).

$$6O_2 + C_6H_{12}O_6 \rightarrow 6CO_2 + 6H_2O \quad (1)$$

Although this is not strictly true (photosynthesis does not make glucose), this is still a useful model. The oxidation of glucose is highly exergonic ($\Delta G = -2823\,\text{kJ}\,\text{mol}^{-1}$), and the reversal of this reaction—the conversion of carbon dioxide to a carbohydrate—is therefore endergonic and depends on an energy input to transform it into a thermodynamically spontaneous, exergonic process. A critical aspect of the operation of photosynthesis as an integrated system is, therefore, energy.

In addition to understanding the physiology of this integration of processes, this chapter will also introduce the models in which physiological knowledge is systematically encoded. Apart from integrating our knowledge of photosynthesis, these models allow the process to be realistically simulated in ways that allow the larger scale effects of photosynthesis on, for example, crop growth and yield or biosphere carbon fluxes to be understood. Realistic photosynthesis models are also indispensable for identifying options for improving the process and then projecting the result of any such improvements on larger scale phenomena, such as crop productivity.

The mechanistic description of photosynthesis is usually done at the molecular level; given that the process is an aggregation of chemical and physical processes, this is natural. Some of these processes, such as excitation transfer, redox reactions, or the molecular transformations of metabolism, function at the nanoscale within chloroplasts, which are themselves distributed within mesophyll cells, while other processes, such as the diffusion of carbon dioxide, take place within the larger scale organisation of the leaf's structure of air spaces and mesophyll cells. Understanding leaf photosynthesis, therefore, depends on being able to upscale from the nanoscale of the molecular engines of photosynthesis—such as the photosystems and other protein complexes of the thylakoids and stroma—to the level of the leaf, the canopy or biome. The effective, combined activity of these components depends on them being coordinated, which requires the control of their activity. There is abundant control of photosynthesis, and this seems to be targeted upon balancing the needs of maximal assimilation of carbon dioxide—maximising the light-use efficiency of photosynthesis—against the risk of damage due to reactive oxygen species and the loss of water from photosynthesising leaves.

1.2 The environment

1.2.1 *Carbon dioxide*

It is useful to have some values for the amount of carbon dioxide in the atmosphere both current and in past, as well as the spectrum and irradiance of light at the Earth's surface. Carbon dioxide is the one of the substrates for photosynthesis, and the global average atmospheric carbon dioxide level for 2018 was 407 ppm. The recent pre-industrial carbon dioxide level was about 280 ppm and began rising due to human activity at about 1800 (Post et al., 1990); this increase of about 130 ppm is significant for photosynthesis. Note that the standard CO_2 concentration routinely used in measurements of leaf photosynthesis has increased in the last decades; in the 1970s, it was only 330 ppm, while today 410 ppm would be more appropriate; this nearly 25% increase in CO_2 mole fraction used as a standard will result in increased assimilation rates, etc.

In the longer term, atmospheric carbon dioxide levels have fluctuated tremendously (Chapter 11). In the Archean period (4–2.5 GYa), when prokaryotic photosynthesis evolved, carbon dioxide levels are thought to have been in the range of 4×10^3–1×10^6 ppm and free oxygen levels were less than 10^{-6} of current levels (Catling and Zahnle, 2020). In the more recent geological past since land plants colonised the Earth's surface (beginning in the Ordovician period), CO_2 levels have been estimated by indirect means to have been typically around 1000 ppm, only decreasing to modern levels during the Tertiary period (Witkowski et al., 2018; Mills et al., 2019). During the previous 400,000 year period during which atmospheric composition could be measured from air bubbles in ice cores (such as the Vostok core), CO_2 levels have fluctuated from 190 ppm to 290 ppm. These CO_2 levels are important in understanding photosynthesis from an evolutionary point of view (for example, the currently poor catalytic properties of rubisco) but also illustrating something of the flexibility of photosynthesis with regards to the availability of its principle substrate, CO_2.

1.2.2 *Irradiance*

The photosynthetically active irradiance is conventionally defined as extending from 400 nm to 700 nm and quantum sensors designed to measure the photosynthetically active quantum flux (commonly used units $\mu mol\ m^{-2}\ s^{-1}$ and conventionally abbreviated to PAR, PAQF, or PPFD (photosynthetic photon flux density)) will measure in this wavelength range. When measured in units of $W\ m^{-2}$, the PAR band comprises about 45% of the solar spectrum. Note, however, that wavelengths outside this range are photosynthetically active—carbon dioxide fixation has been measured in leaves illuminated with 720 nm radiation or more, and below 400 nm (McCree, 1972; Inada, 1976; Hogewoning et al., 2012). The spectrum of light at the surface of the Earth is very variable, depending on the height of the sun above the horizon, the presence of clouds, and shade cast by leaves. In full sunlight with the sun high above the horizon, the spectrum is broad (Fig. 1), while within or under a canopy wavelengths in the range of 400–690 nm are strongly diminished relative to the far-red wavelengths (Fig. 2).

The spectrum of irradiance at the Earth's surface may have been different in the past. In the Archean period, the solar spectrum at the surface of the Earth may have been dominated by wavelengths in the orange region of the spectrum, with a peak at 550–600 nm and falling off more strongly at longer or shorter wavelengths than currently (Arney et al., 2016). This difference in spectrum is due to the different composition of the atmosphere compared to now, and when considering the evolution of photosynthesis it is important to bear in mind that the spectrum and intensity of the irradiance, and other environmental factors, may have been very different in the past compared

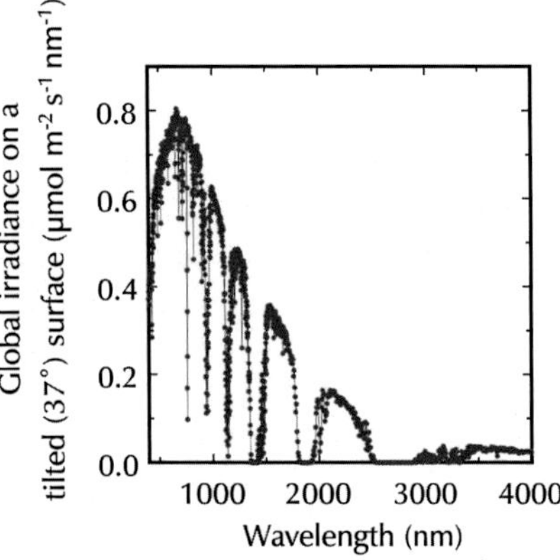

FIG. 1 The relative spectral irradiance on a quantum flux basis at the surface of the Earth in the 48 contiguous states of the USA. This irradiance is a modified form ASTMS spectral irradiance transformed to a quantum flux and was developed for use in solar power modelling (hence the 37 degree angle). This spectrum is itself a modelled spectrum for an air mass of 1.5. This spectral irradiance, while developed for use in the USA, is widely available and detailed and thus is widely used as a standard solar spectrum.

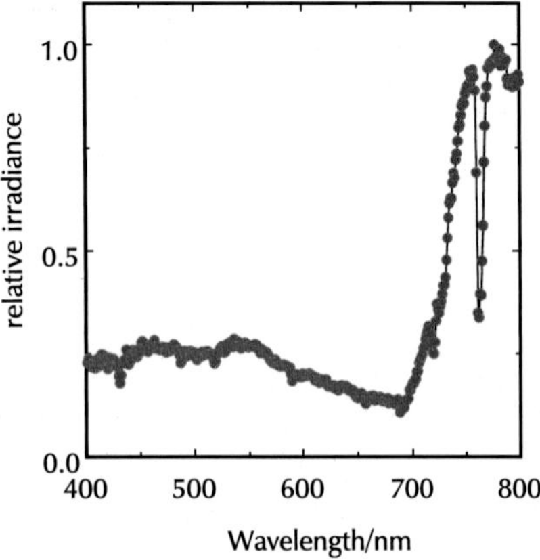

FIG. 2 The relative spectral distribution of the quantum flux in the 400–800nm range under a dense forest canopy.

to those in modern or more recent geological times. The maximum PAR encountered at sea level is normally taken to be 2000 μmol m^{-2} s^{-1} though higher irradiances occur, especially when the sky is partly clouded as the clouds reflect light.

1.3 Light absorption, photosynthetic pigments, and their organisation

Light is the source of energy upon which photosynthesis depends, and the first step in harvesting this energy is the absorption of light by photosynthetic pigments. The spectral dependency of light absorption for photosynthesis by leaves and other photosynthetic tissues, such as algal thalli, depends on the light absorption properties of photosynthetic pigments (usually summarised using the extinction coefficient), the concentration of the pigments, the scattering of light by the tissue, and the presence of non-photosynthetic pigments (Merzlyak et al., 2009). The light absorption maxima of a normal leaf are in the blue (400–500nm) and red (670–690nm) spectral regions, with a minimum at about 560nm (Fig. 3), and the depth of the 560nm minimum relative to the maxima in the blue and the red depends on leaf chlorophyll content (Björkman and Demmig, 1987; Merzlyak et al., 2009). Normal green leaves absorb about 80%–90% of the photon flux within the 400–700nm region (Ehleringer, 1981; Björkman and Demmig, 1987), so they are in general good traps for radiation in the spectral range of 400–700nm. The presence of hairs, scales, etc., however, has a major effect on leaf light-absorption (e.g. Ehleringer, 1981). It is important to note that there is no single value for leaf-light absorption for any wavelength in the 400–750nm range nor for any range of wavelengths in the photosynthetically active range. Non-photosynthetic pigments, such as flavonoids, also absorb in the photosynthetically active range of 400–700nm. In some cases, these result in leaves with conspicuous colouration, for example, anthocyanins and carotenoids formed as a result of stress or senescence, but they may also be present in unstressed healthy leaves in garden forms of plants and in some wild types (e.g. *Setcreasea purpurea*). Additionally, leaves have less conspicuous non-photosynthetic pigments that while absorbing

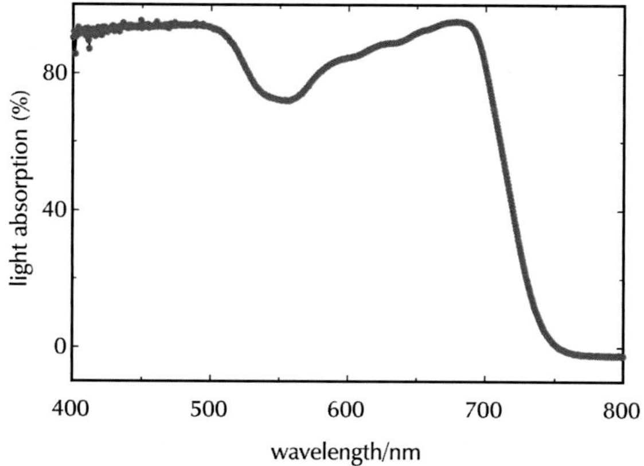

FIG. 3 The spectral dependency of light absorption by a tomato leaf in the 400–800 nm range.

most strongly at wavelengths less than 400 nm still absorb in the region of 400–500 nm or longer (Cerovic et al., 2002). The presence of these non-photosynthetic pigments can have an impact on the light-use efficiency of photosynthesis depending on their location within the leaf; the blue-light absorbing non-photosynthetic pigments, for example, do have a marked effect on photosynthetic light-use efficiency in the 400–500 nm spectral region (Inada, 1976; Hogewoning et al., 2012). In addition to this non-photosynthetic pigment problem, light absorbed by carotenoids, which is in the 400–540 nm region, is used less efficiently for photosynthesis (around 70% compared to light absorbed by chlorophylls (Hogewoning et al., 2012)). The spectral dependency of leaf photosynthetic light-use efficiency is, therefore, complex.

In the relation to the total solar spectrum, only a fraction of the available spectrum is used by leaves (compare Figs. 1 and 3) though non-oxygenic photosynthetic organisms can make use of wavelengths up to about 1000 nm for photosynthesis (Rutherford et al., 2012). Red, green, and brown thalloid algae absorb light over a similar spectral range to land plants, but within that range, the red and brown algae have a different spectral dependency of absorption because of their different pigment composition (Lüning and Dring, 1985).

The photosynthetic pigments generally found in plant leaves are chlorophyll *a*, chlorophyll *b*, neoxanthin, lutein, violaxanthin (this pigment is partly reversibly converted to antheraxanthin and zeaxanthin in the light), β-carotene and, in some plants, α-carotene. The total amount of chlorophyll (a + b) in a leaf varies between species and the history and developmental stage of the leaf, but is typically around 500 μmol m^{-2} (e.g. Fig. 4).

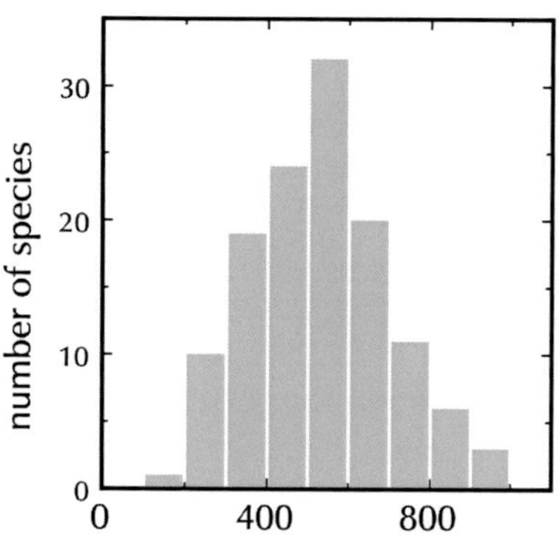

FIG. 4 The distribution of leaf chlorophyll content in neotropical forest plants. *Data taken from Matsubara, S., Krause, G.-H., Aranda, J., Virgo, A., Beisel, K.G., Jahns, P., Winter, K., 2009. Sun-shade patterns of leaf carotenoid composition in 86 species of neotropical forest plants. Funct. Plant Biol. 36, 20–36.*

In vivo, these pigments are highly organised by being bound to specific membrane-bound proteins that are largely specifically associated with one of two photosystems, photosystem I and photosystem II. The detailed organisation of the photosystems is increasingly well understood for cyanobacteria and higher plants (see Chapter 1). Each photosystem type has its own reaction centre type and these play a critical role in photosynthetic electron transport as they are the sites in which the energy of light absorbed by the pigment of bed of either PSI or PSII is converted to chemical energy, making use of the fact that an excited chlorophyll molecule is a stronger reducing agent than a ground state chlorophyll molecule.

1.4 Photosynthetic metabolism and the fixation of carbon dioxide

1.4.1 The Calvin-Benson-Bassham (CBB) cycle

The predominant function of photosynthesis is the fixation of carbon dioxide, and the operation of photosynthesis is moulded to the chemistry of this process. It, therefore, makes sense to describe the chemistry of the assimilation of carbon dioxide fixation before any other functionality of photosynthesis (Chapter 3). Before doing that, however, it is important to stress that while photosynthetic metabolism is dominated by chemistry associated with the Calvin-Benson-Bassham pathway or cycle (CBB; Fig. 5), there is much greater diversity of metabolism associated with the chloroplast than just that emerging from the CBB (Rolland et al., 2012). So while the CBB might be quantitatively the most important pathway, there is more to the chloroplast than just the CBB even though the total activity of these other pathways is not usually quantified or taken in account when trying to understand the relationship between photosynthetic electron transport and metabolism.

Cyanobacterial photosynthesis and thus eukaryotic photosynthesis use the CBB. It is convenient to see this cycle as beginning and ending with ribulose-1,5-bisphosphate (RuBP; a 5-carbon phosphorylated carbohydrate) as this is the substrate that reacts with carbon dioxide in a reaction catalysed by the enzyme ribulose-1,5-bisphosphate carboxylase oxygenase (rubisco). This carboxylation reaction produces two 3 carbon products (two molecules of phosphoglycerate (PGA)), hence the name 'C3', and is the only reaction in the carboxylation phase of the CBB (Fig. 5). The reduction of each PGA, in the reduction phase, to from the triose phosphate glyceraldehyde-3-phosphate (GAP) requires 1 NADPH and 1 ATP. This three carbon sugar phosphate is the first sugar formed in the process of carbon assimilation. GAP is equilibrated with another triose phosphate, dihdroxyacetone phosphate (DHAP), and the GAP + DHAP pool is often collectively referred as 'triose phosphate'. The conversion of PGA to GAP/DHAP comprises the reduction phase of the CBB. The bulk of the CBB is dedicated to regenerating the 5-carbon RuBP from the 3-carbon products of carboxylation and reduction and involves rearrangements involving 3-to-6, 6-to-4, 4-to-7, and 7-to-5 conversions. The final step, the phosphorylation of ribulose-1-phosphate to RuBP, requires an ATP. The CBB and CO_2 assimilation, therefore, requires 2 NADPH and 3 ATP for each CO_2 fixed—a ratio that is of critical importance for the operation of the light reactions.

The export of carbon from the CBB is important as this allows metabolites to feed into other metabolic pathways leading, in particular, to the synthesis of sucrose and starch. This export also allows the release of inorganic phosphate (Pi) from phosphorylated intermediates as they further metabolised; this Pi is essential for ATP synthesis and the continued operation of photosynthetic electron transport. At steady-state for each cycle of the CBB one sixth of the carbon can be exported while 5/6 should remain in the cycle; if too much is exported the metabolite pools of the cycle will crash decreasing

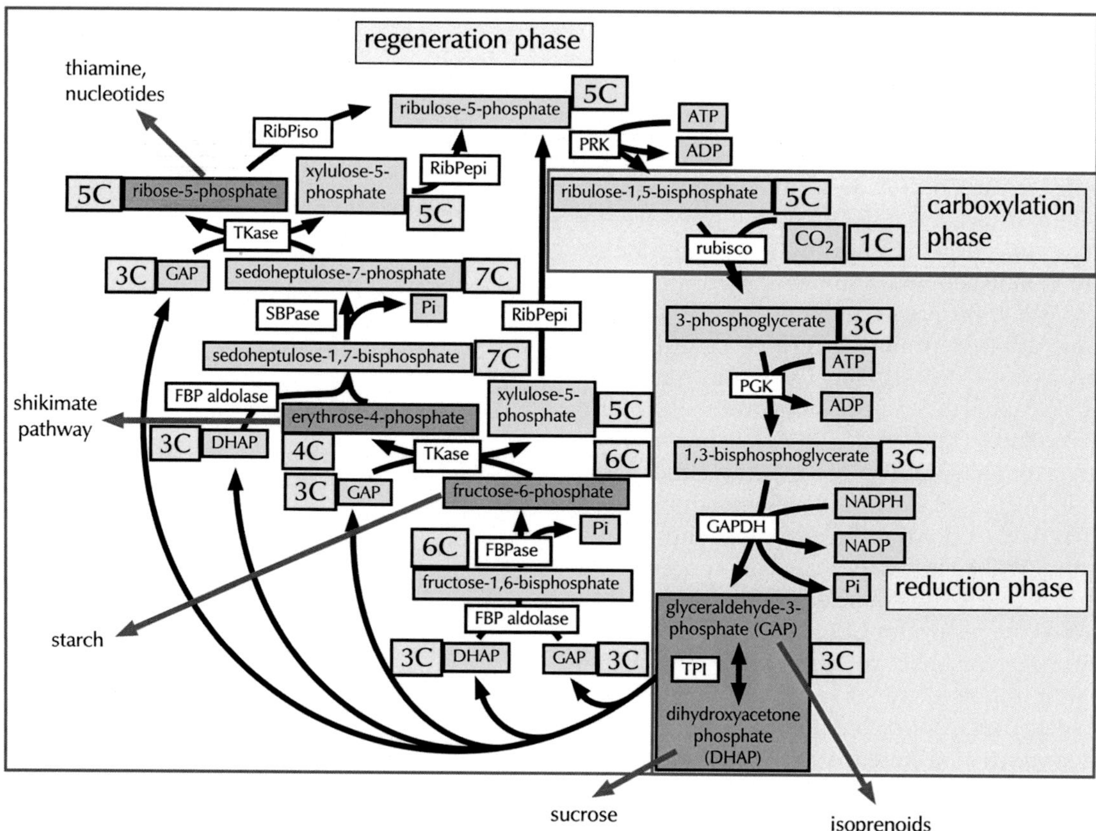

FIG. 5 The Calvin-Benson-Bassham cycle showing the metabolic intermediates, substrates or products of the cycle *(green or purple boxes)*; the number of carbons in each intermediate *(yellow boxes)*; the enzymes *(white boxes)*; the major points of export for various biosynthetic pathways *(purple boxes)*; and three main phases of the pathway (carboxylation, reduction and regeneration). The abbreviations for the enzymes are: *rubisco*, ribulouse-1,5-bisphosphate carboxylase oxygenase; *PGK*, phosphoglycerate kinase; *GAPDH*, glyceraldehyde phosphate dehydrogenase, *FBP aldolase*, fructose-1,6-bisphosphate aldolase; *FBPase*, fructose-1,6-bisphosphatase; *TKase*, transketolase; *Ribepi*, ribulose-5-phosphate epimerase; *SBPase*, sedoheptulose-1,7-bisphosphatase; *RibPiso*, ribose-5-phosphate isomerase; *PRK*, phosphoribulokinase.

assimilation, while if too little is exported the chloroplast Pi pool is depleted. Reduced carbon in the form of DHAP and fructose-6-phosphate is removed from the CBB to make sucrose and starch, respectively. Sucrose synthesis takes place in the cytosol, which requires that DHAP destined for sucrose synthesis be exported from the chloroplast in exchange for Pi released from DHAP and other sugar phosphates in the cytosol as part of the sucrose synthesis pathway. Starch synthesis takes place in the chloroplast and also involves the release of Pi. The phospho-groups added via ATP in the turnover of the CBB are therefore released by the metabolism of sugar phosphates to make sucrose and starch. The turnover of the CBB also regenerates NADP from NADPH. It is worth noting here that some other CBB intermediates are the starting points of further chloroplast metabolic pathways (Fig. 5).

1.4.2 *Rubisco—A problem enzyme*

General characteristics of rubisco and the carboxylation of RuBP

The enzyme rubisco is important for several reasons in addition to its function as the main route for carbon dioxide assimilation in the biosphere. Rubisco is a relatively big enzyme—the eukaryotic form is comprised of 8 large subunits and 8 small subunits and has an mass of 540 kD, it is slow (turnover number 1.4–4.1 (Galmés et al., 2014; Hermida-Carrera et al., 2016)) and has a poor affinity for CO_2 (for plants K_m 6.4–15.5 μmol dm^{-3} (Galmés et al., 2014; Hermida-Carrera et al., 2016) (excluding bryophytes)). The solubility of CO_2 at an atmospheric mole fraction of 400 ppm is 13 μmol dm^{-3} but in an actively photosynthesising leaf the concentration will be less than that. This concentration drop is due to the reliance of C3 leaves on diffusion as the means for CO_2 transport (Chapter 3). Diffusive transport occurs from a site of high concentration to one of low concentration, with the rate of transport (the flux) depending on the concentration difference and the conductance for diffusion. CO_2 diffuses from the surrounding air to the chloroplast, so the concentration in the chloroplast must be lower than the equivalent concentration in the surrounding air and by an amount that will depend on the flux of CO_2 and the total diffusive conductance of the leaf. The main diffusion limitations in a leaf are usually due to the stomata and the mesophyll, with the latter being due to the gaseous intercellular spaces and the aqueous diffusion path across the cell walls and the cell to the chloroplast. As a result of the diffusive limitation acting on the transport of CO_2 to the chloroplast from the free air surrounding the leaf, the actual CO_2 concentration in a C3 chloroplast is usually equivalent to a mole fraction of 200 ppm or a concentration of only 6 μmol dm^{-3}. In an actively photosynthesising C3 leaf, therefore, rubisco will be in a sub-saturating CO_2 concentration and will not be operating at its maximum turnover number. The large size of rubisco coupled with the low turnover number results in rubisco comprising up to 50% of the soluble protein in leaves of plants like wheat, and while the concentration active sites in a rubisco crystal is 8.75–10 mmol dm^{-3}, their concentration in the chloroplast stroma is c. 4 mmol dm^{-3} (Harris and Königer, 1997).

Oxygenation of RuBP by rubisco and photorespiration

The relatively poor catalytic properties of rubisco with regards to the carboxylation of RuBP are compounded by the fact that rubisco also catalyses a competing reaction in which O_2 oxygenates rubp (for plants the K_m of the oxygenation reaction is 183–565 μmol dm^{-3} (Galmés et al., 2014), and the solubility of O_2 is around 265 μmol dm^{-3}). The oxygenation of RuBP produces one molecule of PGA and one of phosphoglycolate (PGO), a two carbon compound. The PGA can be further metabolised by the CBB, while the PGO is further metabolised via the photosynthetic carbon oxidation pathway (PCO; also called the photorespiratory pathway). This spans the chloroplasts, cytosol (especially the peroxisomes), and mitochondria and returns to the chloroplast a molecule of glycerate, which is then phosphorylated to PGA. For every two PGO processed by the PCO, one CO_2 is released and one glycerate produced. Oxygenation of RuBP, therefore, has two negative effects; first, it uses RuBP, which needs to be re-synthesised via the CBB and the PCO, and second, it results in the evolution of CO_2, an act that directly antagonises the assimilation of CO_2. While the ATP/NADPH stoichiometry of the CBB is 1.5, that of the PCO is 1.75.

The oxygenation activity of rubisco occurs at a significant rate at normal in vivo O_2 and CO_2 concentrations; at about 23°C, eliminating photorespiration will produce an approximately 50% increase in the assimilation rate. Overall, it has been calculated that in a spinach leaf (a typical C3 leaf), the poor catalytic properties of rubisco with regards to the carboxylation of RuBP on the one hand and the rather effective catalysis of the oxygenation of RuBP on the other hand results in a carboxylation rate of only $1.2\,s^{-1}$ per active site and an oxygenation rate of $0.4\,s^{-1}$ per active site (so for every 3 carboxylation reactions there is one oxygenation), resulting in a net carboxylation rate of only $1\,s^{-1}$ per active site (Flamholz et al., 2019). The poor catalytic properties of rubisco have considerable ecological and agricultural consequences that we will come back to later.

While it is illustrative to give some typical, broad-brush values for the relative rates of the oxygenation and carboxylation reactions, these are, however, not fixed but will depend on the specific catalytic properties of rubisco (these vary even within genotypes of a species), leaf temperature, and the concentration of CO_2 and O_2 in the chloroplast. High temperature favours the oxygenation reaction for two reasons; first, the effective activation energy of the oxygenation reaction is higher than that of the carboxylation reaction, so oxygenation responds more strongly to increasing temperature, and second while increasing temperatures decrease the solubility of both gases in water, the relative decrease in the solubility of CO_2 with increasing temperature is greater than that for O_2, again favouring the oxygenation reaction. The concentration of CO_2 and O_2 in the chloroplast also has an effect on the rates of carboxylation and oxygenation and consequently the ratio of these rates. Increasing the CO_2 concentration will decrease the rate of oxygenation and increase the rate of carboxylation, so modern greenhouses typically use CO_2 mole fractions of 1000 ppm to essentially eliminate the oxygenation reaction and photorespiratory losses. Reducing the O_2 mole fraction will decrease the rate of oxygenation and an O_2 mole fraction of 2% is commonly used in research to eliminate the oxygenation reaction. On the other hand, any restriction in the supply of CO_2 to the chloroplast from the air surrounding the leaf will diminish the CO_2 concentration in the leaf. The stomatal and the mesophyll conductances (see Chapter 3) exert such a limitation and both conductances are dynamic, adjusting in response to environmental and physiological factors.

An important consequence of the temperature and CO_2/O_2 sensitivity of rates of carboxylation and oxygenation is that the NADPH/ATP demands of metabolic activity associated with the CBB depends on the CO_2 and O_2 concentrations in chloroplast and leaf temperature. The O_2 concentration is assumed to be the same as that of the surrounding air, so it is the CO_2 concentration that will vary in a way that depends on diffusive limitation or the leaf environment. Both leaf temperature and effects of diffusive limitation are dynamic and can change rapidly (see Section 2, this chapter). As a result, the ratio of the metabolic demand for ATP and NADPH can vary between 1.5 and 1.75 in the seconds to minutes time range, depending on leaf temperature and the CO_2/ O_2 ratio in the chloroplast (Fig. 6), with the latter being strongly influenced by the gaseous composition of air around the leaf (important in horticulture and research) and changes in diffusional limitations, arising from, for example, changes in stomatal opening.

Rubisco and its activation and inactivation

In addition to being a catalytic underachiever, rubisco also suffers from another problem—that of inactivation. This inactivation can occur via three main routes and the enzyme rubisco activase is required for its reversal. The catalytic activation of rubisco depends on the carbamylation of a lysine residue in the active site of rubisco—this bound CO_2 is not a substrate for the carboxylation reaction. If a RuBP binds to active site before carbamylation, it is tightly bound and cannot be carboxylated, so creating an inactive rubisco (Bracher et al., 2017). Rubisco is also susceptible to being inhibited by variants of the normal carboxylation and oxygenation reactions of RuBP forming various tight binding products (e.g. xylose bisphosphate and penotdiulose bisphosphate (Pearce and

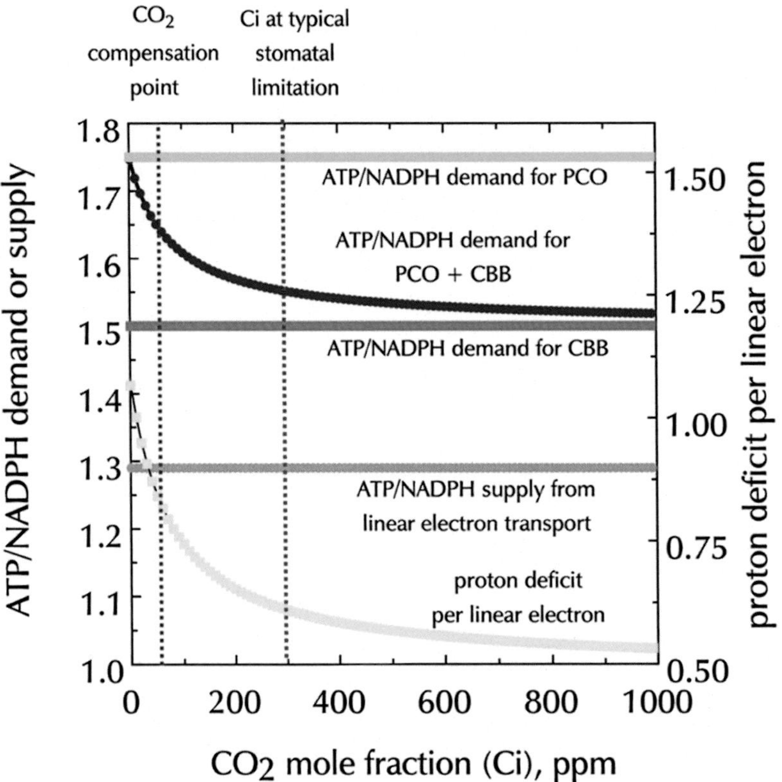

FIG. 6 The dependency on Ci, the CO_2 mole fraction in the substomatal cavity, of the ATP/NADPH for the required for the combined activities of the CBB and PCO, and the proton deficit for each electron passing along the LET to the NADPH that will be used for the PCO and CBB. Also shown are the ATP/NADPH for the PCO (1.75) and the CBB (1.5). Note that the estimation of ATP/NADPH for the PCO+CBB at any Ci will depend on the specificity factor of Rubisco.

Andrews, 2003)). Finally, in darkness or low light, carboxyarbinitol is phosphorylated to form carboxyarbinitol-1-phosphate (CA1P), which is a tight binding rubisco inhibitor. In the light free, CA1P is dephosphorylated and thus inactivated as an inhibitor (Parry et al., 2008). Reactivating these inhibited forms of rubisco requires the action of a helper protein, rubisco activase, which removes tightly bound inhibitors from the active site of rubisco. Rubisco activase exists in two forms, α and β. This regeneration of the rubisco active site requires ATP and, in addition, the activase is a regulated by ADP/ATP [Mg^{2+}] and by reduction of an —S—S— group by thioredoxin, though the regulation of the α and β forms differs and not all species have both forms. In *Arabidopsis*, the α form is regulated by redox state (via thioredoxin) and by ADP, while the β form is redox insensitive and relatively ADP insensitive. The Solanaceae have only the β form, though this is more ADP sensitive than is the β form of *Arabidopsis*. The temperature sensitivity of the activity of rubisco activase has been shown to correlate with the temperature sensitivity of assimilation (Parry et al., 2012). The cause of this sensitivity of activase to relatively moderate temperatures (less than 40°C) is not fully understood but may be due to temperature-induced changes to cofactors (redox state, ADP) involved in activase activity or be due to the effects of even moderate temperature on the activase itself, an enzyme that is required to have a high conformal flexibility to fulfil its function (Parry et al., 2013). The activation of rubisco is thought to play a role in limiting the response of assimilation to a sudden increase in irradiance (see Section 2 of this chapter) but is otherwise not thought to limit assimilation except at temperatures high enough to inhibit the activase.

1.4.3 Regulation of the CBB

The regulation of the CBB is required for two main reasons. It should respond proportionately to the supply of reducing power and ATP from the photosynthetic electron transport chain, which is to say that ideally metabolic demand of the CBB and associated metabolic pathways should be active enough not to limit electron transport, but not so active that they deplete the chloroplast ATP and NADPH pools. Regulation is also needed to prevent energetically wasteful metabolic cycling in the chloroplast in the dark (Heldt et al., 2005). The CBB, therefore, needs to be inactive in the dark and active in the light, and its activity in the light needs to be proportionate to the capacity of the electron transport chain to provide ATP and NADPH. The regulation of the CBB can be divided into coarse control or activation, and fine control. Activation of the CBB occurs in the light and arises from the pH and/or [Mg^{2+}] sensitivity of several of the enzymes in the pathway (phosphoglycerate kinase, fructose-1, 6-bisphosphatase, the transketolases (FBPase), and sedoeheptulose-1,7-bisphosphase (SBPase)). Once illuminated, the pH of the stroma will fall and the [Mg^{2+}] will increase, and this will broadly activate the CBB, serving as a master on/off switch for the cycle (Leegood, 1990). Once activated, the operation of the CBB can be understood via a mix of thermodynamics and kinetics. Most steps in the CBB are not far from equilibrium (i.e. $\Delta G \sim 0$; Fig. 7), and these are not subject to substantial regulation (which is not to say they do not exercise any limitation on the CBB). The main kinetic control of the pathway is exercised at a small number of steps that have a large negative standard $\Delta G^{o\prime}$ (i.e. are strongly exergonic; FBPase, SBPase, and rubisco) or that involve the consumption ATP or NADPH (phosphorubulokinase (PRK) and glyceraldehyde phosphate dehydrogenase (GAPDH)). Some of these controlled steps are subject to complex multilayered control. In the case of rubisco, this control is largely achieved by the control of rubisco activase, which is described earlier. As with the activase, control of FBPase, SBPase, PRK, and GAPDH is achieved partly via reductive activation by the thioredoxin system. The significance of reductive activation is that it allows the photosynthetic electron transport chain, which supplies reductant, to activate critical steps in the CBB via a feedforward mechanism. In some species, the activation of PRK and GAPDH is further complicated by them forming a complex with a small protein CP12. While in this complex GAPDH and PRK are inactive, and CP12 is itself subject to regulation by thioredoxin—when reduced the CP12-PRK-GAPDH complex dissociates and the enzymes can become active.

The four CBB redox activated enzymes (so excluding CP12 and rubisco activase) act in different ways to limit the CBB. FBPase and SBPase do not use either ATP or NADPH but have a large negative $\Delta G^{o\prime}$ because of the hydrolysis of a phosphate group. In *Chlorella pyrenoidosa,* their working ΔG is more negative than the $\Delta G^{o\prime}$, so the reactions are well away from equilibrium and the resulting ratios of FBP/F6P and SBP/S7P are 5.5 and 2.2, respectively, lower than would be expected if they were at equilibrium. Recent developments in metabolomics have allowed for a more comprehensive analyses of CBB metabolite levels under a range conditions. These analyses show that at steady state, the levels of SBP and FBP remain steady with increasing irradiance, while the levels of S7P and F6P are both higher than those of SBP and FBP (i.e. consistent with the *C. pyrenoidosa*

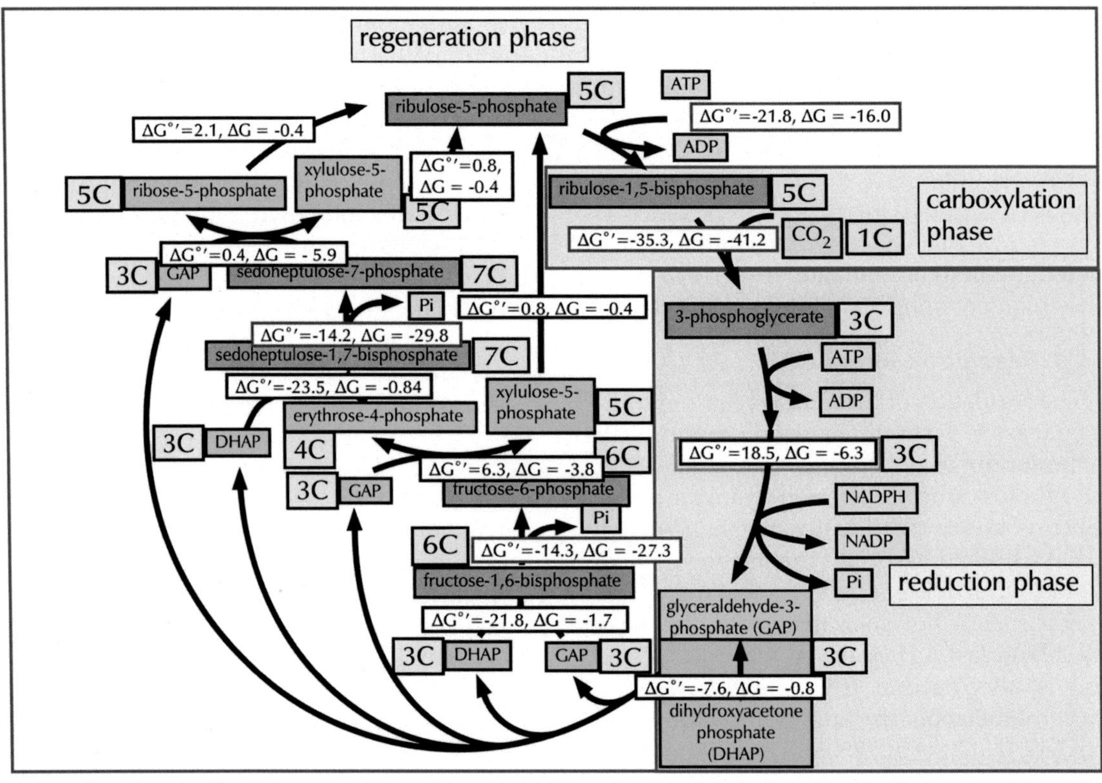

FIG. 7 As for Fig. 5, but showing the energetics of the CBB. The standard biological ($\Delta G^{o\prime}$) and actual (ΔG) free energies of each step are shown *(white boxes)*, with the steps furthest from equilibrium shown with *red frames*. Note that when at equilibrium $\Delta G = 0$. *Credit: Data from Bassham, J., Krause, G., 1969. Free energy changes and metabolic regulation in steady-state photosynthetic carbon reduction. Biochim. Biophys. Acta Bioenerg. 189, 207–221.*

results) and increase with increasing irradiance (Borghi et al., 2019). The activation of FBPase has also been shown to increase with irradiance (Harbinson et al., 1990). In addition to being subject to redox activation, SBPase and FBPase are activated by their substrates and inhibited by S7P and F6P, respectively, providing an another layer of rapid control—the regulation of enzyme activity by thioredoxin is rather slow (c. 2 min half-time (Sassenrath-Cole and Pearcy, 1994)). PGK and GAPD use ATP and NADPH, respectively, and together reduce PGA to triose phosphate, the key reduction step in the CBB. This two enzyme step is considered to be close to equilibrium under many conditions (Dietz and Heber, 1984; Heber et al., 1986) even though GAPDH is regulated by thioredoxin and in some species also via the interaction with CP12; GAPDH activity increases with irradiance (Howard et al., 2008). Being at near-equilibrium would allow the concentration of triose phosphate to increase in response to increases in the ATP/(ADP + Pi) and (NADPH + H^+)/NADP ratios, so pushing the CBB forward to FBPase and SBPase. The $\Delta G^{o\prime}$ of PGA reduction to GAP is positive so for this reaction to move in the direction of net triose phosphate synthesis requires a high mass reaction ratio (i.e.

PGA × ATP × NADPH × H^+)/(GAP × ADP × NADP × Pi) or (PGA × ATP × NADPH × H^+)/(DHAP × ADP × NADP × Pi) (Dietz and Heber, 1984). The steady-state PGA/DHAP ratio, which is overall greater than 1, decreases slightly with increasing irradiance and therefore flux (Borghi et al., 2019) suggesting that the PGA/triose phosphate reaction can become slightly limiting at high fluxes (Dietz and Heber, 1984). PRK commits the final step in RuBP regeneration, so replenishing the RuBP consumed by carboxylation and oxygenation. In addition to being feed-forward activated by thioredoxin (as is rubisco by the α form of rubisco activase), it is inhibited by ADP and PGA, so a build-up of PGA will slow the formation of RuBP. Note that an increase in PGA will arise from a decrease in either of the ATP/(ADP + Pi) and (NADPH+H^+)/NADP ratios so any decrease in the rate of triose phosphate formation due to a decrease in electron transport will feedback on the rate of RuBP synthesis via both the increases in PGA and ADP.

1.4.4 *The thermodynamic efficiency of the CBB*

Despite the kinetic imperfections of rubisco, carboxylation by the CBB is thermodynamically very efficient. For each CO_2 fixed, 3 ATP and 2 NADPH are required. Using the standard free energies of NADPH and ATP production (220.0 kJ mol^{-1} and 31.9 kJ mol^{-1}, respectively (Bell, 1985)), and the standard free energy for fixation of one mole of carbon dioxide (470 kJ mol^{-1}), the overall efficiency of the CBB 88%. The actual free energies of the reactants and products in vivo will not be the standard free energies, but this gives an idea of the efficiency of the C3 CO_2 fixation process. The problem with the CBB lies therefore not substantially with its basic energetic efficiency but with the poor kinetic and catalytic properties of rubisco that result in an enzyme with a low turnover number, is prone to inhibition, and catalyses an unproductive reaction (the oxygenation of RuBP) at an appreciable rate at normal atmospheric conditions.

1.5 Photochemistry, and subsequent electron and proton transport

The CO_2 fixing chemistry of the CBB and the associated photosynthetic carbon oxidation pathway (photorespiration) need NADPH and ATP. In addition, other metabolic activities take place in the chloroplast, and while these activities are quantitatively generally nowhere near as important as CO_2 fixation and photorespiration, they are not insignificant and these other pathways will have their own demands for ATP and reductant as either reduced ferredoxin or NADPH. The role of photosynthetic light reactions is to provide these cofactors and to do so in a way that is safe for the chloroplast and cell. The problem with the light reactions is O_2, the gaseous by-product of photosynthesis. Normal ground state O_2 is a reactive gas that can be converted by either chemical reduction to superoxide or physical transformation from the triplet ground state of O_2 to more reactive and damaging forms of oxygen. These transformations can occur as part of the operation of the light-reactions and the operation and regulation of the light-reactions is a balance between the efficient synthesis or ATP and reducing power on the one hand with operational constraints that avoid the excessive production of reactive oxygen species on the other.

1.6 A summary of photochemistry and the electron transport system

1.6.1 *The general organisation of the light-harvesting and electron and proton transport*

The two photosystems, PSII and PSI, share a common ancestor and therefore have basic similarities, but they are now comprised of distinct populations of proteins (albeit with features shared between the photosystems) and cofactors, most of which are now specific to each photosystem. Some cofactors, chlorophylls a and b, and carotenoids (most importantly lutein, β-carotene, and neoxanthin) are pigments so are key to photosynthetic light-absorption and are bound to specific sites in the pigment binding proteins of PSI and PSII. The amount of PSII or PSI reaction centres in a typical leaf is about 1 μmol m^{-2} per photosystem type while total leaf chlorophyll content is approximately 0.5 mmol m^{-2} (Osmond et al., 1980; Schöttler and Tóth, 2014). Note that these are approximate values. The absorption of a photon in the range of PAR by a photosynthetic pigment results in an electron being excited from the ground state to a higher energy level; if this energy level is higher than the first excited state (the S1 state), it will relax to the S1 state within approximately 10^{-14} s. These excited states can migrate from one pigment molecule to another, though in the case of the carotenoids and chlorophyll *b* migration to chl a is energetically downhill so excited states formed on carotenoids and chl b end up on chl a. Once on chl a, the excited state is free to migrate from chl a to chl a via either Forster resonance or as a delocalised exciton.

Within each PSII and PSI, there is a specialised structures called reaction centres that have the ability to convert the some of the energy of the excited state on chl a to chemical energy; reaction centres of PSII and PSI differ in their energetics and the detail of the operation, but they share many features, consistent with them having a common ancestor. Despite the complexities of both photosystems, they have a quite basic function; in both cases, they take an electron from a relatively weaker reducing agent and donate it to a relatively stronger reducing agent. This is would normally be an endergonic process, but it is driven (i.e. made exergonic) by the energy of the excited state of chl a. This process of charge separation leading the generation of reductants and oxidants in the reactions centres of photosystems I and II provides the energetic impulse that drives most of the life in the biosphere. The quantum efficiencies of the photosystems are high; for PSI it thought to be about 0.99, while for PSII, it is probably around 0.9, which is higher than that typically measured using chlorophyll fluorescence (about 0.82), the difference being due to the presence of weak fluorescence from PSI in the total fluorescence signal.

The two types of photosystem are connected to differently to the photosynthetic electron (and proton) transport network. Electron transport is dealt with in detail in Chapter 2, so in this chapter, we deal only with those aspects of the operation of electron transport that are relevant to the operation and regulation of electron and transport proton as part of bigger operation and regulation of photosynthesis. Photosynthetic electron transport pathways are divided into two main classes: the first is linear electron transport through PSII and PSI (LET), while the second is comprised of variants of cyclic electron transport passing through PSI (CET) (Fig. 8). All types of CET involve the plastoquinol pool, the cytochrome b_6/f complex, PSI, and ferredoxin, while all types of LET involve PSII (with its oxygen evolving complex; OEC), the plastoquinol pool, the cytochrome b_6/f complex, PSI, and ferredoxin. The cytochrome b_6/f complex and ferredoxin are important nodes in electron transport as a whole. Photosynthetic electron transport is associated with proton release into the aqueous thylakoid lumen. The lumen is physically separated from the stroma by a membrane that is impermeable to the passage of protons, so the addition of these protons to the lumen creates a proton potential (a proton energy difference, or proton motive force ($\Delta\mu_{H+}$; in kJ mol^{-1})) between the lumen and the stroma, which is comprised of voltage ($F\Delta\psi$ where F is the Faraday; protons carry a positive charge) and concentration (2.3RTΔpH; the lumen becomes more acidic) components (Eq. 2).

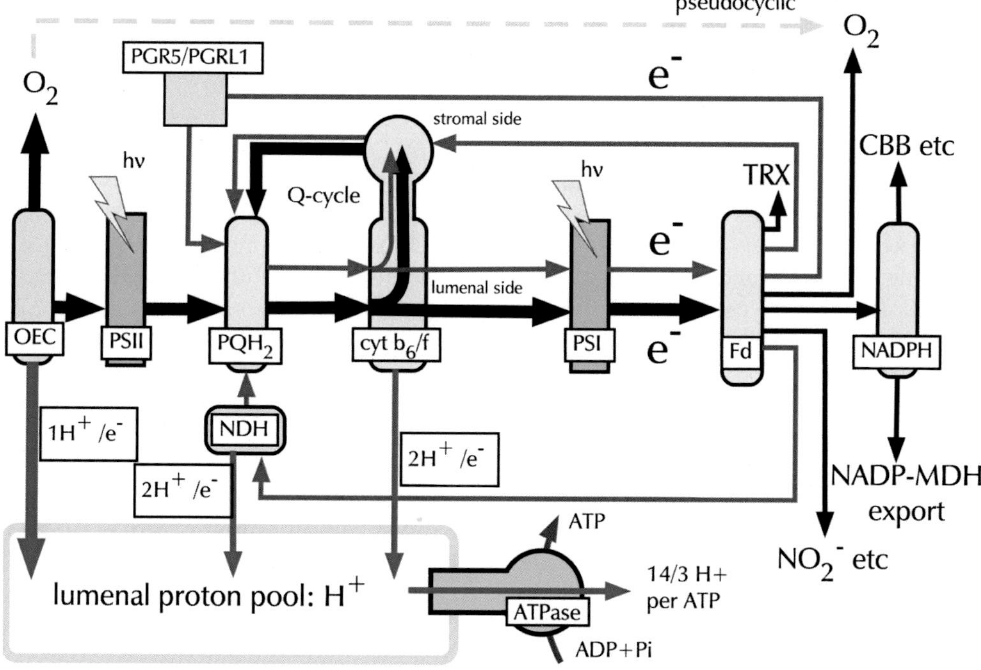

FIG. 8 A summary diagram of the photosynthetic electron transport system. Points of light energy input (PSII and PSI) are shown in *bright green*, major distributive nodes are in *yellow*, the linear electron path (including the Q-cycle) is indicated by the *black arrows* and the cyclic electron paths are shown the *red arrows*. The *blue arrows* indicate the points where protons are released into the thylakoid lumen, building up the proton motive force that drives ATP synthesis. The abbreviations are as in the text and TRX is thioredoxin.

$$\Delta\mu_{H^+} = F\Delta\psi + 2.3RT\Delta pH \qquad (2)$$

The pmf between the stroma and lumen is used by the chloroplast ATPase to drive the formation of ATP, and based on the structure of the ATPase, it is believed that the proton/ATP stoichiometry is 14/3 or c. 4.67.

1.6.2 Linear and cyclic electron transport have different capacities to form reductant and ATP

LET runs from water to the terminal electron acceptor of the pathway, which is usually defined as either ferredoxin or NADP+. Ferredoxin, in particular, is a node from which electrons—reducing power—are supplied to various chloroplast metabolic pathways, including, via NADPH, to the CBB. In addition to the CBB, the major destinations can be summarised as (Fig. 8):

1. Use by nitrite reduction and other reductive metabolic pathways in the chloroplast
2. The reduction of oxygen to superoxide (also called the Mehler reaction) by reduced ferredoxin and subsequent reduction of that superoxide to water
3. The reduction of oxaloacetate to malate by NADPH via NADP-malate dehydrogenase—the malate can be exported from the chloroplast, effectively removing reductant from the chloroplast, and
4. The reduction of thioredoxin (quantitatively minor but very important in control of photosynthetic metabolism)

For each electron moving along the LET $3H^+$ are released into the thylakoid lumen. Considering stoichiometries of ATP formation compared to ferredoxin or NADPH formation, with a ratio of 4.67 protons per ATP, LET results in 0.64 ATP per electron, or 1.29 ATP per NADPH. There are however some variants to this stoichiometry depending on what happens to the reduced ferredoxin or NADPH.

LET leading to O_2 reduction (option 2; the Mehler reaction) is often called 'pseudocyclic electron transport' because it shares some important characteristics with CET. Like CET, the pseudocyclic path does not result in the net formation of reductant; O_2 and $4H^+$ are released into the lumen by the oxidation of water, but O_2 is reduced to O_2^- by reduced ferredoxin in the stroma. This O_2^- is then reduced to water via superoxide dismutase and hydrogen peroxide, which requires another reduced ferredoxin equivalent (the actual reductant is reduced glutathione, which is reduced by NADPH) in addition to that oxidised by O_2 (see Chapter 5), so overall the pseudocyclic form of LET releases protons into the lumen but does not result in the net formation of reductant.

The use of NADPH by NADP malate dehydrogenase (NADP-MDH; a thioredoxin activated enzyme) to reduce oxaloacetate to malate (option 3) results ultimately in the export of reductant from the chloroplast. The malate is exported from the chloroplast in exchange for oxaloacetate, with the malate being oxidised by NAD malate dehydrogenase in the cytosol forming NADH (Scheibe, 2004). Overall, the series of reactions associated with NADP-MDH allows photochemically formed reductant to be exported, and thus removed, from the chloroplast but 3 protons are still deposited in the lumen. So overall electron transport to oxaloacetate results in ATP formation with only the transitory formation of reductant in the chloroplast. This removal of reductant from the chloroplast by the export of malate is commonly called the 'malate valve'.

In addition to LET, there is also cyclic electron transport around PSI. In plants, there are 2, and possibly 3, CET options. By definition in a cyclic electron path there is no net formation of reduction, and as all PSI cyclic routes pass via the plastoquinol pool and the cytochrome b_6/f complex, they all release at least $2H^+$ into the lumen per electron. The route via plant plastid complex I (NDH) (Fig. 8) was first associated with the phenomenon of chlororespiration and in C3 plants is believed to support only a low cyclic flux (e.g. Feild et al., 1998) due to the low concentration of the NDH complex (Burrows et al., 1998). An advantage of the NDH path is that electron transport through NDH is protonmotive because it releases and extra $2H^+$ into the lumen per electron in addition to those released by the cytochrome b_6/f complex (Strand et al., 2017). The second CET path is that attributed to a complex of PGR5 and PGRL1 (Hertle et al., 2013) that reduces PQ to PQH_2 and results in the release of $2H^+$ per electron as PQH_2 is oxidised. Finally, a cyclic pathway involving plastoquinone reduction via the cytochrome b_6/f complex has recently been suggested (or revived) (Nawrocki et al., 2019); in this pathway, PQ is reduced by the cytochrome b_6/f complex on the stromal side of the complex as part of the reducing function of the Q-cycle, and the oxidation by the cytochrome b_6/f complex of PQH_2 so formed will release $2H^+$ per electron. The cyclic and pseudocyclic pathways overall result in no net formation of reductant, but they are coupled to the formation of $\Delta\mu_{H+}$ and so they drive ATP synthesis.

Finally, it is important to note that not all LET generated reductant used in chloroplast metabolism (i.e. reductant not used for O_2 reduction and the malate valve) is used by the CBB and PCO pathways. Reduced ferredoxin is the reductant for required for various other metabolic pathways in the chloroplast, such as nitrite reduction, fatty acid synthesis, sulfite reduction, etc. (path 'NO_2^-, etc'. Fig. 8). Of these, nitrite reduction is potentially the most quantitatively important but even then carbon assimilation is 10–20 times greater than nitrogen assimilation (Robinson, 1988) and often less. Some plants take up NH_4^+ and therefore do not need to reduce NO_3^-, and NO_3^- reduction also occurs to a variable extent non-photosynthetically in the roots (Andrews, 1986; Bloom, 2011), so chloroplast NO_2^- reduction is variable in extent and may even be absent. Nitrite reduction and assimilation requires 8 reduced ferredoxin but only one ATP (10 reduced ferredoxin if cytosolic nitrate reduction is included) (Noctor and Foyer, 1998), while sulfite reduction requires 6 reduced ferredoxin and 2 ATP equivalents. These are very different ratios of demand for ATP and reductant compared to the CBB or PCO, and nitrite or sulphite reduction results in a surplus of ATP relative to the ratio of ATP/reductant formation by LET.

How the supply of ATP and reductant (NADPH or reduced ferredoxin) is balanced with the demand for reductant and ATP by the metabolic processes of the stroma remains a challenge in terms of accounting for all the fluxes and understanding the regulation of these fluxes. Overall, the coordination of electron transport processes with metabolism needs flexibility in two ways: first, the ratio of supply of ATP and reductant, and second, the rate of their supply and the regulation of electron transport and light-harvesting in the event that the capacity for supply exceeds needs or capacity of metabolism.

1.6.3 Meeting the ratio of demand for ATP and NADPH by the CBB and PCO

The LET electron transport chain translocates $3H^+$ per electron that passes from water to ferredoxin (or NADP), enough to make 3/4.67 (0.64) ATP, giving an ATP/reductant ratio of 0.64 if calculated in terms of reduced ferredoxin or 1.29 if calculated in terms of NADPH. The CBB has an ATP/NADPH demand of 1.5, while the PCO has a demand of 1.75, both in excess of 1.29. The actual ATP/NADPH demand of the combined activity of the CBB and PCO will vary with CO_2 mole fraction in the leaf and leaf temperature (Fig. 6), increasing with decreasing mole fraction and increasing temperature, both of which increase the relative rate of oxygenation, compared to carboxylation, of RuBP. This results in an ATP or lumen H^+ deficit associated with LET, at least as far as the CBB + PCO are concerned. For a leaf with a Ci of 300 ppm, a typical value for a leaf photosynthesizing in air with an ambient CO_2 mole fraction of around 400 ppm, the combined ATP/NADPH demand of the CBB + PCO is about 1.55. This is based on specificity factor of rubisco of 2720 (for tobacco and calculated using the value of the CO_2 compensation point (Γ^*) estimated using a finite mesophyll conductance (Table 1.3; von Caemmerer, 2000). An ATP/NADPH demand of 1.55 results in a shortfall of ATP of 0.26 per NADPH, or $0.61H^+$ per electron passing along the electron transport chain. The lowest ATP/NADPH demand is when there is only carboxylation and the ATP/NADPH demand is then 1.5, which still results in an ATP shortfall of 0.21 per NADPH. Clearly photosynthesis happens, so the question is where do the extra ATP, and thus the extra lumenal H^+, come from? The answer can lie both with that LET activity that is not generating NADPH used by the CBB + PCO (e.g. pseudocyclic electron transport) and cyclic electron transport around PSI—these are often collectively referred to as 'alternative electron transport pathways' to distinguish them from LET that produces reductant (i.e. NADPH) that is used by the CBB + PCO. The unifying feature of these alternative electron transport pathways is that they can make extra ATP either without the net formation of reductant in the chloroplast (cyclic, pseudocyclic paths, and the malate valve) or with the formation of reductant in the chloroplast that is coupled to processes that does not require much or any ATP. Overall, therefore, alternative electron transport pathways can provide the ATP missing from 'normal' LET in terms of the needs of CBB +PCO.

If we take the ATP/NADPH of 1.55, with a shortfall of $0.61H^+$ per single electron passing along the LET, the proton pumping, and thus the ATP synthesising, capacity of another 0.2 LET is necessary to balance the NADPH generated by that 1 LET. This extra LET-coupled, ATP synthesising capacity could be accounted for by pseudocyclic flow, the Malate valve, or NO_2^- reduction. In terms of CET, this extra $0.61H^+$ could from 0.3 cycles of a PGR5/PGRL1-type CET, or 0.15 cycles of NDH CET. Some means by which regulation of alternate electron transport could be achieved are also apparent. If the rate of ATP formation is insufficient compared to reductant, the pool of stromal reductant will be become more reduced, and the rate of CET has been shown to increase in response to an increase in stromal reduction, as will the rates of the Mehler reaction and the activity of the malate valve. While it is relatively easy to account for the possible options by which an ATP shortfall for the CBB + PCO could be met, it is much harder to describe what actually happens in vivo because it impossible or difficult to measure alternate electron transport activities with the resolution required. One investigation on arabidopsis revealed that that at low light CET did not respond to changes in demand for ATP/reductant arising from changes in the form of nitrogen supply and the amount of photorespiration, implying that changes in the Mehler reaction or malate valve were adjusting the supply of ATP relative to NADPH, but at high light CET became more important as a tool for adjusting the supply of ATP versus reductant (Walker et al., 2014). Accounting for the tuning of the ratio of supply of ATP and reductant remains a challenge and work in progress.

1.7 Assimilation and its response to irradiance in leaves

Photosynthesis is a flexible system, which responds to increasing irradiance by increasing the rates of CO_2 fixation (along with oxygenation of RuBP), but only up to a limit. This is shown in **light-response curves** that show, on the x-axis, irradiance, and on the y-axis, either gross or net CO_2 fixation (Fig. 9). When calculated as **net CO_2 fixation**, no correction is made for leaf respiration so at low irradiances CO_2 fixation will be negative because the rate of respiratory CO_2 release exceeds that of photosynthetic uptake. The irradiance at which rate of net photosynthetic CO_2 uptake is zero because the rate of photosynthesis equals the rate of respiration is known as the **light compensation point**. If the rate of CO_2 fixation is adjusted (i.e. increased) to remove the effect of respiration it is called the **gross photosynthetic rate**; this will be zero in the dark when there is no photosynthesis and be positive when there is any CO_2 fixation.

A light response curve (Fig. 9, in this case, a net assimilation rate curve) can be divided into at least three phases that represent different limitations to photosynthesis. Region A is the strictly light-limited part of the response curve. This occurs at low irradiances and in this phase assimilation responds linearly to increasing irradiance and the light-use efficiency of photosynthesis (i.e. the quantum yield of photosynthesis) is maximal. Photosynthesis is then limited by the capacity for photosynthetic light-absorption and the limiting (i.e. maximum possible) quantum efficiency with which absorbed photons are transformed into proton and electron transport, and how the ATP and NADPH produced by this electron and proton transport is translated to CO_2 fixation. As irradiance is increased further, the light-limited linear phase is succeeded by the curvilinear region (B), the gradient of which decreases progressively with increasing irradiance. In this region the light-use efficiency of assimilation, also referred to as the quantum yield, decreases with increasing irradiance. This **quantum yield (Φ_{CO2})**, or efficiency, is normally calculated as the (gross rate of photosynthesis)/(irradiance producing that gross rate)—the irradiance is either the incident irradiance (most common) or the absorbed irradiance. The Φ_{CO2}

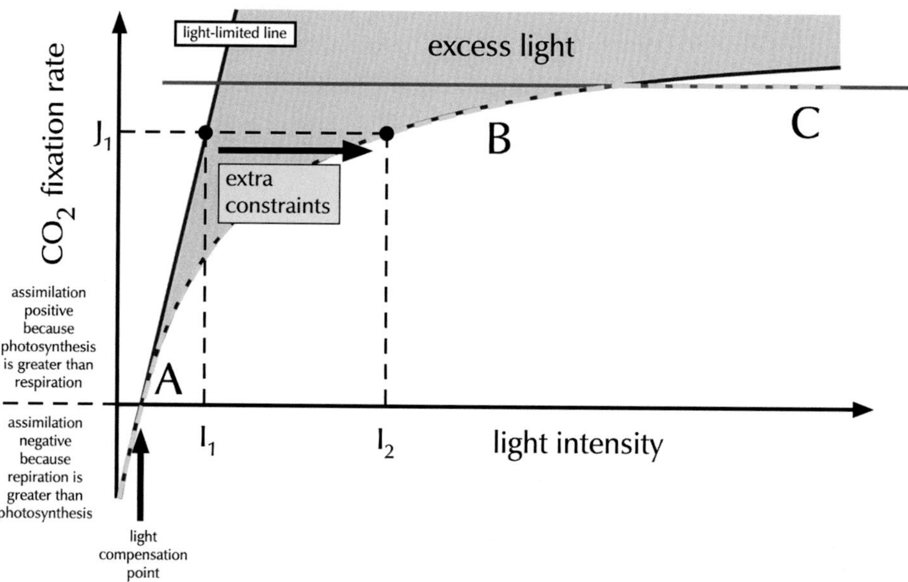

FIG. 9 A summary of a photosynthetic light-response curve, in this case a net photosynthesis response curve. A shows the phase where photosynthesis is strictly light-limited, B the phase of increasing light-saturation where additional constraints act to reduce the quantum efficiency of photosynthesis below the maximum quantum efficiency achieved in phase A (and its continuation the 'light-limited line') and C a phase that may develop in some leaves where assimilation transitions to being limited by a process that causes an increase in the kinetic limitation on electron transport (e.g. limitation by rubisco or triose phosphate utlilisation).

measured using incident irradiance will be lower than that measured using absorbed irradiance. With increasing irradiance assimilation will eventually become light-saturated and insensitive to further increases in irradiance—this is the **light-saturated rate** or the **maximum rate of assimilation (Amax, Pmax**, etc.).

If not light-saturated, assimilation in region B can still respond to an increase in irradiance, though the increase per unit increase in irradiance is lower than in region A. Photosynthesis in region B therefore faces constraints in addition to those experienced in region A. Further, the constraints limiting the response of assimilation to increasing irradiance beyond the light-limiting phase of the response curve may change as the assimilation rate increases (Fig. 9). It is possible that assimilation could enter a third phase (Fig. 9, region C) where it experiences further limitation that places a ceiling on assimilation.

1.7.1 Assimilation in the light-limited part of the irradiance response curve

Variation in the light-limited slope, or light-limited quantum efficiency, of the assimilation response curve is dominated by three main factors.

1. The absorption of photosynthetically wavelengths of light

This is dealt with in detail in the section 'Light absorption, photosynthetic pigments and their organisation' above. In the case of hairy or scaly leaves or leaves with a low photosynthetic pigment content, low light absorption can have a significant effect on the light-limited quantum efficiency of assimilation if this is quantified on an incident rather than on an absorbed light basis. Adaptations to reduce light-absorption are found most commonly in plants adapted to high light, high stress environments (e.g. desert plants, epiphytes).

2. The efficiency of use of photosynthetically absorbed wavelengths for electron transport

This will depend primarily on the quantum efficiency for charge separation (or electron transport) in open reaction centres (i.e., those in a state where charge separation can take place). For PSII, this is probably about 0.9 (about 0.82 is commonly measured maximum obtained using chlorophyll fluorescence but this is an underestimate) and for PSI probably about 0.99. These yields this will affected by the accumulation of photodamage, which can occur to either PSI or PSII but is more commonly encountered in PSII, or to slowly reversible down-regulation of PSII, which is often mistaken for PSII photodamage (see Chapter 4). Accumulated photodamage (or downregulation) can reduce substantially (a reduction of 80% or more can occur) the quantum efficiency of PSII electron transport, so this is often a significant effect. In linear electron transport what goes through PSI has to first go through PSII so any damage to PSII (or PSI) will have an identical effect on linear electron transport and the light use efficiency of assimilation. This loss of quantum efficiency due to photodamage or slowly reversible downregulation is a major result of plant stress, due, for example, to cold, drought or high light. In addition to photodamage, the quantum efficiency of linear electron transport, and thus assimilation, will be diminished by an imbalance in the potential rates of electron transport by the two photosystems (note that some PSI will participate in cyclic electron transport). If potential electron transport by one photosystem exceeds that of the other the efficiency of the system excited in excess will be reduced until the rates of electron transport through both photosystems are equal. This effect, which is known only from controlled laboratory measurements but can be expected in the field, is normally small, typically no more than 10%–15% under extreme conditions, and state transitions act to at least partly rebalance these imbalances (see Chapter 1).

3. The efficiency with which electron (and proton) transport is translated to assimilation

This is strongly dominated by the competition between oxygenation and carboxylation of RuBP. Critical to this are the relative concentrations of CO_2 and O_2 at the site of carboxylation in the chloroplast, the kinetic properties of rubisco and temperature, and are described in more detail in the section 'Oxygenation of RuBP by rubisco and photorespiration'. The effect of oxygenation of RuBP and the subsequent PCO cycle on assimilation is substantial; on an incident light basis in the absence of oxygenation (2% O_2), the light-limited quantum yield of CO_2 fixation is about 0.08, while in an atmosphere of 21% O_2 that yield varies from about 0.07 at about 12°C to about 0.04 at 40°C. It is because of this major effect of oxygenation on the light-use efficiency of assimilation that a typical modern greenhouse used for tomato production will normally use a CO_2 mole fraction of about 1000 ppm—this largely suppresses the oxygenation reaction.

1.7.2 What is the maximum efficiency for the quantum efficiency of assimilation on an absorbed light basis?

The maximum theoretical quantum efficiency of assimilation on an absorbed light-basis is often reported as 0.125, and this is based on a very idealised model of 2 NADPH required per CO_2 fixed (see above), so 4 e^-, which requires 8 photons, 4 by PSII and 4 by PSI, and with both photosystems operating with a quantum efficiency of 1. The less than ideal efficiency of PSII for electron transport (maximum quantum yield ~0.9) would reduce that efficiency from 0.125 to 0.119 (a factor of 0.95). There are few measurements of the light-use efficiency for CO_2 fixation on an absorbed light basis, but at a wavelength that optimally balances the excitation of PSI and PSII, efficiencies of about 0.093 have been reported (in 2% O_2, 400 ppm CO_2 so practically no oxygenation of RuBP). This is considerably less than the efficiency of 0.119 obtained after correcting for the low maximum photochemical yield of PSII. The value of 0.119, like that of 0.125, is however based on the assumption that the quantum yield of CO_2 fixation is limited by NADPH formation. Assuming NADPH limitation means no account is taken of the shortfall in ATP formation compared to NADPH formation by LET (see above) compared to the demands CBB (or the CBB + PCO (see above)). The activity of the alternative electron transport pathways needed to make this ATP will result in a loss of the light-use efficiency for assimilation. This can be illustrated using a scenario in which a leaf is photosynthesising in a 2% O_2 atmosphere with 400 ppm CO_2, so no photorespiration and the same conditions under which the yield of 0.093 was measured and in which pseudocyclic electron transport provides the extra ATP for the CBB. Under these conditions, the ATP/NADPH demand for the CBB is 1.5, but the ATP/NADPH production ratio of the LET is only 1.28; a shortfall of 0.22 ATP per 2 electrons moving through the LET chain, or 0.11 ATP per single electron. To meet the demands of the CBB, we would need the NADPH produce by 4 LET and the ATP produced by 4.68 LET so 0.68 LET would need to operate pseudocyclic electron transport and produce only ATP. Combining this scenario with a maximum average photochemical efficiency of 0.95 would result in a maximum quantum yield for assimilation of 0.102. Using purely cyclic electron flow to produce the extra ATP would result in a higher yield. This scenario illustrates the need to consider the energetic realities of running the CBB or CBB + PCO when estimating maximum light-use efficiencies and the 0.125 yield so widely encountered in text books and lectures is in practice not realistic. Fully accounting for the observed quantum efficiencies of light-limited assimilation remains, however, 'work in progress'.

1.7.3 The loss of light-use efficiency as irradiance increases beyond the light-limited region; Pmax and excess light

The decrease in light-use efficiency of assimilation as irradiance increases above the light-limited region results in ultimately in the saturation of assimilation and Pmax. Overall there is great variation in Pmax and therefore the sweep of the light response curve (Fig. 10), with some C3 desert winter annuals achieving net assimilation rates of over 60 $\mu mol\,m^{-2}\,s^{-1}$ (and this was not their Pmax) at an irradiance of 1800–2000 $\mu mol\,m^{-2}\,s^{-1}$, while some shade adapted herbs are light-saturated at an irradiance of only 50 $\mu mol\,m^{-2}\,s^{-1}$, which is less than the light compensation point of a typical crop plant, with a gross Pmax of about 1.0 $\mu mol\,m^{-2}\,s^{-1}$.

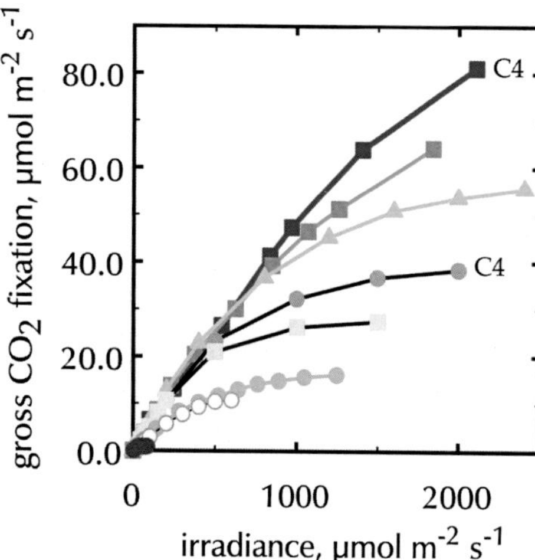

FIG. 10 A range of light response curves of gross assimilation rates from C3 plants with different photosynthetic capacities and two examples of response curves from C4 plants. The C3 plants are: ■ *Camissonia brevipes* (data from Mooney et al., 1976), ▲ *Hirschfeldia incana*, ■ *Oryza sativa*, ● *Pisum sativum*, ○ *Arabidopsis thaliana* grown at 200 molm-2s-1 (a typical growth irradiance for this species under controlled environment conditions) and ● *Begonia luzonensis.* The C4 species included for comparison are ■ *Amaranthus palmeri* (the highest published assimilation rate) and ● *Zea mays*.

The light response curve is a compromise between on the one hand the costs of building a photosynthetic system that can achieve high assimilation rates, most prominently the costs of acquiring the nitrogen needed for rubisco, and the costs in terms of transpired water of acquiring CO_2, and on the other hand, the benefit of the extra assimilation that arises from each increment of photosynthetic capacity. To realise those benefits depending on the availability of irradiance, the source of the energy upon which photosynthesis depends, and the presence of a high metabolic, electron transport and CO_2 diffusive transport capacities, which allow high photosynthetic light-use efficiencies at high irradiances. The optimum compromise between the costs and benefits of photosynthesis varies from habitat to habitat, from niche to niche, and seasonally within a habitat. In particular, the poor catalytic properties of rubisco mean that large quantities of this enzyme are needed to support high rates of photosynthesis, which translates to a need for large amounts of mineral nitrogen. This is the reason why agriculture demands so much nitrogenous fertiliser and why the Haber-Bosch is so important, consuming 1%–2% of the World economy's energy supply. In fact, within a species Pmax, is proportional to leaf rubisco content. In nature, the availability of mineral nitrogen (typically as NO_3^- or NH_4^+) is limiting in many habitats resulting in low Pmax for plants adapted to them. Many climax tree species, for example, growing in an environment where nutrients like mineral nitrogen are scarce, have a relatively low Pmax, in the order of only 10 μmol m^{-2} s^{-1} and will be light saturated by an irradiance of 1000 μmol m^{-2} s^{-1} or less despite having leaves present above the canopy in full sunlight where irradiances will reach 2000 μmol m^{-2} s^{-1}. On the other hand, plants that grow in disturbed sites (in the ecological sense of disturbed), with higher soil nutrients and a high irradiance, will have higher assimilation rates—normally over 20 μmol m^{-2} s^{-1}: many of our crop plants are of this plant type. So while Pmax in C3 plants can be high this performance is only rarely encountered because the cost–benefit outcome of assimilation usually favours a much lower assimilation rate because of other limiting factors, such as a nutrients, water, or light. Understanding photosynthesis in the agriculture or nature is all about disentangling limiting factors and making sense of photosynthetic shortcomings. This also implies that while considering the limitations acting on photosynthesis it is important to remember that the short-term physiological limitations measured experimentally are superimposed upon a substrate of longer term evolutionarily determined optimisations that have shaped the properties and limitations of photosynthesis.

The limitations acting on photosynthesis give rise to the physiological constraints that result in the loss of light-use efficiency as irradiance is increased above the light-limited part of the irradiance response curve (phase B Fig. 9). The impact of the loss of light-use efficiency can be seen in this example by comparing the irradiance needed to achieve the assimilation rate J1 (Fig. 9); in the case of the light-limited line, this is achieved by irradiance I1, while for curve ABC, it requires an irradiance of I2. This loss of light-use efficiency results in the phenomenon of excess irradiance; light which is excess of the needs of assimilation in curve B compared to curve A. Note that this excess light is an absorbed flux of quanta which will produce excited singlet states of chlorophyll *a*—the driving force for reaction centre charge separation and photosynthetic electron transport. If these excited singlet states of chlorophyll *a* are not successfully dissipated by photochemistry, there is the risk that they will result in damaging side reactions connected to O_2 within the photosystems or the photosynthetic pigment beds (see Section 1.9 and Chapter 5). Understanding the regulation of photosynthesis in the excess light conditions that characterise region B of the light-response curve is closely entwined with moderating the damage cause by these side reactions. The chemistry of O_2 is pivotal in this.

1.7.4 *The light-use efficiency of photosystems I and II declines with increasing irradiance*

The decreasing light-use efficiency of assimilation that results from increasing irradiance is correlated with a loss of light-use efficiency for electron transport by photosystem II (Fig. 11A), which in turn is correlated with the loss of light-use efficiency for electron transport by photosystem I (Fig. 11B). The loss of light-use efficiency is therefore largely a system-wide loss rather than a diversion of NADPH, or reduced ferredoxin, and ATP to other fates. This also implies that excess light is not only in excess of the needs of assimilation but also of electron transport, and that a consequence of the loss of photochemical efficiency is that some excited singlet chlorophylls are not going to be dissipated photochemically. That the efficiency of PSI declines in parallel with that of PSII also implies that linear electron transport is the dominant electron transport pathway in C3 plant thylakoids.

The loss of PSI efficiency for electron transport

Most commonly the loss of PSI efficiency *in folio* is due to the formation of oxidised P700 in the reaction centre of PSI—in fact, measurement of the amount of $P700^+$ is the means by which PSI efficiency is normally measured. The $P700^+$ that accumulates under conditions of donor side limitation is relatively long-lived with a half-life of about 3 ms or more and its presence in vivo implies two important things. First, the paramount limitation acting on photosynthetic electron transport acts before PSI (on its donor side), rather than after it (on its acceptor side); if the limitation was on the acceptor side, a loss of PSI efficiency would be due to the absence of electron acceptors and P700 would not become oxidised in the way that it does. Second, $P700^+$ is as good a quencher of excited chlorophyll singlet states as is P700, though quenching by $P700^+$ does not result in photochemistry. The effectiveness of quenching by $P700^+$ can be shown by direct measurements of the excited state lifetime (the fluorescence lifetime) of PSI (see Chapter 1). The

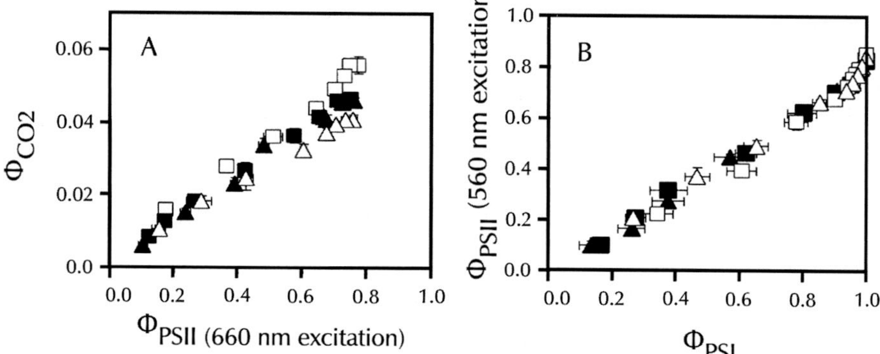

FIG. 11 The parallel decreases with increasing irradiance in leaves of *Lycopersicum esculentum* of (A) the light-use efficiencies of carbon dioxide fixation (Φ_{CO2}) and PSII electron transport (Φ_{PSII}), and (B) the light-use efficiencies for PSII electron transport and PSI electron transport (Φ_{PSI}). In this case the light-use efficiencies for PSII electron transport were measured in slightly different ways in figures (A) and (B). The method used to measure the light-use efficiency of PSI is based on a light-induced absorbance change measurement that looks through the leaf while the actinic light that drives photosynthesis was from red (660 nm) LEDs, a wavelength that is strongly absorbed by the leaf. To match the measurement profile in the leaf for the light-use efficiency of carbon dioxide fixation (A) chlorophyll fluorescence was excited by a 660 nm light, and to the match the measurement profile for the light-induced absorbance-change (B) chlorophyll fluorescence was excited with a 560 nm, more penetrating excitation. The leaves of *Lycopersicum esculentum* were from plants grown with normal nutrition and low nitrogen, and high and low irradiances. See de Groot et al. (2003) for more details.

quenching of excited chlorophyll singlet states in PSI is therefore not affected by the loss of PSI quantum efficiency for electron transport when that occurs via the formation of P700$^+$; the problem of excess light in PSI is therefore avoided even though the quantum efficiency of PSI is decreased.

There are exceptions to this rule that the predominant loss of PSI efficiency is due P700$^+$-formation following a donor side limitation. During the early stages of photosynthetic induction (the start-up of photosynthesis in dark-adapted leaf), the formation of P700$^+$ is conspicuously limited by acceptor side limitation (Harbinson and Hedley, 1993; Klughammer and Schreiber, 1994), and rapid changes in irradiance, for example, may lead to an increase in the limitation of PSI electron on the PSI acceptor side.

PSII: The loss of efficiency arises from QA reduction and the qE component of NPQ

The loss of PSII quantum efficiency occurs via a different mechanism and is rather more complex. While electron transport through PSI is predominantly limited on the donor side of PSI, that of PSII is predominantly limited on its acceptor side, with the loss of efficiency being due to the accumulation of reduced Q_A. As a result the accumulation of oxidised PSII reaction centres does not occur in vivo as it does for PSI. The loss of the quantum efficiency of electron transport by PSII due to Q_A^- accumulation is quantified by means of the chlorophyll fluorescence-derived parameter q_P. The value of q_P is non-linearly related to the fraction of PSII with a reduced Q_A but it is a linear measure of the impact of this reduction on PSII efficiency. It might sound strange to propose that something is non-linearly related to the amount of something but linearly related to the effect it has, but this can be explained by the interconnectivity of PSII and the fact that PSII with Q_A^- is not a quencher of excited states of chlorophyll. When an excited state arrives at a closed PSII trap it therefore is not quenched but can migrate to another PSII trap, so when the relative number of PSII with Q_A^- is small, the impact of PSII efficiency is small and only becomes significant when the number of closed traps increases. The effect of trap closure on PSII efficiency is therefore not directly due the closure of traps but to the increased time it takes for the excited state to find an open trap. This increase in time increases the likelihood that the excited state will be quenched by fluorescence or other non-photochemical quenching processes active in the PSII pigment bed.

With increasing irradiance and decreasing PSII efficiency the value of q_P decreases (Fig. 12), but not by enough to account for the fall in PSII efficiency. To fully account for this loss of PSII efficiency, it is necessary to also include the effect on PSII efficiency of another mechanism that dissipates or quenches excited states in the PSII pigment bed: the q_E component of NPQ (see Chapter 4). The effect of q_E on the light-use efficiency of PSII is quantified another chlorophyll fluorescence-derived: Fv′/Fm′,

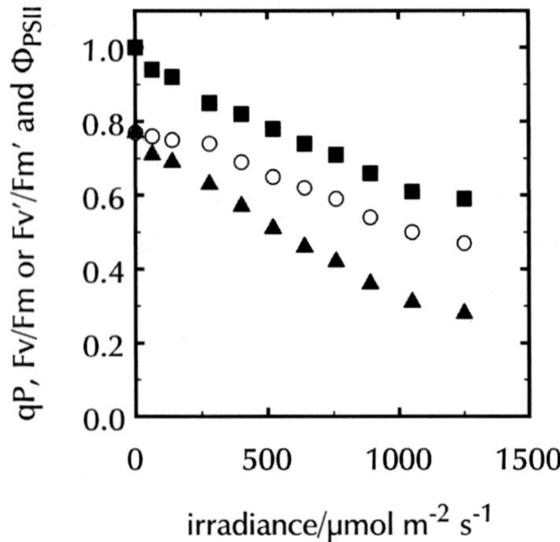

FIG. 12 An example of the irradiance dependency of the light-use efficiency for PSII electron transport (ΦPSII; ▲), the dark-adapted Fv/Fm or Fv′/Fm′, which same parameter as Fv/Fm only measured in the light or while not in a dark-adapted state (○) and qP (▪). These data were obtained from a leaf of *Juanulloa aurantiacum*.

which is a relative measure of the quantum efficiency for electron transport by open PSII reaction centres (i.e. with Q_A oxidised) in the light. Fv′/Fm′ is entirely comparable to the Fv/Fm parameter, with the exception that Fv/Fm is measured leaves that have been dark-adapted for 15–30 min to allow the relaxation of q_E. Fv/Fm is therefore the maximum relative quantum yield for PSII electron transport in a leaf in the absence of all q_E that can relax in 15–30 min. There is also some slowly reversible q_E, which can be mistaken for PSII photodamage, which—as the name suggests—relaxes more slowly (Chapter 4). Essentially, the q_E mechanism creates sites in the PSII pigment bed that compete with the PSII reaction centres in the dissipation or quenching of excited singlet states of chlorophyll (Chapter 4). When the excited state is quenched by the reaction centre, it produces photochemistry, which leads to electron transport, while if it is dissipated by a q_E centre, it is converted entirely to heat. The end result is that the more q_E is switched on the lower becomes Fv′/Fm′. The total quantum efficiency of PSII electron transport is the product of q_P and Fv′/Fm′ and as irradiance increases both q_P and Fv′/Fm′ decrease so the loss of PSII light-use efficiency is due to both the reduction of the PSII acceptor side, in particular the reduction of Q_A as this is related to the loss of q_P, and the increase in q_E, which results in a decrease in Fv′/Fm′ (Fig. 12).

The limitation of electron transport and the importance of lumen pH

In an illuminated thylakoid, the PSII acceptor side gets reduced and the PSI donor side gets oxidised and the crossover point of this reduction and oxidation occurs at the point where PQH_2 reduces the Rieske FeS complex of the cytochrome b_6/f complex, the first step in the oxidation of PQH_2 (Chapter 2). This oxidation is the rate limiting step of the intersystem photosynthetic electron transport chain and it is a pH sensitive reaction, slowing as the thylakoid lumen acidifies below pH6.5; in plants the pH optimum of this reaction is from 6.5 to 8.0. Lumen pH, part of the transthylakoid proton motive force (pmf) that drives ATP synthesis, therefore plays a critical role in regulating the electron transport that creates the pmf. If the pmf grows, lumen pH falls and electron transport slows. The limitation of electron transport at $PQH_2/cytb_6/f$ step results in the reduction of Q_A and thus the decrease in q_P, and the accumulation of $P700^+$. Decreases in lumen pH are also important in the activation of q_E and thus the decrease of Fv′/Fm′ via the protonation of PsbS and the activation of violaxanthin de-epoxidase, with latter resulting the in the formation of zeaxanthin whose presence catalyses the effect of PsbS on q_E (see Chapter 4). Lumen pH is therefore a pivotal component in the operation and regulation of photosynthesis at the level of the thylakoid influencing as it does electron transport, ATP synthesis, and the efficiency of light-harvesting in PSII. Electron transport beyond PSI reduces ferredoxin and then thioredoxin so activating photosynthetic metabolism and thus the demand for ATP and NADPH, and this will be dependent on the efficiency of PSI and thus the accumulation of $P700^+$. Reduction of stromal electron acceptors will also activate some alternative electron transport pathways, so increasing proton accumulation in the lumen, and decreasing lumen pH and increasing pmf.

It is relatively easy to measure the limitation of electron transport imposed by pH at the level of the cytochrome b_6/f complex by measuring the kinetics of $P700^+$ reduction. Measurements of this kind show that for most leaves under non-stressed conditions the limitation on electron transport does not change with increasing irradiance (Fig. 13). This implies that assimilation in these cases is either electron transport limited or that the balance between limitation and activation is very fine, resulting no measurable change in the limitation of electron transport with increasing irradiance. If the irradiance response curve was to become significantly limited by some process outside electron transport, such as by rubisco or some other part of

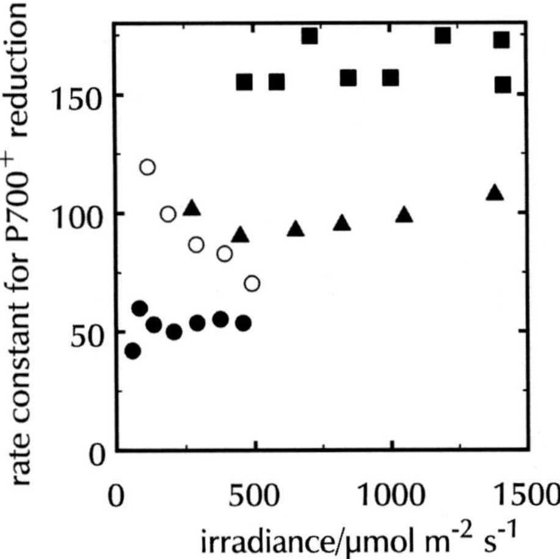

FIG. 13 The irradiance dependency of the first-order rate constant for P-700+ reduction, a measure of the kinetic limitation imposed electron transport. In most cases for unstressed healthy leaves this rate constant varies only little with changing irradiance. Data are shown measured on *Pisum sativum* (▪), *Lycopersicum esculentum* (▲), *Juanulloa aurantiacum* (●; a tropical woody epiphyte), and *Stephanotis floribunda* (○; a tropical woody climber). *S. floribunda* is unusual in showing a large change in the rate constant with increasing irradiance. In all cases the highest irradiances used would have resulted in saturation or near-saturation of assimilation irradiance response.

the CBB, or by triose phosphate utilisation (shown by phase C in Fig. 9), then electron transport would be need to downregulated and this would be detectable via a decrease in the kinetics of $P700^+$ reduction. While most leaves show no measurable slowing in the kinetics of $P700^+$ reduction with increasing irradiance there are exceptions to this and such an exception (that of the tropical climber *Stephanotis floribunda*) is also shown in Fig. 13. In addition to being able to measure the kinetics of electron transport from PQH_2 via the cytochrome b_6/f to $P700^+$ it is possible to measure the kinetics of proton efflux via the ATPase that is coupled to ATP formation. Those kinetics are also often independent of irradiance in normally functioning, unstressed leaves, paralleling the results obtained from measurements of electron transport.

1.8 Can metabolism control electron transport?

To answer to that is a definite 'yes!'. It is easy to show metabolism acting to slow down electron transport and this seems to occurs by slowing electron transport at the level of the cytochrome b_6/f complex, so via decreasing lumen pH. This can be shown when assimilation is slowed by decreasing the CO_2 mole fraction in the air around a leaf under non-photorespiratory conditions, which allows the metabolic demand for ATP and NADPH to be simply controlled by the CO_2 supply. Under these conditions and with a constant light-intensity, the light-use efficiencies for electron transport by photosystems I and II decrease along with the light-use efficiency of CO_2 fixation and in parallel with decreases in the kinetics of the reduction of $P700^+$ (Fig. 13). In many ways, this is a similar story to that obtained with increasing irradiance; the light-use efficiencies of the photosystems decline, but in this case, irradiance is constant and the loss of the light-use efficiencies of the photosystems occurs because of an increased restriction on electron transport—in the case of increasing irradiance, the limitation on electron transport is usually constant, but the light-use efficiencies of the photosystems declines as the irradiance increases.

1.9 Why are electron transport and q_E controlled; the significance of reactive oxygen species

The most widely accepted reason for controlling electron transport is to reduce of the risk of damage from reactive oxygen species (ROS). q_E is believed to be required for similar reasons. The chemistry of molecular oxygen (O_2, dioxygen) results in the photosynthetic machinery, in particular the reaction centres and the photosynthetic electron transport chain, forming reactive forms of oxygen or reactive oxygen species (ROS). These ROS can damage the cell in general and the photosynthetic machinery in particular. There are two forms of ROS generated by the photosynthetic machinery: superoxide (O_2^-) and singlet oxygen ($^1O_2$).

1.9.1 *Superoxide and electron transport*

Superoxide (O_2^-) is formed by the reduction of O_2 and oxygen is a very good single electron acceptor with an $E^{o'}_{m7}$ of −330 mV resulting in an E_m − 373 mV at 21% O_2 and 1 mmol dm^{-3} O_2^- (Fig. 14); the choice of 1 mmol dm^3 for the concentration of superoxide was based on a reasoned guess. The acceptor side of PSI extending to the ferredoxin (Fig. 14) and many of the stromal enzymes, like nitrite reductase, that have reducing cofactors, can reduce O_2 to superoxide (Robinson, 1988; Rutherford et al., 2012). When this reduction occurs in the PSI reaction centre, it is associated with photodamage to PSI. If it occurs in the stroma, where ferredoxin is believed to be the main reducing factor, it gives rise to pseudocyclic electron transport, which is useful but some superoxide can escape from the pseudocyclic path posing a hazard to chloroplasts, so superoxide formation needs to be controlled. The uncontrolled over-reduction of the PSI acceptor side leading to damage to PSI and the excessive formation of ROS in the stroma are therefore believed to be a major reasons why electron transport is limited on the donor side of PSI.

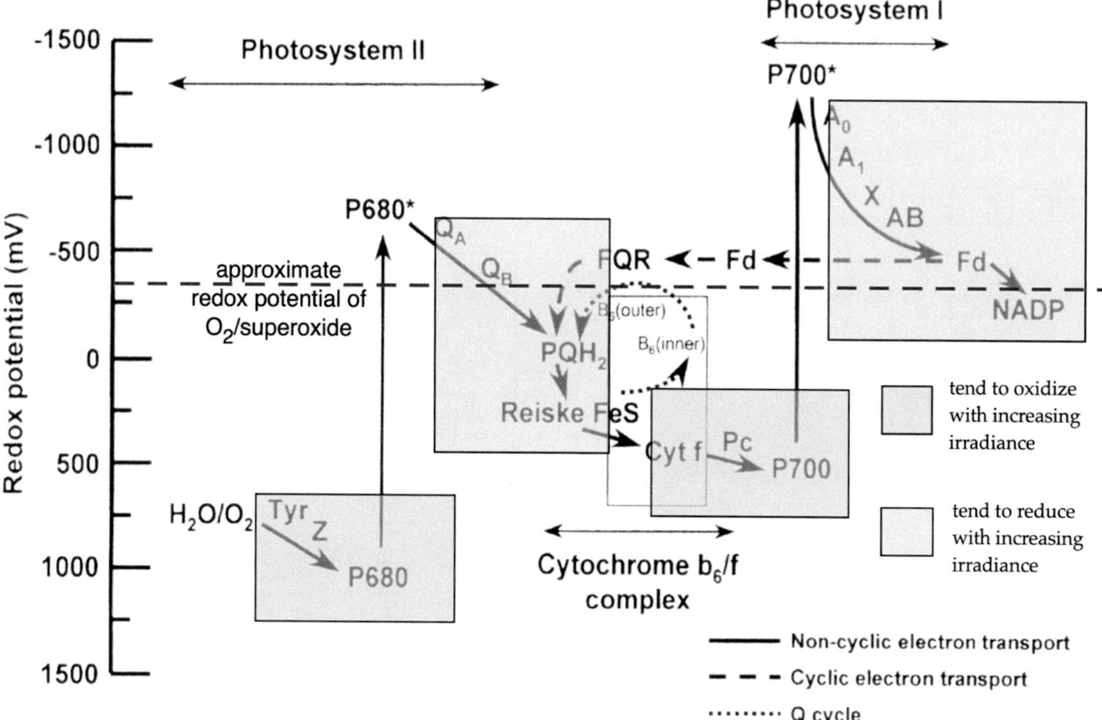

FIG. 14 The Z-scheme for photosynthetic electron transport showing those parts that tend to get reduced or oxidised with increasing irradiance. The approximate redox potential of superoxide is shown, calculated assuming a concentration of about 1 mmol dm^{-3}. For reasons of clarity cyclic electron transport is represented by one generalised path—that of ferredoxin quinone reductase (FQR). *Redox potentials in the Q cycle are not shown accurately.*

This however still leaves PSII at risk, in this case from the other major ROS—singlet oxygen.

1.9.2 *Singlet oxygen and PSII*

O_2 is an unusual molecule because it has a triplet ground state (i.e. it is actually $^3O_2$) with the two highest energy electrons having parallel spins and each singly occupying two π antibonding orbitals (Fig. 15). In contrast, nearly every other molecule has a singlet ground state; in the singlet state, a molecule with an even number of electrons the number of electrons with parallel and antiparallel spins is the same. In the case of singlet ground state molecules, a triplet state can form but it will have a higher energy than the singlet ground state (this is the case with chlorophyll). The triplet state in this case is therefore an excited state. In the case of O_2 two different singlet states can be formed, which according to spectroscopic notation are referred as $^1\Delta_g$ and $^1\Sigma_g^+$—these differ in the whether the electrons are paired in one orbital or occupy two different orbitals (Fig. 15). The $^1\Delta_g$ form has an energy $94\,kJ\,mol^{-1}$ higher than the $^3O_2$ ground state while $^1\Sigma_g^+$ has an energy of $157\,kJ\,mol^{-1}$ above the ground state. If a singlet molecule reacts with $^3O_2$, the triplet state is conserved resulting in a relatively energetic triplet product. As a result, $^3O_2$ is relatively unreactive, but when transformed to a singlet state this constraint is removed and O_2 becomes more reactive. Only the $^1\Delta_g$ singlet state of O_2 ($^1O_2$) can

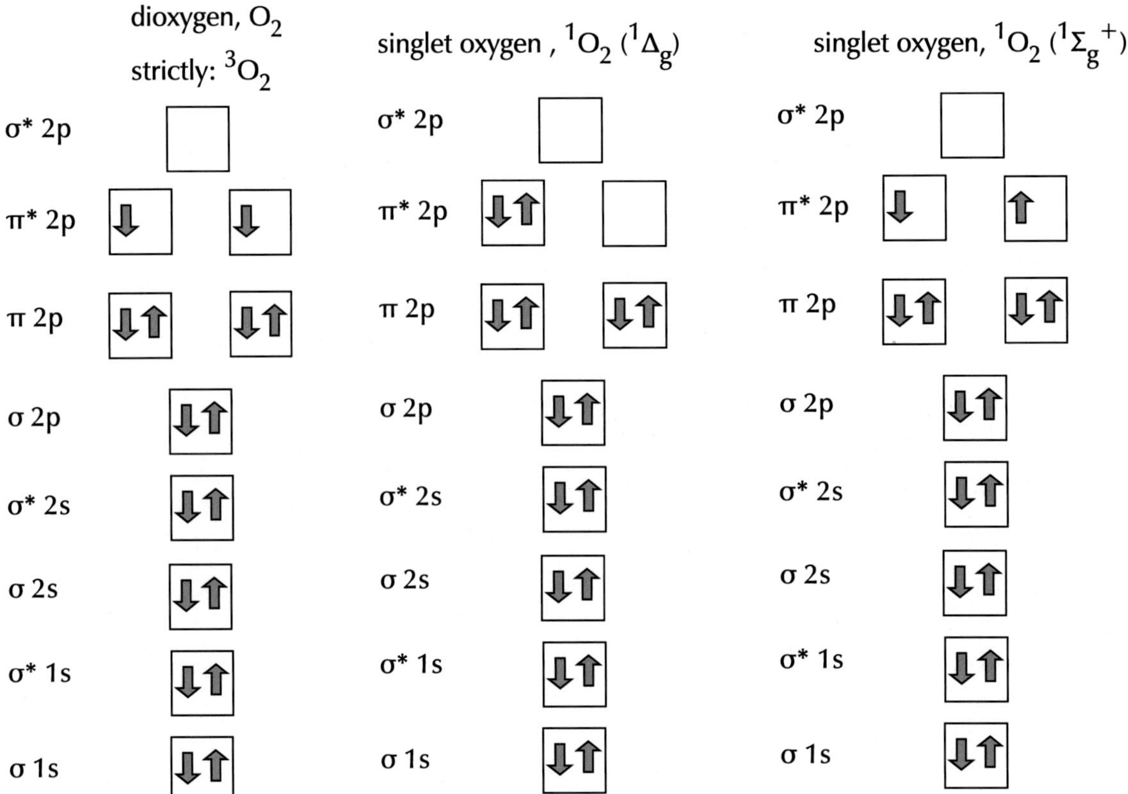

FIG. 15 The electronic configuration of dioxygen (O_2) in the triplet ground state, the lowest energy singlet state (1Δg) and the higher energy singlet state (1Σ+g).

form in photosynthesis so '$^1O_2$' will from now be used as a synonym for the $^1\Delta_g$ singlet state.

$^1O_2$ can be formed when $^3O_2$ physically reacts with another triplet molecule with the result that both molecules becomes singlet. As the lowest singlet state of O_2 has an ΔG 94 kJ mol^{-1} higher than the triplet state this conversion can only occur if the other triplet state molecule has a $\Delta G > 94$ kJ mol^{-1} higher than its singlet ground state. In the case of chl a, the lowest triplet state has a ΔG of 126 kJ mol^{-1} so ^{3}chla* can convert $^3O_2$ to the more reactive $^1O_2$. The problem for photosynthesis is that ^{3}chla* is readily formed from ^{1}chla* either by relaxation of ^{1}chla* or as a result of back-reactions in reaction centres (Rutherford et al., 2012). In vitro only about one third of ^{1}chla* relax to the ground state via fluorescence decay while about two thirds of ^{1}chla* decay to form ^{3}chla*—formation of triplet chlorophyll is twice as likely as fluorescence and chlorophyll is known as a fluorescent molecule. In vivo the pigment beds of PSI and PSII have carotenoids adjacent to nearly all chlorophylls—these carotenoids can accept a triplet state from a ^{3}chla* to form a triplet carotenoid which decays promptly and harmlessly to the singlet ground state. This is the so-called 'carotenoid valve', and it provides good overall protection against ^{3}chla* formed in the pigment beds of PSI and PSII. The lifetime of ^{1}chla* in PSI is short and almost independent of the state of the PSI reaction centre so the formation of ^{3}chla* in PSI is not thought to be major risk. PSII, on the other hand, has a longer ^{1}chla* lifetime (and has more fluorescence as a consequence) and ^{1}chla* lifetime can increase substantially upon closure (Q_A reduction) of the PSII reaction centre. PSII is therefore thought to be at greater risk of ^{3}chla* formation than PSI and the carotenoid valve is not 100% effective so the risk of $^1O_2$ formation is not eliminated. The thylakoid membranes are also at particular risk of damage from $^1O_2$ because they contain lipids with highly unsaturated fatty acids. These are vulnerable to a free-radical chain reactions initiated by the attack by the $^1O_2$ on a double bond of the fatty acids. Protection against these free-radical chain reactions is achieved by various antioxidants, such as tocopherol, but some lipid oxidation still occurs (Triantaphylidès et al., 2008).

In addition to forming in the pigment bed, ^{3}chla* can also form in the reaction centres of PSII and PSI as a result of back-reactions or, in the case of PSII the partial relaxation of the excited radical pair formed between ^{1}chla* and phaeophytin in the PSII reaction centre (see Chapters 1 and 2). The formation of ^{3}chl* from the radical pair is enhanced when Q_A is reduced, which will occur routinely in response to increasing irradiance. The problem for the PSII reaction centre is the lack of any protective carotenoids that would normally quench any ^{3}chl* formed; the oxidised primary donor in PSII formed by photochemistry would oxidise any adjacent carotenoids, destroying them, so the reaction centre of PSII lacks the protection of the carotenoid valve. Unlike the ^{1}chl* state, which is mobile and can migrate out of the closed reaction centre, the ^{3}chl* state can only move via the Dexter mechanism which occurs only between molecules that are in van der Waal's contact and so the ^{3}chl* state is relatively immobile. It is believed that $^1O_2$ formed by the interaction between $^3O_2$ and ^{3}chl* is the cause of the most of the damage that occurs in PSII in the light. $^1O_2$ formation is therefore a particular risk in PSII because of ^{3}chl* formation in the pigment bed and the reaction centre. The solution to the PSII $^1O_2$ problem is believed to be q_E and other quenching mechanisms that work in PSII. q_E activates as irradiance reaches above the light-limited phase when the limitation of the electron transport chain would be expected to result in the accumulation of reduced Q_A. This would result in both more ^{3}chl* in the PSII reaction centre and due to an increased fluorescence lifetime more ^{3}chl* formation in the PSII pigment bed. This would lead to more $^1O_2$ damage either in the reaction centre or the thylakoid membrane more generally. q_E quenches ^{1}chl* in

PSII well enough that it competes with PSII photochemistry and it results in the reduction of Fv/Fm to the lower values of Fv′/Fm′. As a result of q_E the steady state chlorophyll fluorescence yield in leaf is only slightly higher than the dark-adapted Fo level, which is much lower than the yield at Fm—the yield corresponding to complete Q_A reduction in the absence of q_E.

2 Photosynthetic responses to fluctuations in irradiance

2.1 Measurement of photosynthesis transients: The new normal

The natural environment is hardly ever stable, and plants have to cope with this environmental variability. Most notable of all environmental factors, irradiance can change rapidly and at large amplitudes, shifting photosynthesis rate from irradiance-limiting to saturating, and back. Irradiance in the field fluctuates because wind moves leaves and clouds, and because the sun's elevation angle additionally changes continuously. Wind-induced leaf flutter can induce rapid fluctuations in the subsecond range, whereas cloud movement adds slower, more gradual fluctuations in the range of minutes. At the same time, photosynthesis has mainly been studied under stable environmental conditions in the laboratory. This makes sense when characterizing the effects of irradiance, CO_2 concentrations, or temperatures on steady-state photosynthesis rate, but often, the rate of change of photosynthesis after a change in a given environmental factor is ignored. Until recently, irradiance fluctuations were only considered to be important for the carbon budget of heavily shaded plants growing under the forest canopy. However, in the past decade, a number of influential publications have brought about a paradigm shift: the relatively slow response of photosynthesis to changes in irradiance is now being recognized as a limitation to diurnal photosynthesis in its own right, and is under strong genetic control. In many new studies characterizing photosynthesis, data of photosynthesis transients after a shift from, for example, low to high irradiance are added to the more established characterization of steady-state irradiance and CO_2 response curves. Characterization of nonsteady state, or dynamic photosynthesis, is becoming the new normal.

2.2 Regulation of photosynthesis in fluctuating irradiance

As its name implies, photosynthesis is light-driven, and the activity of many of its components respond to irradiance either because of more or less direct mass action effects or by more indirect regulation of their activity. These components include electron and proton transport within and around the thylakoid membrane, Calvin Benson Bassham cycle (CBB) metabolite turnover rates and enzyme activation states, as well as stomatal conductance, and are described in greater detail in this and other chapters. Importantly, these components are activated and deactivated with different time constants following a step change in irradiance, which is why their effect on dynamic photosynthesis depends on the frequency of irradiance fluctuations.

When a leaf that had been shaded for long enough to ensure complete inactivation of photosynthesis (e.g. 60 min) is suddenly exposed to high irradiance, its photosynthesis rate increases near-exponentially, until it reaches a new steady-state; this process of activation of photosynthesis is called photosynthetic induction. A leaf's photosynthetic induction state is a measure of its readiness to increase its photosynthetic rate upon exposure to a higher irradiance. After a long exposure to shade, it takes relatively long (30–120 min) until full photosynthetic induction under high irradiance is attained. If that same leaf was shaded for a short time (e.g. 2 min), photosynthesis rate would increase much more quickly upon subsequent exposure to high irradiance. This is so because deactivation of the system in low irradiance is not instantaneous and after only brief exposure to shade, the leaf's photosynthetic induction state will still be relatively high. Considering this temporal dependency (rate of activation and deactivation of the system under a given set of conditions) is important for modelling the system, when designing experiments for studying dynamic photosynthesis, and when interpreting data derived from such experiments.

2.2.1 *Electron and proton transport regulation*

Redox state and pH changes along the thylakoid membrane rapidly respond to changes in irradiance. Within milliseconds after an increase in irradiance, the increase in photochemistry powers an increase in LEF that powers the CBB and alternative electron sinks. Protons passing from the stroma into the lumen rapidly increase lumen pmf. Changes in the relative strength of alternative electron sinks and CEF change the ratios of NADPH and ATP, while the activity of several ion transporters affects pmf partitioning into $\Delta\Psi$ and ΔpH. These mechanisms are important under conditions where metabolic demands for ATP and NADPH change, such as during irradiance fluctuations. Also, because lumen pH has a regulatory role in slowing linear electron transfer from PSII to PSI via the cytochrome b_6f complex, as well as in qE, rapid adjustments of the pmf through alternative electron flows are important for optimal functioning of the system under rapidly changing conditions. Mutants lacking the protein PROTON GRADIENT REGULATION 5 (PGR5) grow under stable but die under fluctuating irradiance, even though they can perform qE. This lethality in *pgr5* mutants under fluctuating irradiance stems from severe damage to photosystem I, which is believed to occur as a consequence of a low pmf and a lack of pH-dependent regulation of cytochrome b_6f activity.

After a transition from high to low irradiance, qE can take several minutes to relax, and a transiently high NPQ consequently competes with LEF, reducing diurnal photosynthesis under fluctuating irradiance. In *Arabidopsis thaliana* overexpressing K^+ efflux antiporter proteins, the ΔpH component of the pmf was increased such that qE induction and relaxation were much faster than in wildtype plants, which diminished the transient reductions in LEF and CO_2 fixation after drops in irradiance (Armbruster et al., 2014). In tobacco, when several components of NPQ (PsbS, violaxanthin de-epoxidase and zeaxanthin epoxidase) were overexpressed simultaneously, the rate of NPQ relaxation after drops from high to low irradiance was increased substantially and growth in the field was increased by 14%–20% (Kromdijk et al., 2016). However, in Arabidopsis, in which the same set of NPQ components was overexpressed, growth under controlled irradiance fluctuations in the climate chamber was reduced, despite faster NPQ relaxation after reductions in irradiance (Garcia-Molina and Leister, 2020). These contrasting findings suggest that our understanding of the limitations of photosynthesis and growth under fluctuating irradiance is incomplete.

2.2.2 *Enzyme activation and metabolite turnover*

The activation state of several enzymes in the CBB depends indirectly on irradiance, such that it is low in the shade and progressively increases as irradiance increases. The regulation of the activation state of several CBB enzymes occurs through thioredoxins (TRX), of which chloroplasts typically contain several isoforms for increased flexibility and functionality. Some TRX isoforms are activated by reduction through ferredoxin, while others depend on NADPH for reduction. While ferredoxin reduces TRXs at higher irradiance, NADPH can also reduce TRX at lower irradiances. In this way, a matching of CBB and LEF activity is ensured over a range of irradiances, and the activity of several enzymes that require high activation in high irradiance can be maintained in the shade. Of the different TRXs, especially the chloroplastic f-type TRX isoforms are important under fluctuating irradiance, as they affect the activation states of FBPase, SBPase, and Rubisco activase (Rca), all of which can regulate photosynthetic CO_2 fixation under fluctuating irradiance. Oxidized FBPase in darkness maintains a basal activity of 20%–30%, while the oxidized form of SBPase is completely inactive. The activation state of PRK has additionally been shown to be under the control of TRX m and f. Apart from regulation of the CBB enzyme activation states, the two TRX systems are also involved in the regulation of the ATP synthase, starch synthesis, the malate valve and cyclic electron flow. Thus, they broadly coordinate responses across the complete system to rapid irradiance fluctuations. The activity of CBB enzymes is additionally regulated by stromal pH (which increases due to light-driven proton movement from the stroma to the lumen) and by ADP/ATP ratios. The CBB enzymes SBPase, FBPase and PRK activate and deactivate relatively quickly, with time constants in the range of 1–3 min (activation) and 2–4 min (deactivation), respectively. In a shade-adapted leaf that is suddenly exposed to a high irradiance, these enzymes have been implicated in slowing down RuBP regeneration, and are assumed to be the main limitation to CO_2 fixation within the first minute after a switch from low to high irradiance.

With capacity of RuBP regeneration restored, the limitation that takes over is typically that of Rubisco activation. The activation state of Rubisco under a range of irradiances resembles a irradiance response curve of photosynthesis, although even in leaves adapted to darkness or shade, a basal Rubisco activation state of 30%–50% has been measured. In other words, it seems that a relatively large fraction of Rubisco never deactivates. When irradiance increases, the remainder of inactive Rubisco is activated, at time constants of 3–5 min; this means that incomplete Rubisco activation can be a substantial limitation to CO_2 fixation during the first ~10 min of photosynthetic induction. Activation of Rca after a low-to-high irradiance transition has in spinach been shown to occur with a time constant of approx. 4 min. In many genotypes, Rca is present in two isoforms: the longer α isoform and the shorter β isoform. The α-isoform is redox regulated, whereas the β-isoform is regulated by the α-isoform. Taking away the α-isoform by means of transgenics has shown to keep Rca in a constitutively high activation state, leading to faster photosynthetic induction. Modification of the composition of Rca isoforms or of their concentration relative to that of Rubisco might be a means to improve photosynthesis under fluctuating irradiance. Some of the triose phosphates synthesized by the CBB are transported out of the chloroplast and converted into sucrose. This process frees up phosphate, which is transported back into the chloroplast in exchange for triose phosphates and used for ATP synthesis at the thylakoid membrane. One of the enzymes facilitating triose phosphate conversion into sugars, sucrose phosphate synthase (SPS), also increases slowly in activity after transitions from low to high irradiance. So far, however, the slow increase in SPS activity has only been shown to limit photosynthesis during induction under elevated CO_2 concentrations.

After a drop from high to low irradiance, large metabolite pools remaining in the CBB can lead to a transient enhancement of photosynthesis that can last several seconds—this phenomenon is referred to as postillumination CO_2 fixation. At the same time, enhanced rates of turnover of glycine in the photorespiratory pathway can transiently decrease photosynthesis rates after a drop in irradiance; this phenomenon is termed postillumination CO_2 burst, and can take up to approx. 1 min.

2.2.3 CO_2 *diffusion*

In low irradiance, with reduced demand for CO_2 diffusion, stomata typically reduce their aperture, while after a switch to high irradiance, stomata open. Stomatal opening is relatively slow, and additionally shows very large genotypic variation (time constants in the range of 4–29 min have been reported). Because of their initially small aperture and their slow opening, stomata can transiently limit CO_2 uptake during photosynthetic induction. After a drop in irradiance, on the other hand, stomatal closure is very slow compared to the near-immediate drop in CO_2 uptake, leading to a transient reduction in water use efficiency.

Several anatomical features on the leaf surface affect the rate of stomatal movement: guard cell and subsidiary cell type, as well as stomatal density and size. Two types of guard cell exist: dumbbell-shaped and kidney-shaped. Dumbbell-shaped guard cells are mainly found in grasses, ferns, and gymnosperms; these guard cells generally exhibit much faster movement than kidney-shaped guard cells, due to a larger cell membrane to surface volume ratio. Stomatal density and stomatal size are typically negatively correlated, such that leaves of a given genotype either have few large or many small stomata. Among closely related genotypes of the same genus, it has been demonstrated that smaller stomata confer faster rates of stomatal movement, probably again due to a larger surface to volume ratio. Lastly, guard cells are surrounded by a variable number of subsidiary cells, which exchange water, ions, and hormones with the guard cells and form a functional unit—the stomatal complex—with the guard cells. There is large variability in the number and patterning of subsidiary cells surrounding the guard cells, and their role in dynamic photosynthesis is not well defined. However, in the grass *Brachypodium distachyon*, in which subsidiary cells usually flank guard cells, a lack of subsidiary cells leads to extremely slow stomatal movement (Raissig et al., 2017), showing that at least in some species, subsidiary cells are required for rapid adjustments of stomatal opening.

Manipulating stomata for faster reactions to changes in irradiance may increase both photosynthesis and water use efficiency under fluctuating irradiance. Indeed, two recent examples in guard cells of transgenic Arabidopsis show that such manipulation is possible: Papanatsiou et al. (2019) expressed the synthetic, blue irradiance activated BLINK1 potassium channel protein in guard cells, while Kimura et al. (2020) used a *proton ATPase translocation control 1* (PATROL1) gene overexpressor line. In both cases, stomata opened faster upon increases and closed faster upon decreases in irradiance, and plants showed improved growth as well as water use efficiency under fluctuating irradiance.

Mesophyll conductance (g_m) reduces the CO_2 available to Rubisco even further. It has been known that steady-state g_m reduces CO_2 diffusion to a similar extent as does g_s. Also, g_m can be variable under dynamic irradiance, and some of the components determining g_m are flexible and dependent on light intensity. Indeed, g_m was recently shown to increase during photosynthetic induction in leaves of Arabidopsis and tobacco (by ~0.3 mol m^{-2} s^{-1}, a similar extent as is often observed in g_s; Sakoda et al., 2021). However, diffusional limitation due to g_m was found to be less severe than that imposed by stomata (Sakoda et al., 2021). In conclusion, stomatal conductance and kinetics seem to be the most promising targets to improve on transient diffusional limitations to photosynthetic induction, and on instantaneous water use efficiency under field conditions.

2.3 Environmental factors affect the rapidity of photosynthetic responses to fluctuating light

As described in the previous sections, transient limitations to dynamic photosynthesis include RuBP regeneration state, Rubisco activation state, stomatal conductance, and NPQ. These components are not only affected by irradiance (see previous paragraphs), but also by environmental factors such as CO_2 concentration, temperature, air humidity, soil salinity, drought stress and the spectral composition of irradiance. Given this, it is not surprising that the rate with which photosynthesis reacts to changes in irradiance is also affected by these environmental factors.Especially CO_2 concentration ([CO_2]) has been shown to have a strong effect on the rate of photosynthetic induction: the higher [CO_2], the faster the rate of induction. Also, the loss of photosynthetic induction in the shade was shown to be slowed down under increased [CO_2]. In tomato leaves, doubling [CO_2] from 400 to 800 ppm resulted in an increase of 12%–17% in dynamic photosynthesis rates (Kaiser et al., 2017), and this increase was additional to positive effects of elevated [CO_2] on steady-state photosynthesis. Reasons for this increase in dynamic photosynthesis under high [CO_2] are not entirely clear, but may include (a) reduced diffusional limitations due to a steeper concentration gradient between [CO_2] in the air surrounding the leaf and [CO_2] in the chloroplast, (b) faster activation and slower deactivation of Rubisco, and (c) smaller postillumination CO_2 burst due to reduced rate of RuBP oxygenation.

A large number of free-air CO_2 enrichment (FACE) experiments in numerous habitats have demonstrated that growth in most plants responds positively to an increase in atmospheric [CO_2], as would be expected from plants grown in modern greenhouse production systems as well as from knowledge of steady-state and dynamic responses of leaf photosynthesis to elevated [CO_2]. This positive response of plant growth is termed the CO_2 fertilization effect. In FACE experiments, typically, a number of plants is surrounded by a ring that releases pure CO_2 into the air, and the direction as well as dosing rate depend on wind speed and direction. Because wind speed and direction change constantly, fluctuations in elevated [CO_2] are inevitable. A recent study has shown that fluctuations in [CO_2] depress growth relative to constant [CO_2] of the same average, elevated concentration (Allen et al., 2020). The physiological responses underlying this reduction in growth are not clear, but an important implication is that the CO_2 fertilization effect on plant growth in a CO_2-rich future climate will likely be stronger than what is suggested by results from most FACE experiments.

Temperature affects the kinetics and equilibrium of all biochemical reactions. This is also true for several processes related to enzyme activation and deactivation and hence, photosynthetic induction. For example, the action of Rca is temperature dependent; the rate of Rubisco activation is thus faster the higher the temperature, albeit with an optimum. The rate of photosynthetic induction seems to follow an optimum response with regard to temperature, but data underpinning this conclusion is still scarce.

Several factors have direct effects on stomatal conductance and thus rate of CO_2 diffusion into the leaf, but do not affect Rubisco activation state (at least, in the short term): air humidity, salinity, and drought stress. These factors have been shown to exacerbate the effects of low stomatal conductance on photosynthesis under fluctuating irradiance. As the air and soil get dryer, and the soil becomes more saline, stomata close. This typically reduces steady-state g_s at both low and high irradiance and the rate of stomatal closure after a transition from high to low irradiance, but not necessarily the rate of opening after a switch from low to high irradiance. In fact, examples from leaves at low air humidity show that the rate of stomatal opening (i.e. how quickly g_s transitions from one steady-state to another) can be increased in dryer air. Nevertheless, the net effect of these stresses is that they reduce time-integrated dynamic photosynthesis rates, as they exacerbate stomatal closure.

Irradiance spectrum is another factor that can potentially affect dynamic photosynthesis. For example, blue irradiance (400–500 nm), which is a potent effector of stomatal opening in the presence of red irradiance (600–700 nm), can potentially modulate the rate of stomatal movement after a change in irradiance and therefore overall photosynthesis rates. However, results from tomato leaves have so far not proven blue irradiance to have an effect on dynamic photosynthesis. Far-red irradiance (700–800 nm) preferentially excites photosystem I, and as such can increase photosynthesis after a shift from high to low irradiance, as in that situation it can improve the flow of residual electrons through the linear electron transport chain; this effect has recently been shown to increase photosynthesis in fluctuating irradiance (Kono et al., 2019).

2.4 Genotypic variation in dynamic photosynthesis

Apart from various environmental factors, which can modulate the rate of response of photosynthesis to a change in irradiance, dynamic photosynthesis is also strongly affected by plant genetics. Indeed, substantial genotypic variation [calculated as percentage genetic variation (PGV)] in photosynthetic induction has recently been shown to exist for a range of important crops. For example, substantial variation for integrated net photosynthesis rate during the first 5 min of photosynthetic induction has been documented in rice (PGV: 102%, $n = 14$ genotypes; Acevedo-Siaca et al., 2020), soybean (PGV: 159%, $n = 37$; Soleh et al., 2017), and cassava (PGV: 91%, $n = 13$; De Souza et al., 2020). Also, in wheat, the time to reach 95% of full Rubisco activation showed a PGV of 62% among 10 genotypes (Salter et al., 2019). These differences in the photosynthesis response to changes in irradiance were linked to differences in the rate of Rubisco activation, initial g_s in low irradiance as well as the rate of stomatal opening. Furthermore, substantial variation in a range of other traits defining transient photosynthesis, chlorophyll fluorescence and g_s responses to fluctuating irradiance has recently been shown in banana (Eyland et al., 2021), barley (Salter et al., 2020), canola (Liu et al., 2021), rice (Qu et al., 2016), sorghum (Pignon et al., 2021), and wild relatives of modern wheat (McAusland et al., 2020). Generally, steady-state g_s at any irradiance has been shown to vary between genotypes by up to 250% PGV, and rates of stomatal opening as well as closure have additionally been shown to be strongly genotype dependent. An important conclusion from these studies is that variation of traits related to dynamics of photosynthesis and stomatal conductance is typically larger than the variation of steady-state traits, potentially enabling the breeding for faster transient responses to fluctuations in irradiance. Importantly, also, there are several components for which genotypic variation has not yet been demonstrated, such as rates of RuBP regeneration and Rubisco deactivation as well as NPQ relaxation upon shifts to low irradiance. These processes are less easily observable with the tools that are readily available. It will be exciting to see what insights the development of new (rapid) phenotyping tools will bring in terms of the discovery of genes underlying photosynthetic responses to irradiance fluctuations.

3 Modelling leaf photosynthesis as a system

3.1 Steady-state models of photosynthesis

3.1.1 The FvCB model

A wide range of mathematical models of photosynthesis are derived from the model by Farquhar et al. (1980), hereafter the FvCB model. The FvCB model is a model of photosynthesis that focuses on the concept of limiting factor as the key concept for simplification and yet it can reproduce accurately the steady-state responses of CO_2 assimilation to CO_2, light intensity, and temperature. The fundamental idea in this model is that photosynthesis cannot go faster than the rate at which rubisco fixes CO_2 but also not faster than the rate at which the substrate for carboxylation (i.e. RuBP) is being regenerated by the Calvin Benson Bassham cycle. Furthermore, the system is assumed to be so well regulated that it can move from one limitation to the other without any losses in efficiency. This assumption is not valid for photosynthesis under a fluctuating environment but is good enough to capture the behaviour under steady-state conditions. Many variants of this original model have been developed to incorporate additional processes and address specific cases (von Caemmerer, 2000).

In the original FvCB model, RuBP regeneration was assumed to be limited by the rate at which NADPH or ATP was being produced (the difference between the two alternatives resides on the stoichiometry assumed in the equations, but otherwise produce very similar results). However, nowadays there is empirical evidence that, at least under certain conditions, RuBP regeneration may be limited by the activity of enzymes in the Calvin Benson Bassham cycle, such as SBPase, though the mathematical formulation of this limitation remains the same. Furthermore, many of the variants derived from the FvCB model use a third limitation, denoted as triose phosphate utilisation (TPU), which corresponds to the rate at which phosphate is regenerated through the export of triose phosphates into the cytosol.

In a modern variant of the FvCB model, the rate of CO_2 assimilation (A) is calculated independently for each limiting factor and the minimum of these is taken:

$$A = \min\left(A_{Rubisco}, A_{RuBP}, A_{TPU}\right) \tag{3}$$

Each term in Eq. (3) is described by an empirical expression, except $A_{Rubisco}$, which is based on a model of rubisco kinetics n. The most common equation assumes rubisco to be fully active and considers the competition between O_2 and CO_2:

$$A_{Rubisco} = \frac{(C_C - \Gamma_\star)V_{c,\max}}{C_C + K_{mC}\left(1 + \frac{O_2}{K_{mO}}\right)} - R_d, \tag{4}$$

where C_C is the CO_2 concentration inside the chloroplast, $\Gamma_\star$ is the value of C_C at which the first term in Eq. (4) becomes zero (i.e. CO_2 compensation point), $V_{c,\,max}$ is the maximum rate of carboxylation and assumed proportional to the amount of rubisco, K_{mC} and K_{mO} are the Michaelis–Menten constants with respect to CO_2 and O_2, respectively, and R_d is the rate of mitochondrial respiration in the light. The parameter $V_{c,\,max}$ is widely used as a measure of photosynthetic capacity.

In order to couple electron transport to CO_2 assimilation, stoichiometries for production and consumption of NADPH or ATP are used, based on the canonical description of the Calvin cycle, photorespiration and electron transport. For example, if one assumes that the rate of RuBP regeneration is limited by the rate of NADPH production, the rate of CO_2 assimilation limited by RuBP regeneration is modelled as

$$A_{RuBP} = \frac{(C_C - \Gamma_\star)J}{4C_C + 8\Gamma_\star} - R_d, \tag{5}$$

where J is the linear electron flux. Depending on the context, J may be calculated from measurements (e.g. chlorophyll fluorescence) or calculated empirically as a function of light intensity. The most common model to calculate J as a function of light intensity is a non-rectangular hyperbola:

$$J = \frac{J_{max} + \alpha \cdot I - \sqrt{(J_{max} + \alpha \cdot I)^2 - \alpha\theta J_{max} I}}{2\theta}, \tag{6}$$

where J_{max} is the maximum electron flux, α is the efficiency at which light is converted into electrons at low light (i.e. the initial slope of the hyperbola) and θ is an empirical factor that determines the curvature of the hyperbola. Finally, the rate of CO_2 assimilation limited by TPU is most often modelled as

$$A_{TPU} = 3TPU - R_d, \quad (7)$$

where TPU is the maximum rate at which triose phosphate are used.

Eqs. (4)–(7) are the most common in modern variants of the FvCB model, though alternative exists. For example, in recent years, alternative equations have been used for the TPU limitation as CO_2 assimilation may decrease with CO_2 concentration under this limitation, though the exact nature of such decrease is still under debate. Also, stoichiometries in the denominator of Eq. (5) may differ, as authors may assume limitation by ATP instead of NADPH or may consider the effects of alternatives forms of electron transport.

The most common procedure to parameterize the FvCB model or modern variants is to measure the rate of CO_2 assimilation as a function of intercellular CO_2 at saturating or high light intensity (Sharkey et al., 2007) producing the so-called 'A/Ci curve'. Typically, this type of response curve will allow estimating, at least, $V_{c,\,max}$, J, and TPU (Fig. 16). The value of J_{max} can then be derived from Eq. (6) by making assumptions about the other parameters or using a light response curve to estimate them. By allowing to estimate key parameters related to the underlying (photo-)biochemistry from gas exchange, the FvCB model and modern variants have achieved great popularity in photosynthesis research.

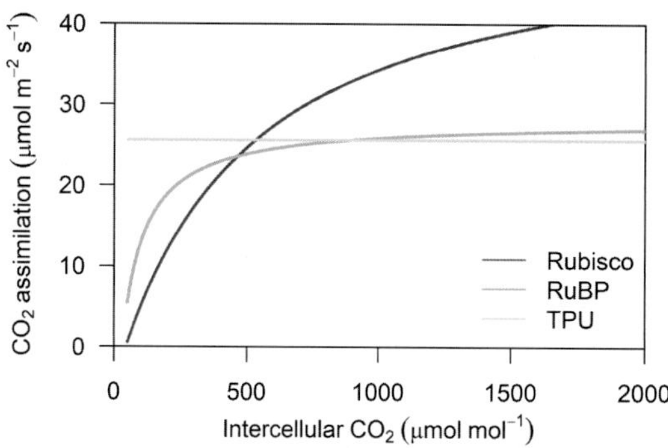

FIG. 16 Example of CO_2 assimilation as a function of intercellular CO_2 molar fraction at high light intensity and for different limiting factors in a modern variant of the FvCB model including limitations due Rubisco, RuBP regeneration and triose phosphate utilisation (TPU). The model assumes an infinite mesophyll conductance.

3.1.2 Scaling up from chloroplasts

The relatively ease with which the FvCB variants can be parameterized and their simple analytical form make these equations also popular in multiscale models where photosynthesis is being scaled up to the level of leaves or canopies. In order to successfully scale up these models, three additional processes need to be considered:

1. CO_2 diffusion from air to chloroplasts.
2. Distribution of light intensity within leaves or canopies.
3. The distribution of photosynthetic capacity within leaves or canopies.

The diffusion of CO_2 from the air surrounding the leaves into the chloroplast is generally modelled in analogy to Ohm's law. That is, each diffusional barrier is described as a conductance (g) such that the flux of CO_2 (F_C) can be modelled as

$$F_C = g\,\Delta C, \quad (8)$$

where ΔC is the difference in CO_2 concentrations across the barrier. The two most common barriers considered are the stomatal pores (i.e. stomatal conductance) and all diffusional barriers at the cellular level lumped into a single term (i.e. mesophyll conductance). For each barrier, an equation like Eq. (8) is used and combined with Eq. (3) leading to a coupled system of equations. This system can often be solved analytically resulting in polynomials of second or third degree that, when solved, yield the rate of CO_2 assimilation.

Light distribution within leaves and canopies has traditionally been modelled with an statistical approximation that approximates the leaf or canopy with a turbid medium (Baldocchi and Amthor, 2001). According to this approach, the intensity of a beam of light that travels through a leaf or canopy decreases exponentially, with the exponential coefficient dependent on chlorophyll content (within leaves) and leaf area density and angle distribution (within canopies). When this approach is applied in one dimension (i.e. the vertical direction), one obtains an exponential profile of light intensity and this is the most common approach. However, at the canopy level, it is also possible to extend the concept of turbid medium to two or three dimensions to take into account complex geometries (e.g. row crops or sparse forest canopies).

Detailed models of cellular and leaf anatomy have been developed in order to capture the effect of anatomy and sub-cellular processes on CO_2 diffusion and light distribution (Berghuijs et al., 2016). These models may represent diffusion in one, two or three dimensions and may be limited to the cell or the whole leaf. Also, these models may use approximate geometries to represent the different sub-cellular elements and tissues or actual measurements of anatomy from microscopy and other optical techniques.

3D models of leaf anatomy and plant architecture have been coupled with Monte Carlo physics-based ray tracers in order to compute the amount of light absorbed by each cell or organ in a more realistic fashion (Townsend et al., 2018). These models are computationally more expensive to run and much harder to parameterize due to the level of detail at which anatomy and architecture are described, but they allow for a higher degree of realism.

The distribution of photosynthetic capacity is generally represented as spatial changes in $V_{c,max}$, J_{max}, and TPU, whereas the rest of parameters are often assumed to be uniform. Also, these parameters are often assumed to be linearly related to nitrogen content. The problem is then reduced to measuring the coefficients of these linear relationships and to model the spatial distribution of nitrogen content itself. This has been achieved by either by (i) measuring the empirical distribution of nitrogen within the leaf or canopy and the linear relationships between the parameters and nitrogen content or (ii) calculating the distribution of nitrogen and the linear coefficients that maximises photosynthesis at a particular spatial and time scale, depending on the problem (Yin et al., 2018). These optimisation studies show that, both within leaves and canopies, the optimal distribution of nitrogen is proportional to light intensity and hence decreases exponentially with the canopy or leaf (Farquhar, 1989). However, experiments show that real distributions tend to be more uniform than expected from optimisation theory.

3.1.3 Reaction-based models

Despite the success of the FvCB model and its variants, there are limitations to the use of those models. Most notably, the lack of an explicit description any biochemical reactions (except for rubisco) means that such models cannot be used to analyse the importance of the different reactions in photosynthesis or gain insights on how metabolic regulation is achieved. Therefore, alternative models have been constructed, often to address specific research questions and with much lower levels of reuse and adoption than the FvCB model and its variants. Broadly, these models fall into three categories: (i) models of the electron transport chain (Morales et al., 2018b), (ii) models of the Calvin Benson Bassham cycle and associated metabolic pathways (Zhu et al., 2007) and (iii) models of the whole photosynthetic system (Zhu et al., 2013).

Reaction-based models may also be classified depending on whether they consider the kinetics of reactions explicitly or only the stoichiometries. Kinetic models describe each individual reaction according to generic rate equations that are generally parameterized from published literature. Stoichiometric models allow simulating large networks of reactions without the need to parameterize its kinetics and are used to analyse the flow of energy and matter through the network (Luo et al., 2009). In all cases, a fundamental component of these models has been the need for realistic boundary conditions as photosynthesis is highly integrated with cellular metabolism (e.g. exchange of redox power between chloroplasts and cytosol) and it is sensitive to several environmental factors.

Reactions catalysed by enzymes that occur in liquid media (e.g. the stroma of the chloroplast) are generally modelled according to Michaelis–Menten (a.k.a. hyperbolic) kinetics. The assumption of this approach is that, at the scale of the simulations, the reactions are in quasi-steady state and the rate at which substrates/products are consumed/generated can be calculated from (i) the concentration of the enzyme, (ii) the catalytic constant of the enzyme, (iii) the Michaelis–Menten constants with respect to substrates and products, (iv) the equilibrium coefficient of the reaction, and (iv) activation and inhibition constants with respect to other metabolites that affect the enzyme (i.e. allosteric regulation). Sometimes enzymes may also be separated into active and inactive pools and the activation process modelled explicitly (e.g. activation of enzymes in the Calvin Benson Bassham cycle by thioredoxins). A generic equation for a reversible reaction with two substrates and products (i.e. $A+B \rightarrow C+D$) in the absence of allosteric regulation is

$$V = V_m \frac{[A][B] - \frac{[C][D]}{K_{eq}}}{K_{mA}K_{mB}\left(1 + \frac{[A]}{K_{mA}} + \frac{[B]}{K_{mB}} + \frac{[C]}{K_{mC}} + \frac{[D]}{K_{mD}}\right)}, \tag{9}$$

where V_m is the maximum rate of the reaction (product of the concentration of enzyme and catalytic constant), square brackets ([]) indicate the concentration of a substrate or product, K_{eq} is the equilibrium coefficient of the reaction, and K_m stands for Michaelis–Menten constant. As can be inferred, a large number of parameters are required to fully parameterized any photosynthetic pathways, and these parameters are generally taken from published results on in vitro studies.

Once all reactions of the pathways of interest are modelled, a system of differential equations is then built by applying mass balance to the pool of each metabolite, that is,

$$\frac{dA}{dt} = V_{in,A} - V_{out,A}, \tag{10}$$

where $V_{in,\,A}$ and $V_{out,\,A}$ are the rate of the reactions where A is a product and substrate, respectively. The resulting system of equations is then solved with numerical algorithms that can compute the rates of all reactions and concentrations of all metabolites as a function of environmental conditions, either in the steady-state or as a function of time.

When modelling the electron transport chain, equations, such as Eq. (9), cannot be used to represent most of the redox reactions responsible for electron transport as these are (i) not catalysed by enzymes and (ii) electron donors and receptors are not freely diffusing in a medium. Instead, reversible second order kinetics are used. For example, a generic rate equation for electron transfer between A and B (i.e. $A^- + B \rightarrow A + B^-$) is

$$V = [A^-][B]k - [A][B^-]k/K_{eq}, \tag{11}$$

where k is a second order rate constant and K_{eq} is the equilibrium coefficient of the reaction, calculated from the difference in midpoint redox potentials of A and B. In practice, modelling of protein complexes in the electron transport chain can be rather complex as (i) a protein complex may have multiple components engaged in electron transfer that cannot be modelled independently and (ii) the redox state of a component may affect the kinetics a reaction where it is not involved, through changes in the electrical field arising from its charge affecting the energy of a reactant in another reaction. These constraints result in models where the transitions between all possible states of a protein complex (define by all the combinations of redox states of its components) have to be modelled explicitly. For example, if a protein complex is modelled as being composed of three components, and each component can be in two states (reduced and oxidised), there are eight possible states for the protein complex and therefore 8 equations implementing mass balance (i.e. like Eq. 9) would be required. In reality, protein complex can have numerous components and this leads to models with a large number of equations (Lazár and Jablonský, 2009).

3.2 Dynamic models of photosynthesis

Whereas much of the research on photosynthesis has been (and still is) done under steady-state controlled conditions, there is increasing concern that much of the natural behaviour of photosynthesis in its natural context is not being captured by this type of research. This had led to an increasing interest in researching and understanding photosynthesis under fluctuating environmental conditions, with an emphasis on fluctuating light. In turn, this has led to an increasing number of models of photosynthesis that incorporate processes necessary to capture the dynamic responses of photosynthesis to these fluctuations.

Kinetic reaction-based models can easily be used to study dynamic photosynthesis, though the extent to which they are useful will depend on whether relevant processes are captured. For example, although many kinetic models of the Calvin-Benson-Bassham cycle have been developed in the past decades, these models are not used to study dynamic photosynthesis as they lack a description of rubisco activation or dynamic regulation of the electron transport chain, both of which can have a great impact on dynamic photosynthesis. However, new models have been developed recently, most of them focused on the dynamic regulation of the electron transport chain (Morales et al., 2018b).

As alternative to kinetic reaction-based models, a group of models have also been developed that extend the FvCB model by adding processes known to limit dynamic photosynthesis during light transients (Morales et al., 2018a). The advantage of these models is that (i) they reuse parameters from the FvCB model and its modern variants and (ii) many of the new parameters introduced by these extensions can be estimated from gas exchange measurements during and following irradiance transitions (e.g. induction curves). These models are constructed by starting with a steady-state model and assuming that parameters associated to photosynthetic capacity and CO_2 diffusion vary in time.

The dynamic changes in these models are implemented in a two-step approach:

1. Calculate the value of a parameter in the steady-state model as function of the environmental conditions being simulated (i.e. the value of a parameter once steady-state conditions is reached).
2. Assume that the actual value of the parameter changes towards the calculated value at a given rate.

Most commonly, first order kinetics are assumed, meaning that the rate of change of a parameter p (i.e. dp/dt) is calculated as.

$$\frac{dp}{dt} = \frac{p_c - p}{\tau_p}, \tag{12}$$

where p_c is the value calculated in step 1 and τ_p is the time constant (i.e. inverse of a rate constant) of change in p. In some cases, more complex kinetics may be assumed to capture experimental results (e.g. stomatal conductance). Also, the time constant for increases and decreases in a parameter may differ (e.g. stomatal conductance, rubisco activity). Typical parameters modelled through this approach include stomatal conductance, the relativity activity of enzymes (i.e. a factor between 0 and 1) including rubisco and other enzymes involved in the regeneration of RuBP or the quantum yield of PSII (which is a component of the parameter α in Eq. (6)).

3.3 About the use and parameterisation of models of photosynthesis

3.3.1 *The different uses of models*

Mathematical models are used for different purposes and choices during model development are strongly determined by its intended use. These choices often include simplifications that may be valid or innocuous under the original intended use but may be inadequate for other applications. Therefore, when considering the use of an existing photosynthesis model (or when judging the merits of such model) it is of upmost importance to consider the purpose of the model. Within photosynthesis, models have been used for a variety of purposes including (i) parameterisation of photosynthesis, (ii) estimation of unknown quantities from indirect measurements, (iii) predictions about CO_2 uptake or productivity, and (iv) identification of potential strategies to improve CO_2 uptake or productivity.

Models have become commonplace in the analysis of experimental results in photosynthesis research, to the point that these activities will often not be perceived as 'modelling'. However, not acknowledging the role of models in such procedures can result in unnecessary confusion. As an example, we may explore the meaning of the parameter $V_{c,\max}$ from Eq. (4). Whenever such symbol is used in a study on photosynthesis, it will correspond to a parameter in a model and its value would have been obtained by fitting the model to the data. However, many different models may include a parameter called $V_{c,max}$, such as models of in vitro rubisco kinetics, chloroplast metabolism, or leaf CO_2 assimilation (with all the variations within these domains). Furthermore, the procedure by which $V_{c,max}$ is estimated given a model and dataset may vary from study to study and also affects its value. Thus, when comparing values of $V_{c,max}$ from different studies, the researcher should consider whether such values were obtained with the same models and fitting procedure. If not, then the differences in reported values may not entirely correspond to differences in biology. Unfortunately, there is no widely accepted standard model and fitting procedure for obtaining most parameters in photosynthesis research.

As predictive tools, models of photosynthesis may be used on their own or, more commonly, as components of a model of plant growth or CO_2 cycle, at the level of plant, ecosystem, or planetary level. Predictions may be of interest in themselves, for example, to forecast yields or estimate carbon balances, but they are often used in the context of scenario analysis, sensitivity analysis, and thought (or in silico) experiments, as a complement to experimental research. For example, the potential benefits of increasing the amounts of SBPase or accelerating the relaxation of NPQ in low light on photosynthesis were demonstrated with metabolic models before, or in parallel with, experiments with transgenic plants. Similarly, mathematical models of photosynthesis have a played a pivotal role in exploring strategies for improving plant photosynthesis through the use of CO_2 concentrating mechanisms as the proposed solutions remain theoretical.

3.3.2 *Mechanistic vs empirical models and model complexity*

Two common adjectives employed in the description of models are *mechanistic* and *empirical* (sometimes replaced by 'phenomenological'). The term mechanistic tends to have a positive connotation, meaning a model that is truer or closer to reality than an empirical model, but it may scare away potential users as it also tends to be associated with models that are more difficult to parameterize or understand. However, there is no widely accepted criteria to determine whether a model is mechanistic or empirical and different researchers may disagree on how to qualify a model.

A useful criterion that may be used to differentiate between *mechanistic* and *empirical* is to consider as mechanistic such models (or components within models) that predict a phenomenon based on a description of underlying biological, chemical, and physical processes, whereas models (or components within models) are empirical when they describe the phenomenon of interest directly in mathematical form. For example, the steady-state response of photosynthesis to light intensity may be described directly with a non-rectangular hyperbola (empirical model) or be predicted as an *emerging pattern* of the underlying biochemical and photochemical reactions (mechanistic model).

Note that this criterion implies that being mechanistic or empirical is not an intrinsic property of the model but depends on its use and therefore the same model may be qualified as being empirical or mechanistic depending on the context. For example, an ecosystem model that simulates gross primary production by scaling photosynthesis from the chloroplast to the canopy level would be mechanistic, even if the rate of photosynthesis at the chloroplast level is described by the aforementioned non-rectangular hyperbola. Indeed, even in the most complex multi-scale models of photosynthesis, the smallest scales are often empirical descriptions of biochemical or physical phenomena.

Another criterion when considering the use of a model is its complexity. Often, model complexity is associated to the number of parameters in the model. However, complexity may also refer to the practical difficulties in using the model or the tractability of the predictions of the model. For these reasons, analytical models are generally considered less complex than computational models, even when they have the same number of parameters. Similarly, mechanistic models tend to be more complex than empirical models, though that choice may not be relevant in practical terms as their intended use would differ. Indeed, the most advanced mechanistic models of photosynthesis may have hundreds of parameters (Zhu et al., 2013; Morales et al., 2018b) and it would not be feasible to estimate their values from the measurements of any particular experiment. However, this is not an issue in practice as such models are not intended to be fitted to experimental results but rather capture the state-of-the-art quantitative knowledge on a particular topic of photosynthesis research. Regardless of how complexity is defined, there is no threshold above which a model may be considered complex and such judgement would always be affected by the experience of researcher and the intended use of the model.

3.3.3 Estimating parameters from data

The number of parameters of a model is particularly relevant when the model needs to be parameterized from a specific genotype or scenario. In this case, the larger number of parameters that need to be estimated, the more information needs to be collected about the genotype or scenario of interest. However, the problem is also subtle as (i) not all parameters are equally relevant for a particular prediction (which can be quantified through sensitivity analysis), (ii) not all parameters are equally easy to estimate from data (a problem known as identifiability of parameters), (iii) not all parameters vary across genotypes or scenario (e.g. kinetic constants of enzymes are often assumed to be highly conserved), and (iv) different estimation techniques may facilitate the process of parameterisation.

Increasing the amount of data does not necessarily make the problem easier. For example, the number of parameters that can be reasonably estimated from a steady-state CO_2 response (A/Ci) curve using a typical steady-state biochemical model is limited and will be the same regardless of whether the curve consists of 15 measurement points or 100, as long as the measurement points are well distributed throughout the range of CO_2 concentrations and therefore contain the same information about the system. The only advantage of taking a hundred measurement points would be an (approximately) ten-fold decrease in the uncertainty on the parameter estimates.

Regarding the different estimation techniques, no widely accepted standard exists, but different types of approaches may be clearly distinguished. One approach is to determine which parameters from a given model can be reasonably estimated from a specific type of measurement and design a stepwise estimation procedure (Yin et al., 2009). This approach may be called 'divided and conquer' as it often requires dividing the data into subsets that are used to estimate different parameters in a specific order (i.e. one step often requiring parameters estimated in a previous step). If there are parameters that cannot be estimated from the data at hand, they will be fixed to literature values and all analysis should be considered conditional on those values being correct. The advantage of this approach is that, in each step, parameters are often easier to estimate, often with linear regression, and thus remains a popular approach. On the other hand, this type of approach lacks a well-defined statistical framework, making it hard to correctly quantify the uncertainty in the estimates and ignores correlations among parameters, which may result in biased estimates.

A more recent alternative approach is the use of non-linear modelling techniques that simultaneously makes use of all the data as well as prior knowledge, generally under a Bayesian parameter estimation framework (Ogle and Barber, 2008). The advantage of this approach is that it follows an orthodox statistical approach and that all correlations are considered. Also, it can take into account the uncertainty that the user may have on parameter values taken from the literature (i.e. the analysis is no longer conditional on specific values for those parameters being true). However, this approach is substantially harder to implement (requiring more advanced computational techniques) and the results will be sensitive to the prior knowledge by the author. To avoid this dependency on prior knowledge, a special case of this approach is non-linear least squares, where the researcher decides to ignore prior knowledge and instead fixes some of the parameters to literature values, thus making analysis conditional of such values.

An additional approach is the use of mixed models (a.k.a. mixed effect, hierarchical or multilevel models) generally under a Bayesian parameter estimation framework (Patrick et al., 2009). The idea of mixed models is applicable when analogous measurements (e.g. steady-state response curves) have been taken on multiple individuals across treatments, genotypes or as biological replicates. In a mixed model, all the data will be used simultaneously, and the researcher will extend the original photosynthesis model with a statistical description of the grouping structure in the data. The aim is to obtain the parameter estimates for each individual as well as the distribution of parameters across individuals. The advantage of this approach versus analysing each individual separately and them combining the result is that information from different individuals will be partially shared, which in statistics is known as *partial pooling*. This process can add robustness to the estimation procedure, especially when some of the measurements may be of lower quality or some of the parameters are less identifiable.

4 Conclusions

In this chapter we have summarised three important aspects of plant photosynthesis:

1. The operation and regulation of photosynthesis in leaves under steady state conditions.
2. The responses of leaf photosynthesis to fluctuating irradiance.
3. The modelling of leaf photosynthesis.

Our primary aim was to show how photosynthesis in leaves is the result of the coordinated activity of numerous photosynthetic subprocesses, such as stomatal regulation and diffusive transport, light-harvesting and electron transport, and metabolic activity associated with the CBB and photorespiratory (or photosynthetic carbon oxidation) pathway, subprocesses that are individually discussed in more detail in other chapters. Note that a complete understanding of the operation of the subprocesses is not possible without understanding how they combine into the integrated machine that is leaf photosynthesis. For example how can the regulation of q_E be fully comprehended if it is never seen in the context of limitation on PSII electron transport imposed by electron transport, which is itself difficult to understand in isolation from the CBB (etc.). In fact, in one of those 'chicken and egg' arguments, it might be better to describe the integrated activity of photosynthesis before covering in detail the mechanisms of the subprocesses.

A particular and evolving challenge is understanding how the combined activity of photosynthetic subprocesses functions in the natural world where the physical environment (irradiance, temperature, the supply of minerals and water, etc.) is not constant (as it would often be in a growth cabinet) but fluctuates over a range of time scales. While the importance of photosynthetic regulation in response to changes in environment has been recognised for many years (Harbinson and Woodward, 1984; Pearcy, 1990) only recently has this topic become a major topic of photosynthesis research. This new focus on the response of photosynthesis to fluctuations involves not only shorter term physiological responses but longer term changes in structure and composition of the photosynthetic machinery (not a topic dealt with in any detail in this chapter). An understanding of physiological responses to rapid environmental changes—the dynamics of photosynthetic regulation—cannot be achieved unless the more classical understanding of photosynthetic operation and regulation under static or steady-state conditions is understood, though understanding control of any kind leads one naturally to analysing responses to a disturbance of some kind. We see, therefore, the expanding field of photosynthetic responses to fluctuations as a natural and relevant extension to our foundation of the understanding of photosynthesis under steady-state conditions.

This understanding of photosynthetic operation and regulation under steady state and fluctuating conditions is codified by means of models. By means of models, we can bring together the dots of individual physiological discoveries into a unified whole. At one extreme, this can result in large, detailed and mechanistically precise models of photosynthesis that are valuable tools for exploring photosynthesis in silico—something that is, for example, particularly valuable if you wish to improve photosynthesis in some way. Other models, however, such as the FvCB model for assimilation, are valuable not because they are complex but because they condense the essence of a complex process into a simple formula that focuses on the important features of process. These simple models are widely used to model photosynthesis as part of more complex models, such as GECROS or APSIM (McCown et al., 1996; Xinyou and Van Laar, 2005), that are used to simulate crop growth. Which brings us to our final concluding thought. Just as our understanding of leaf photosynthesis depends on an understanding of more elemental photosynthetic subprocesses then we need to recognise that to understand the full value of photosynthesis to an individual plant or canopy then need to upscale from leaf photosynthesis to whole plant or canopy photosynthesis. If we want to understand, for example, the capacity of a mixed forest to sink atmospheric CO_2, or the productive capacity of wheat canopy, then we need to upscale from the single leaf to the collection of leaves and understand not only how that population of leaves affects the physical environment (irradiance, temperature, humidity, etc.) within the population, but how individual leaves respond to that environment in terms of their immediate photosynthetic activity and their longer term acclimation. Even in a canopy of a single species, every leaf is still part of an individual and it is at the level of the individual that optimisation of leaf level photosynthesis (e.g. via the partitioning of nitrogen between leaves) seems to occur. Understanding how photosynthesis upscales to these higher levels of integration (the canopy, the community, the biome, and the planet) is essential to understanding the efficiency of photosynthesis in agriculture and natural systems and therefore the role it plays in providing or food, feed for our animals and raw materials such as wood and fibres, and in the cycling of atmospheric CO_2.

References

Acevedo-Siaca, L.G., Coe, R., Wang, Y., Kromdijk, J., Quick, W.P., Long, S.P., 2020. Variation in photosynthetic induction between rice accessions and its potential for improving productivity. New Phytol. https://doi.org/10.1111/np.

Allen, L.H., Kimball, B.A., Bunce, J.A., Yoshimoto, M., Harazono, Y., Baker, J.T., Boote, K.J., White, J.W., 2020. Fluctuations of CO_2 in free-air CO_2 enrichment (FACE) depress plant photosynthesis, growth, and yield. Agric. For. Meteorol. 284, 107899.

Andrews, M., 1986. Nitrate and reduced-N concentrations in the xylem sap of *Stellaria media, Xanthium strumarium* and six legume species. Plant Cell Environ. 9, 605–608.

Armbruster, U., Carrillo, L.R., Venema, K., Pavlovic, L., Schmidtmann, E., Kornfeld, A., Jahns, P., Berry, J.A., Kramer, D.M., Jonikas, M.C., 2014. Ion antiport accelerates photosynthetic acclimation in fluctuating light environments. Nat. Commun. 5, 1–8.

Arney, G., Domagal-Goldman, S.D., Meadows, V.S., Wolf, E.-T., Schwieterman, E., Charnay, B., Claire, M., Hébrard, E., Trainer, M.G., 2016. The pale orange dot: the spectrum and habitability of hazy archean earth. Astrobiology 16, 873–899.

Baldocchi, D.D., Amthor, J.S., 2001. Canopy photosynthesis: history. In: Terrestrial Global Productivity. Academic Press, New York, NY, USA, pp. 9–31.

Bell, L.N., 1985. Energetics of the Photosynthesizing Plant Cell. Harwood Academic Publishers, Chur, Switzerland; New York, NY, p. 402.

Berghuijs, H.N.C., Yin, X., Ho, Q.T., Driever, S.M., Retta, M.-A., Nicolaï, B.M., Struik, P.C., 2016. Mesophyll conductance and reaction-diffusion models for CO_2 transport in C3 leaves; needs, opportunities and challenges. Plant Sci. 252, 62–75. https://doi.org/10.1016/j.plantsci.2016.05.016.

Björkman, O., Demmig, B., 1987. Photon yield of O 2 evolution and chlorophyll fluorescence characteristics at 77 K among vascular plants of diverse origins. Planta 170, 489–504.

Bloom, A.J., 2011. Energetics of nitrogen acquisition. In: Foyer, C.H., Zhang, H. (Eds.), Nitrogen Metabolism in Plants in the Post-genomic Era. Wiley-Blackwell, Chichester, UK, pp. 63–81.

Borghi, G.L., Moraes, T.A., Günther, M., Feil, R., Mengin, V., Lunn, J.E., Stitt, M., Arrivault, S., 2019. Relationship between irradiance and levels of Calvin–Benson cycle and other intermediates in the model eudicot Arabidopsis and the model monocot rice. J. Exp. Bot. 70, 5809–5825.

Bracher, A., Whitney, S.M., Hartl, F.U., Hayer-Hartl, M., 2017. Biogenesis and metabolic maintenance of rubisco. Annu. Rev. Plant Biol. 68, 29–60.

Burrows, P.A., Sazanov, L.A., Svab, Z., Maliga, P., Nixon, P.J., 1998. Duplicate identification of a functional respiratory complex in chloroplasts through analysis of tobacco mutants containing disrupted plastid ndh genes. EMBO J. 17, 868–876.

Catling, D.C., Zahnle, K.J., 2020. The Archean atmosphere. Sci. Adv. 6, eaax1420.

Cerovic, Z.G., Ounis, A., Cartelat, A., Latouche, G., Goulas, Y., Meyer, S., Moya, I., 2002. The use of chlorophyll fluorescence excitation spectra for the non-destructive in situ assessment of UV-absorbing compounds in leaves. Plant, Cell & Environment 25, 1663–1676.

de Groot, C.C., van den Boogaard, R., Marcelis, L.F.M., Harbinson, J., Lambers, H., 2003. Contrasting effects of N and P deprivation on the regulation of photosynthesis in tomato plants in relation to feedback limitation. J. Exp. Bot. 54, 1957–1967.

De Souza, A.P., Wang, Y., Orr, D.J., Carmo-Silva, E., Long, S.-P., 2020. Photosynthesis across African cassava germplasm is limited by Rubisco and mesophyll conductance at steady state, but by stomatal conductance in fluctuating light. New Phytol. 225, 2498–2512.

Dietz, K.J., Heber, U., 1984. Rate limiting fluxes in leaf photosynthesis 1. Carbon fluxes in the Calvin cycle. Biochim. Biophys. Acta 767, 432–443.

Ehleringer, J., 1981. Leaf absorptances of Mohave and Sonoran Desert plants. Oecologia 49, 366–370.

Eyland, D., van Wesemael, J., Lawson, T., Carpentier, S., 2021. The impact of slow stomatal kinetics on photosynthesis and water use efficiency under fluctuating light. Plant Physiol 186, 998–1012.

Farquhar, G.D., 1989. Models of integrated photosynthesis of cells and leaves. Philos. Trans. R. Soc., B 323 (1216), 357–367. https://doi.org/10.1098/rstb.1989.0016.

Farquhar, G.D., von Caemmerer, S., Berry, J.A., 1980. A biochemical model of photosynthetic CO2 assimilation in leaves of C3 species. Planta 149, 78–90. https://doi.org/10.1007/BF00386231.

Feild, T.S., Nedbal, L., Ort, D.R., 1998. Nonphotochemical reduction of the plastoquinone pool in sunflower leaves originates from chlororespiration. Plant Physiol. 116, 1209–1218.

Flamholz, A.I., Prywes, N., Moran, U., Davidi, D., Bar-On, Y.M., Oltrogge, L.M., Alves, R., Savage, D., Milo, R., 2019. Revisiting trade-offs between Rubisco kinetic parameters. Biochemistry 58, 3365–3376.

Galmés, J., Kapralov, M.V., Andralojc, P.J., Conesa, M.À., Keys, A.J., Parry, M.A.J., Flexas, J., 2014. Expanding knowledge of the Rubisco kinetics variability in plant species: environmental and evolutionary trends. Plant Cell Environ. 37, 1989–2001.

Garcia-Molina, A., Leister, D., 2020. Accelerated relaxation of photoprotection impairs biomass accumulation in Arabidopsis. Nat. Plants 6, 9–12.

Harbinson, J., Genty, B., Foyer, C.H., 1990. Relationship between photosynthetic electron transport and stromal enzyme activity in pea leaves: toward an understanding of the nature of photosynthetic control. Plant Physiol. 94, 545–553.

Harbinson, J., Hedley, C.L., 1993. Changes in P-700 oxidation during the early stages of the induction of photosynthesis. Plant Physiol. 103, 649–660.

Harbinson, J., Woodward, F.I., 1984. Field measurements of the gas exchange of woody plant species in simulated sunflecks. Ann. Bot. 53, 841–851.

Harris, G.C., Königer, M., 1997. The 'high' concentrations of enzymes within the chloroplast. Photosynth. Res. 54, 5–23.

Heber, U., Neimanis, S., Dietz, K.-J., Viil, J., 1986. Assimilatory power as a driving force in photosynthesis. Biochim. Biophys. Acta, Bioenerg. 852, 144–155.

Heldt, H.-W., Heldt, F., Piechulla, B., 2005. Plant Biochemistry., p. 630.

Hermida-Carrera, C., Kapralov, M.V., Galmés, J., 2016. Rubisco catalytic properties and temperature response in crops. Plant Physiol. 171, 2549–2561.

Hertle, A.P., Blunder, T., Wunder, T., Pesaresi, P., Pribil, M., Armbruster, U., Leister, D., 2013. PGRL1 is the elusive ferredoxin-plastoquinone reductase in photosynthetic cyclic electron flow. Mol. Cell 49, 511–523.

Hogewoning, S.W., Wientjes, E., Douwstra, P., Trouwborst, G., van Ieperen, W., Croce, R., Harbinson, J., 2012. Photosynthetic quantum yield dynamics: from photosystems to leaves. Plant Cell 24, 1921–1935.

Howard, T.P., Metodiev, M., Lloyd, J.C., Raines, C.A., 2008. Thioredoxin-mediated reversible dissociation of a stromal multiprotein complex in response to changes in light availability. Proc. Natl. Acad. Sci. U.S. A. 105, 4056–4061.

Inada, K., 1976. Action spectra for photosynthesis in higher plants. Plant Cell Physiol. 17, 355–365.

Kaiser, E., Zhou, D., Heuvelink, E., Harbinson, J., Morales, A., Marcelis, L.F.M., 2017. Elevated CO_2 increases photosynthesis in fluctuating irradiance regardless of photosynthetic induction state. J. Exp. Bot. 68, 5629–5640.

Kimura, H., Hashimoto-Sugimoto, M., Iba, K., Terashima, I., Yamori, W., 2020. Improved stomatal opening enhances photosynthetic rate and biomass production in fluctuating light. J. Exp. Bot. 71, 2339–2350.

Klughammer, C., Schreiber, U., 1994. An improved method, using saturating light pulses, for the determination of photosystem I quantum yield via P-700+ -absorbance changes at 830nm. Planta 192, 261–268.

Kono, M., Kawaguchi, H., Mizusawa, N., Yamori, W., Suzuki, Y., Terashima, I., 2019. Far-red light accelerates photosynthesis in the low-light phases of fluctuating light. Plant Cell Physiol., 1–11.

Kromdijk, J., Glowacka, K., Leonelli, L., Gabilly, S.T., Iwai, M., Niyogi, K.K., Long, S.P., 2016. Improving photosynthesis and crop productivity by accelerating recovery from photoprotection. Science 354, 857–861.

Lazár, D., Jablonský, J., 2009. On the approaches applied in formulation of a kinetic model of photosystem II: different approaches lead to different simulations of the chlorophyll a fluorescence transients. J. Theor. Biol. 257 (2), 260–269. https://doi.org/10.1016/j.jtbi.2008.11.018.

Leegood, R.C., 1990. Enzymes of the Calvin cycle. In: Lea, P.J. (Ed.), Methods in Plant Biochemistry. Vol. 3. Elsevier, pp. 15–37.

Liu, J., Zhang, J., Estavillo, G.M., Luo, T., Hu, L., 2021. Leaf N content regulates the speed of photosynthetic induction under fluctuating light among canola genotypes (*Brassica napus* L.). Physiol. Plant. 172, 1844–1852.

Lüning, K., Dring, M.J., 1985. Action spectra and spectral quantum yield of photosynthesis in marine macroalgae with thin and thick thalli. Mar. Biol. 87, 119–129.

Luo, R., Wei, H., Ye, L., Wang, K., Chen, F., Luo, L., Liu, L., Li, Y., Crabbe, M.J.C., Jin, L., 2009. Photosynthetic metabolism of C3 plants shows highly cooperative regulation under changing environments: a systems biological analysis. Proc. Natl. Acad. Sci. U. S. A. 106 (3), 847–852.

McAusland, L., Vialet-Chabrand, S., Jauregui, I., et al., 2020. Variation in key leaf photosynthetic traits across wheat wild relatives is accession dependent not species dependent. New Phytol. 228, 1767–1780.

McCown, R.L., Hammer, G.L., Hargreaves, J.N.G., Holzworth, D.P., Freebairn, D.M., 1996. APSIM: a novel software system for model development, model testing and simulation in agricultural systems research. Agr. Syst. 50, 255–271.

McCree, K.J., 1972. The action spectrum, absorptance and quantum yield of photosynthesis in crop plants. Agric. Meteorol. 9, 191–216.

Merzlyak, M.N., Chivkunova, O.B., Zhigalova, T.V., Naqvi, K.R., 2009. Light absorption by isolated chloroplasts and leaves: effects of scattering and 'packing'. Photosynth. Res. 102, 31–41.

Mills, B.J.W., Krause, A.J., Scotese, C.R., Hill, D.J., Shields, G.A., Lenton, T.M., 2019. Modelling the long-term carbon cycle, atmospheric CO_2, and earth surface temperature from late neoproterozoic to present day. Gondw. Res. 67, 172–186.

Mooney, H.A., Ehleringer, J., Berry, J.A., 1976. High photosynthetic capacity of a winter annual in death valley. Science 194, 322–324.

Morales, A., Kaiser, E., Yin, X., Harbinson, J., Molenaar, J., Driever, S.M., Struik, P.C., 2018a. Dynamic modelling of limitations on improving leaf CO_2 assimilation under fluctuating irradiance. Plant Cell Environ. 41 (3), 589–604. https://doi.org/10.1111/pce.13119.

Morales, A., Yin, X., Harbinson, J., Driever, S.M., Molenaar, J., Kramer, D.M., Struik, P.C., 2018b. In silico analysis of the regulation of the photosynthetic electron transport chain in C3 plants. Plant Physiol. 176 (2), 1247–1261. https://doi.org/10.1104/pp.17.00779.

Nawrocki, W.J., Bailleul, B., Picot, D., Cardol, P., Rappaport, F., Wollman, F.-A., Joliot, P., 2019. The mechanism of cyclic electron flow. Biochim. Biophys. Acta, Bioenerg. 1860, 433–438.

Noctor, G., Foyer, C.H., 1998. A re-evaluation of the ATP: NADPH budget during C3 photosynthesis: a contribution from nitrate assimilation and its associated respiratory activity. J. Exp. Bot. 49, 1895–1908.

Ogle, K., Barber, J.J., 2008. Bayesian data—model integration in plant physiological and ecosystem ecology. In: Lüttge, U., Beyschlag, W., Murata, J. (Eds.), Progress in Botany. Springer Berlin Heidelberg, Berlin, Heidelberg, pp. 281–311, https://doi.org/10.1007/978-3-540-72954-9_12.

Osmond, C.B., Björkman, O., Anderson, D.J., 1980. Physiological processes in plant ecology: toward a synthesis with Atriplex. Ecol. Stud. 36, 468.

Papanatsiou, M., Petersen, J., Henderson, L., Wang, Y., Christie, J.M., Blatt, M.R., 2019. Optogenetic manipulation of stomatal kinetics improves carbon assimilation, water use, and growth. Science 363, 1456–1459.

Parry, M.A.J., Andralojc, P.J., Scales, J.C., Salvucci, M.E., Carmo-Silva, A.E., Alonso, H., Whitney, S.M., 2012. Rubisco activity and regulation as targets for crop improvement. J. Exp. Bot. 64, 717–730.

Parry, M.A.J., Andralojc, P.J., Scales, J.C., Salvucci, M.E., Carmo-Silva, A.E., Alonso, H., Whitney, S.M., 2013. Rubisco activity and regulation as targets for crop improvement. J. Exp. Bot. 64, 717–730.

Parry, M.A.J., Keys, A.J., Madgwick, P.J., Carmo-Silva, A.E., Andralojc, P.J., 2008. Rubisco regulation: a role for inhibitors. J. Exp. Bot. 59, 1569–1580.

Patrick, L.D., Ogle, K., Tissue, D.T., 2009. A hierarchical Bayesian approach for estimation of photosynthetic parameters of C3 plants. Plant Cell Environ. 32 (12), 1695–1709. https://doi.org/10.1111/j.1365-3040.2009.02029.x.

Pearce, F.G., Andrews, T.J., 2003. The relationship between side reactions and slow inhibition of ribulose-bisphosphate carboxylase revealed by a loop 6 mutant of the tobacco enzyme. J. Biol. Chem. 278, 32526–32536.

Pearcy, R.W., 1990. Sunflecks and photosynthesis in plant canopies. Annu. Rev. Plant Biol. 41, 421–453.

Pignon, C.P., Leakey, A.D.B., Long, S.P., Kromdijk, J., 2021. Drivers of natural variation in water-use efficiency under fluctuating light are promising targets for improvement in Sorghum. Front. Plant Sci. 12, 1–16.

Post, W.M., Peng, T.-H., Emanuel, W.R., King, A.W., Dale, V.-H., DeAngelis, D.L., 1990. The global carbon cycle. Am. Sci. 78, 310–326.

Qu, M., Hamdani, S., Li, W., et al., 2016. Rapid stomatal response to fluctuating light: an under-explored mechanism to improve drought tolerance in rice. Funct. Plant Biol. 43, 727–738.

Raissig, M.T., Matos, J.L., Gil, M.X.A., et al., 2017. Mobile MUTE specifies subsidiary cells to build physiologically improved grass stomata. Science 355, 1215–1218.

Robinson, J.M., 1988. Does O_2 photoreduction occur within chloroplasts in vivo? Physiol. Plant. 72, 666–680.

Rolland, N., Curien, G., Finazzi, G., Kuntz, M., Maréchal, E., Matringe, M., Ravanel, S., Seigneurin-Berny, D., 2012. The biosynthetic capacities of the plastids and integration between cytoplasmic and chloroplast processes. Annu. Rev. Genet. 46, 233–264.

Rutherford, A.W., Osyczka, A., Rappaport, F., 2012. Back-reactions, short-circuits, leaks and other energy wasteful reactions in biological electron transfer: redox tuning to survive life in O_2. FEBS Lett. 586, 603–616.

Sakoda, K., Yamori, W., Groszmann, M., Evans, J.R., 2021. Stomatal, mesophyll conductance, and biochemical limitations to photosynthesis during induction. Plant Physiol. 185, 146–160.

Salter, W.T., Li, S., Dracatos, P.M., Barbour, M.M., 2020. Identification of quantitative trait loci for dynamic and steady-state photosynthetic traits in a barley mapping population. AoB Plants 12, 1–10.

Salter, W.T., Merchant, A.M., Richards, R.A., Trethowan, R., Buckley, T.N., 2019. Rate of photosynthetic induction in fluctuating light varies widely among genotypes of wheat. J. Exp. Bot. 70, 2787–2796.

Sassenrath-Cole, G.F., Pearcy, R.W., 1994. Regulation of photosynthetic induction state by the magnitude and duration of low light exposure. Plant Physiol. 105, 1115–1123.

Scheibe, R., 2004. Malate valves to balance cellular energy supply. Physiol. Plant. 120, 21–26.

Schöttler, M.A., Tóth, S.Z., 2014. Photosynthetic complex stoichiometry dynamics in higher plants: environmental acclimation and photosynthetic flux control. Front. Plant Sci. 5, 188.

Sharkey, T.D., Bernacchi, C.J., Farquhar, G.D., Singsaas, E.L., 2007. Fitting photosynthetic carbon dioxide response curves for C3 leaves. Plant Cell Environ. 30, 1035–1040. https://doi.org/10.1111/j.1365-3040.2007.01710.x.

Soleh, M.A., Tanaka, Y., Kim, S.Y., Huber, S.C., Sakoda, K., Shiraiwa, T., 2017. Identification of large variation in the photosynthetic induction response among 37 soybean [*Glycine max* (L.) Merr.] genotypes that is not correlated with steady-state photosynthetic capacity. Photosynth. Res. 131, 305–315.

Strand, D.D., Fisher, N., Kramer, D.M., 2017. The higher plant plastid NAD (P) H dehydrogenase-like complex (NDH) is a high efficiency proton pump that increases ATP production by cyclic electron flow. J. Biol. Chem. 292, 11850–11860.

Townsend, A.J., Retkute, R., Chinnathambi, K., Randall, J.W.-P., Foulkes, J., Carmo-Silva, E., Murchie, E.H., 2018. Suboptimal acclimation of photosynthesis to light in wheat canopies. Plant Physiol. 176 (2), 1233–1246. https://doi.org/10.1104/pp.17.01213.

Triantaphylidès, C., Krischke, M., Hoeberichts, F.A., Ksas, B., Gresser, G., Havaux, M., Van Breusegem, F., Mueller, M.-J., 2008. Singlet oxygen is the major reactive oxygen species involved in photooxidative damage to plants. Plant Physiol. 148, 960–968.

von Caemmerer, S., 2000. Biochemical Models of Leaf Photosynthesis. CSIRO Publishing, Collingwood, Australia.

Walker, B.J., Strand, D.D., Kramer, D.M., Cousins, A.B., 2014. The response of cyclic electron flow around photosystem I to changes in photorespiration and nitrate assimilation. Plant Physiol. 165, 453–462.

Witkowski, C.R., Weijers, J.W.H., Blais, B., Schouten, S., Sinninghe Damsté, J.S., 2018. Molecular fossils from phytoplankton reveal secular Pco2 trend over the Phanerozoic. Sci. Adv. 4, eaat4556.

Xinyou, Y., Van Laar, H.H., 2005. Crop Systems Dynamics: An Ecophysiological Simulation Model of Genotype-by-Environment Interactions. Wageningen Academic Publishers, Wageningen, The Netherlands, p. 155.

Yin, X., Schapendonk, A.H.C.M., Struik, P.C., 2018. Exploring the optimum nitrogen partitioning to predict the acclimation of C3 leaf photosynthesis to varying growth conditions. J. Exp. Bot. 70 (9), 2435–2447. https://doi.org/10.1093/jxb/ery277.

Yin, X., Struik, P.C., Romero, P., Harbinson, J., Evers, J.B., Van Der Putten, P.E.L., Vos, J., 2009. Using combined measurements of gas exchange and chlorophyll fluorescence to estimate parameters of a biochemical C_3 photosynthesis model: a critical appraisal and a new integrated approach applied to leaves in a wheat (*Triticum aestivum*). Plant Cell Environ. 32, 448–464. https://doi.org/10.1111/j.1365-3040.2009.01934.x.

Zhu, X.-G., de Sturler, E., Long, S.P., 2007. Optimizing the distribution of resources between enzymes of carbon metabolism can dramatically increase photosynthetic rate: a numerical simulation using an evolutionary algorithm. Plant Physiol. 145 (2), 513–526. https://doi.org/10.1104/pp.107.103713.

Zhu, X.G., Wang, Y., Ort, D.R., Long, S.P., 2013. e-Photosynthesis: a comprehensive dynamic mechanistic model of C3 photosynthesis: from light capture to sucrose synthesis. Plant Cell Environ. 36, 1711–1727. https://doi.org/10.1111/pce.12025.

CHAPTER

11

Photosynthesis in action: The global view

Emanuel Gloor, Roel Brienen, and David Galbraith

School of Geography, Faculty of Earth and Environment, University of Leeds, Leeds, United Kingdom

1 The past—Role of photosynthesis for co-evolution of life and earth

The most consequential long-term changes of the earth-environment for life caused by photosynthesis is oxygenation of the oceans and, with some delay, of the atmosphere. Originally, approximately 4 billion years ago, the earth atmosphere is thought to have been 'reducing', i.e., it did not contain oxygen. Oxygenation of oceans and atmosphere is the result of the release of oxygen as a by-product of water-splitting photosynthesis. The first organisms that developed this ability are thought to be cyanobacteria. Cyanobacteria contain light absorbing pigments, including chlorophyll *a*, the same pigment that is used by plants to capture light energy. In the geological record, cyanobacteria occur, amongst others, in layered mat-like colonies, and these strata have been estimated to be up to ~3.5 billion years old. Nonetheless, inferences of the dates of first emergence of photosynthetic cyanobacteria are difficult and have large uncertainties (e.g., Xiong and Bauer, 2002; Demoulin et al., 2019). Layered structures in the geological record resemble similar formations found today in shallow sea water, which are called stromatolites. Cyanobacteria have not only survived as single autonomous bacteria but have, over time, also been incorporated into plant leaves where they evolved into chloroplasts. As such, they are responsible for photosynthesis and thus constitute the base of most food-chains on land. Some cyanobacteria do not only photosynthesize but, unlike most other organisms, are also able to fix atmospheric nitrogen. This is an energetically very demanding process as atmospheric N_2 is a highly inert compound. The ability of cyanobacteria to fix atmospheric nitrogen makes them important for higher life as nitrogen is a central component of amino-acids, the building blocks of proteins.

As the first cyanobacteria lived in the sea, the release of oxygen as a waste product of photosynthesis will have first oxygenated the uppermost parts of the ocean where light is available. At a later stage, oxygen will also have started to build up in the atmosphere. One indicator of oxygen in the upper part of the sea is repeated events of iron deposition at the sea floor visible in sedimentary rocks as red bands of 'oxidised iron' alternating with greyish layers. The accepted explanation for the formation of these banded structures is from Preston Cloud (1972). According to Cloud, upwelling waters in the sea rich in dissolved iron (Fe^{2+}) have been oxidised to Fe^{3+} in the upper zone of the sea by oxygen produced by cyanobacteria. In contrast

Photosynthesis in Action
https://doi.org/10.1016/B978-0-12-823781-6.00002-2

to Fe^{2+}, Fe^{3+} is insoluble and thus 'rained' out in particulate form from the water column to the sediments. He attributed the periodic absence of bands to periods of low oxygenation as a result of cyanobacteria periodically poisoning themselves with the oxygen they released. This cycle ended once cyanobacteria had developed defences to live in oxygenated waters. The history of banded iron formations has been constructed by Isley and Abbott (1999) (Fig. 1) based on geological records. Iron band formation occurred first in the geological record around 3.8 billion years ago, with peak formation rate around 2.8 billion years ago and latest formation event around 1.8 billion years ago.

The timing of the onset of oxygenation of the atmosphere has been estimated in various ways. Many of the approaches have been criticised on methodological grounds (Holland, 2006). There is, however, one very elegant line of evidence that has set a firm date on the onset of atmospheric oxygenation. The inference is based on sulphur isotopes with atomic mass units 33 and 34 (^{33}S and ^{34}S). Isotopes of an element differ in the number of neutrons in the atom's nucleus whilst the number of protons in the nucleus and electrons are the same. During processes like diffusion, evaporation, chemical reactions, or physiological processes, the lighter isotope tends to be transported or react slightly more rapidly than the heavier isotope, and thus, the isotope ratio (the ratio of the abundances, e.g., in units of moles of the two isotopes) of the product of the process (or reaction) will be slightly shifted towards the lighter isotope. An illustrative example is carboxylation, the process by which CO_2 is fixed into precursors of sugars in leaves, during the Calvin cycle. This process slightly discriminates against the heavier isotope of the stable carbon isotopes, ^{13}C and ^{12}C. As a result, the isotopic abundance ratio $R \equiv {}^{13}C/{}^{12}C$, e.g., in units of moles per moles, in leaves is slightly lower than in air, or, i.e., leaves are slightly depleted in the heavier isotope. Processes that lead to slight shifts in isotopic ratios are called fractionation processes. Because differences in isotope ratios as a result of natural fractionation processes tend to be small, and isotope ratios can be measured well with mass spectrometers, but not individual isotope abundances, differences are reported as relative deviations δ with regards to a standard R_{st} multiplied by a factor 10^3, i.e., $\delta \equiv \frac{R_{sample} - R_{st}}{R_{st}} \cdot 10^3$. Most known fractionation processes share an interesting property. Fractionation increases linearly proportionally with the mass difference between isotopes. For example, fractionation of ^{34}S with regards to ^{32}S is twice as large as fractionation of ^{33}S with regards to ^{32}S (Fig. 2). This fractionation process property is called mass-dependent fractionation. There are however processes, occurring particularly in the upper atmosphere, which involve photolysis reactions triggered by high energy solar radiation, that

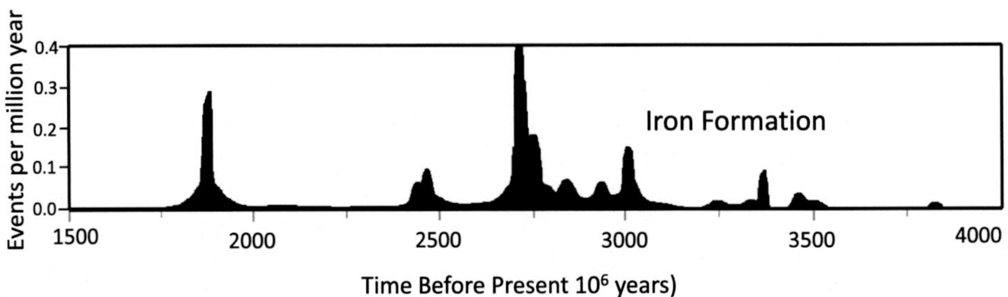

FIG. 1 Sum of Gaussian functions representing the occurrence of banded iron formations from data compiled by Isley and Abbott (1999). *From Isley, A., Abbott, D., 1999. Plume-related mafic volcanism and the deposition of banded iron formation. J. Geophys. Res. 104 (B7), 15461–15477.*

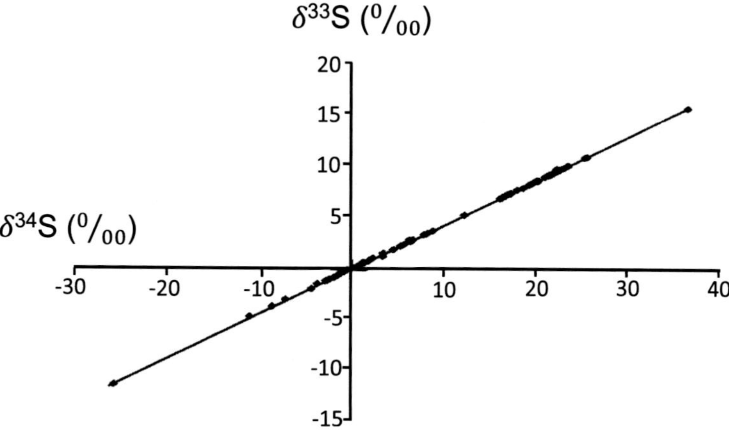

FIG. 2 $\delta^{33}S$ versus $\delta^{34}S$ measured on terrestrial rock samples younger than 100 Ma. *From Farquhar, J., Savarino, J., Airieaua, S., Thiemens, M.H., 2001. Observation of wavelength-sensitive mass-independent sulfur isotope effects during SO2 photolysis: implications for the early atmosphere. J. Geophys. Res. 106(E12), 32829–32839.*

cause mass-independent fractionation. A measure whether mass-independent fractionation processes are occurring is the difference $\Delta^{33}S \equiv \delta^{33}S - 0.5 \cdot \delta^{34}S$, with $\Delta^{33}S \neq 0$ indicating mass-independent fractionation.

How is this related to the oxygenation of the atmosphere? One pathway of dead organic matter decomposition by bacteria in ocean sediments involves formation of pyrite (FeS_2), which occurs in the presence of hydrogen sulphide (H_2S) (conditions which are found today, e.g., at depth in the black sea). Pyrite is a mineral that looks slightly like gold. However, unlike gold, in the presence of oxygen, it decays rapidly. For this reason, it is also known as fool's gold. Now, $\Delta^{33}S$ in pyrite (and baryte) measured across a wide range of rock samples whose ages span the last 4 billion years of the earth history show an intriguing pattern. Before 2.5 billion years ago, there is mass in-dependent fraction, but thereafter, mass in-dependent fractionation disappears (Fig. 3). This pattern can be explained as follows. Mass independent fraction of sulphur compounds can and will occur in an atmosphere void of oxygen, not however in an atmosphere with oxygen that prevents photolysis reactions (Farquhar et al., 2001). The onset of mass-dependent fractionation of sulphur thus coincides with the timing when the atmosphere had reached sufficient levels of oxygen to shield off photolysis reactions of atmospheric sulphur compounds. The oxygenation of the atmosphere is effectively thought of having occurred in steps reflecting various innovations of life as indicated in Fig. 4. Cyanobacteria are thought to have had a substantial influence on atmospheric O_2 from 2.7 billion years ago, elevating atmospheric O_2 to approximately 25% of today's level by 2

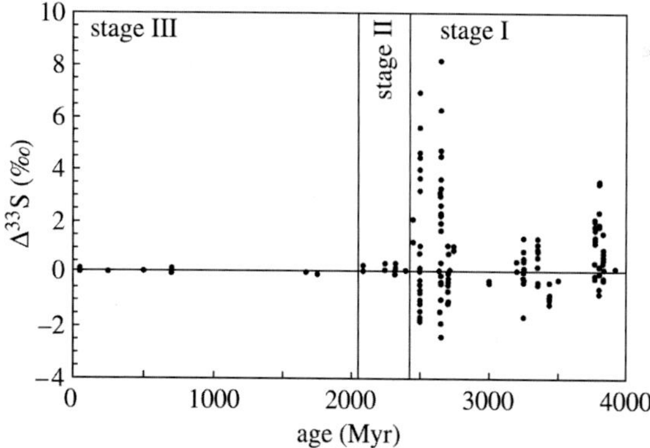

FIG. 3 $\Delta^{33}S$ in rock samples with ages covering the past 4 billion years. *From Holland, H.D., 2006. The oxygenation of the atmosphere and oceans. Philos. Trans. R. Soc. B 361, 903–915. https://doi.org/10.1098/rstb.2006.1838.*

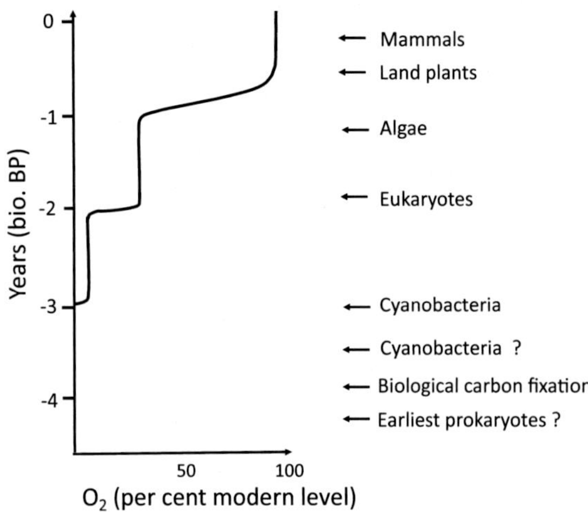

FIG. 4 Schematic of the evolution of atmospheric oxygen over earth's history. *Based on Xiong, J., Bauer, C.E., 2002. Complex evolution of photosynthesis. Annu. Rev. Plant Biol. 53, 503–521.*

billion years ago. First, eukaryotes (organisms whose cells contain a membrane-enclosed nucleus which stores genetic information) appeared around 1.9 billion years ago and led to a further increase of atmospheric O_2 by roughly 1.8 billion years ago. The appearance of algae led to elevation of atmospheric O_2 to contemporary levels (Xiong and Bauer, 2002).

One important effect of the oxygenation of the atmosphere is the formation of the ozone layer in the stratosphere via photolysis of O_2, which splits O_2 into two single oxygen atoms. Ozone O_3 is then formed via combination of O_2 and O. The ozone layer absorbs most of the high energy (ultraviolet) portion of solar radiation, which reaches the earth. This high energy radiation damages DNA, and thus, once the ozone layer had been formed, it provided a protecting shield for life outside the sea. It thus enabled colonisation of land by living organisms. Plants emerged on land in abundance around 0.5 billion years ago.

As already mentioned, oxygenation of the atmosphere is the result of photosynthesis by organisms and subsequent burial of the organisms in sediments after death. Sediments slowly grow via accumulation of new material and thus already buried matter experiences steadily increasing pressure and temperature, which eventually leads to transformation of the organic material (hydrocarbons) into oil and gas. Oil is a mixture of a wide variety of structurally and compositionally quite similar compounds. They consist primarily of the elements carbon (C) and hydrogen (H) following the general formula C_nH_m (where n and m are natural numbers). Structures include C–H chains (C_nH_{2n+2}, alkanes), cyclic (aromatic) compounds, as well as molecules containing both chains and cyclic compounds.

Photosynthesis not only splits water and thereby releases oxygen into the atmosphere but also exerts control on atmospheric CO_2. Burial of organic carbon removes CO_2 from the atmosphere. An illustration of the control by plants of the atmospheric CO_2 is the time-history of atmospheric CO_2 reconstructed based on the carbon isotopic ratio ($^{13}C/^{12}C$), of alkenones (Fig. 5). Alkenones are produced by some algae and are long carbon chains with an oxygen atom towards one end of the chain (e.g., $C_{37}H_{68}O$). The carbon isotopic ratio of alkenones depends on the CO_2 concentration of ocean water, which is related to, and thus an indicator of, atmospheric CO_2 (Pagani et al., 2009). There are two striking features in the reconstructed atmospheric CO_2 record (Fig. 5). CO_2 levels have decreased quite dramatically over the past 50 million years and from roughly 20 million years ago onwards atmospheric CO_2 levels have remained nearly constant, i.e., they have stabilised.

The atmospheric CO_2 decline is thought to be the result of vegetation carbon capture and vegetation mediated enhanced rock weathering, which consumes atmospheric CO_2 and leads to enhanced carbon burial in ocean sediments. The relatively stable CO_2 levels since 24 million years ago indicate the onset of a negative feedback: atmospheric CO_2 is the life ingredient of plants and if it reaches excessively low levels, plants will start to starve and will cease to draw

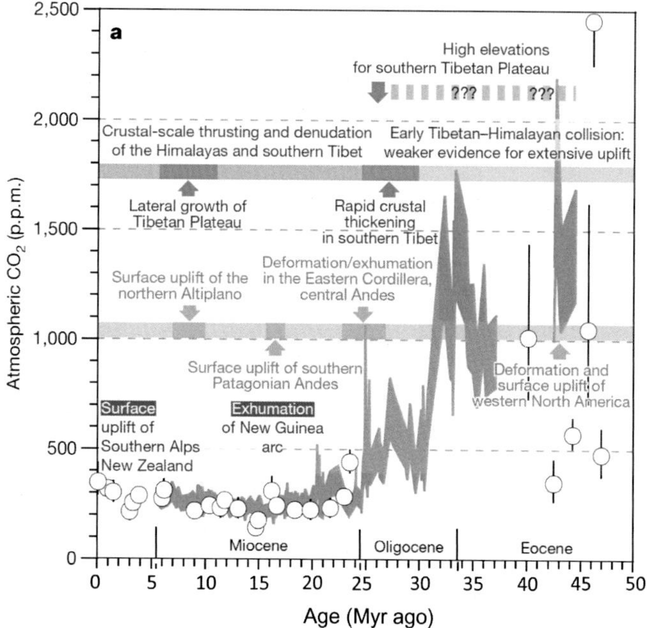

FIG. 5 Reconstruction of atmospheric CO_2 over the past 50 million years based on the $^{13}C/^{12}C$ ratio of alkenones produced by algae. *From Pagani, M., Caldeira, K., Berner, R., Beerling, D.J., 2009. The role of terrestrial plants in limiting atmospheric CO_2 decline over the past 24 million years. Nature 460, 85–88.*

down further atmospheric CO_2 both directly through reductions in photosynthesis and indirectly by slowing rock weathering. They, thus, became the victim of their own success.

Because CO_2 is a greenhouse gas, changes in atmospheric CO_2 levels by plant uptake and burial change the amount of thermal longwave emission from the earth's surface absorbed in the atmosphere, thereby changing the temperature of earth's surface. This process may possibly have contributed to onset of glaciations ~300 billion years ago, at the time of expanded colonisation of land by plants (Berner, 1990).

2 Present—Reversal of long-term natural carbon capture

Since the mid-19th century, humans have understood the full potential of the use of fossil fuels for energy generation and how to extract fossil fuels. Harvesting this resource at ever increasing scale has completely changed the world in which we live. It has enabled revolutions in many aspects of human life like health and longevity (medicine), transport, trade, acquisition of resources (e.g., metals), construction, which together enabled unprecedented exponential growth of the human population but at the same time also widespread destruction of the natural environment, vegetation and fauna. The fulminant reversal of billions of years of carbon capture by living organism is now clearly noticeable in the global records of atmospheric gases like CO_2 and O_2, with atmospheric CO_2 levels in the year 2020 having reached nearly 50% higher levels compared to preindustrial times (AD 1750) (Fig. 6, top panel). Because atmospheric O_2 molar fraction (mol O_2 (mol air)$^{-1}$) is approximately 500 times larger than atmospheric CO_2 molar fraction, relative changes in atmospheric O_2 are much smaller and thus more

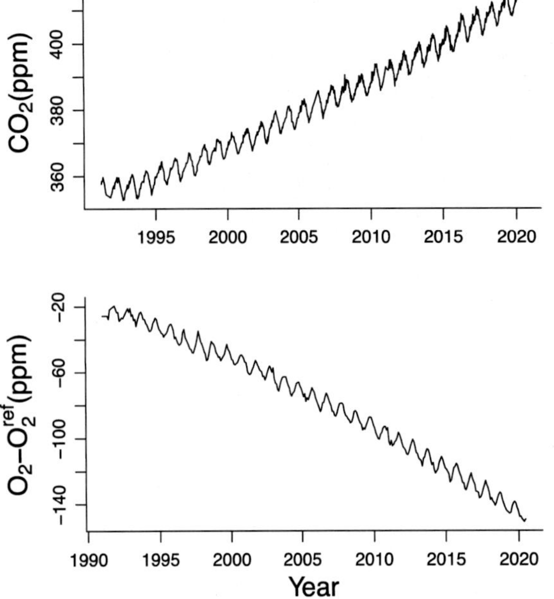

FIG. 6 Atmospheric CO_2 (top panel) and O_2 (bottom panel) measured by R. Keeling and his group at Mauna Loa, Hawaii.

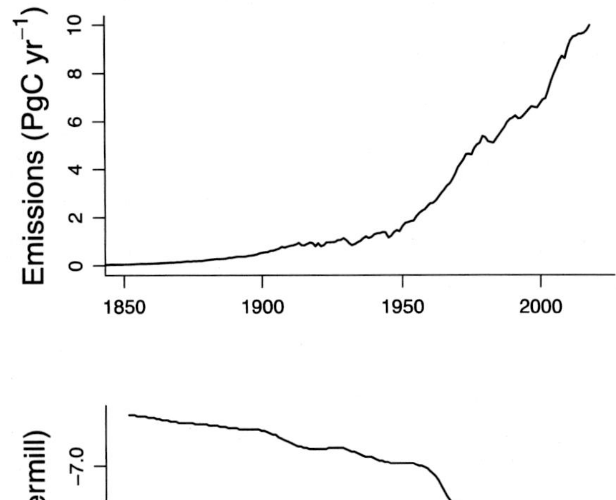

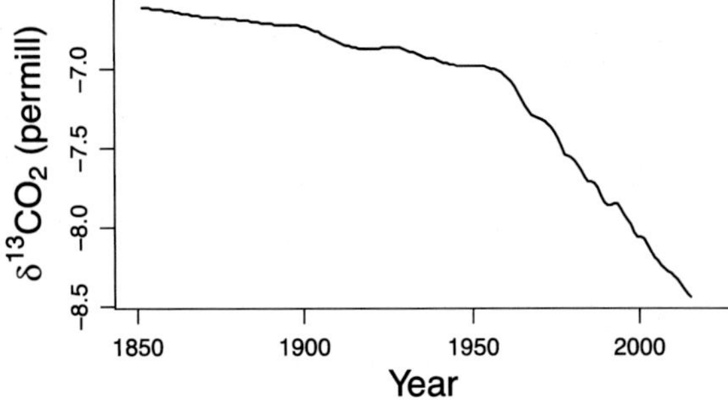

FIG. 7 Fossil fuel carbon emissions estimated from energy statistics (Andres et al., 1999), (top panel), and atmospheric $\delta^{13}CO_2$ data compiled by Graven et al. (2017) (bottom panel).

difficult to measure. In today's atmosphere, oxygen levels are $0.21\,\text{mol mol}^{-1}$ whilst CO_2 levels are approximately $0.0004\,\text{mol mol}^{-1}$. Nonetheless, it has been possible to demonstrate that O_2 levels are indeed decreasing as a result of oxidation (burning) of fossil fuels (Keeling and Shertz, 1992; Fig. 6, bottom panel). Because O_2 levels in the atmosphere are so much larger than CO_2 levels, even if very large amounts of fossil fuels are being burnt, it will have very little impact on the overall quantities of atmospheric O_2. If the atmospheric oxygen decrease, in Fig. 6, bottom panel, was entirely due to fossil fuel burning, the ratio $\frac{\Delta O_2}{\Delta CO_2}$ (mol mol^{-1}) between changes in oxygen and carbon dioxide would be ≈ -1.4 [determined by the mean ratio n/m of C to H of fossil fuels according to $\frac{\Delta O_2}{\Delta CO_2} = -\left(1 + \frac{1}{4}\frac{m}{n}\right)$ (e.g., Randerson et al., 2006)].

A second line of evidence that the increase in atmospheric CO_2 is caused by fossil fuel burning is the time history of atmospheric $^{13}CO_2$. As explained before, photosynthesis by plants and microorganisms slightly favours fixation of the light isotope ^{12}C over the heavier isotope ^{13}C. Thus, organic matter converted to fossil fuels will be slightly depleted in ^{13}C compared to the atmosphere. Therefore, fossil fuel burning will lead to increasingly lighter (i.e., larger ^{12}C fraction) atmospheric CO_2 or, i.e., a continuing decrease in $\delta^{13}CO_2$. This phenomenon is called Suess effect and is indeed observed (Fig. 7, bottom panel).

Whilst the discovery of cheap energy generation has permitted development of the world at an incredible speed, by relying very heavily on ongoing energy provision, humans have manoeuvred themselves into a very difficult situation. This is so because CO_2 is a greenhouse gas and, although the warming potential per molecule is comparably small, given the magnitude of fossil fuel reserves burned so far, and potentially to be burned in the future, it has a substantial effect on earth's surface temperatures (sea and land) directly and indirectly by heating up the oceans. The use of fossil fuel energy and economic stability and high living standards (existing or aspired to) are so closely intertwined that resolving this problem is very difficult. So far, earth's surface temperature has increased by nearly 1.5°C compared to preindustrial times, with warming of the land surface being substantially greater than warming of the sea surface. A doubling of atmospheric CO_2 is likely to lead to warming of the earth's surface by approximately 3°C (Sherwood et al., 2020). Known fossil fuel reserves, if all burned, are more than sufficient to reach such levels. Eventually, fossil fuel reserves will run out, nonetheless recovery of the climate system to preindustrial levels will be slow on time-scales relevant to humans, lasting hundreds to thousands, to ten thousands of years (like, e.g., ice sheet regrowth).

Currently, both the land vegetation and oceans moderate the human made 'carbon problem' by taking up each roughly 25% of CO_2 entering the atmosphere as a result of fossil fuel burning and cement manufacture (Friedlingstein et al., 2019). CO_2 uptake by the oceans is primarily a physical process. The human-caused excess atmospheric CO_2 partial pressure in the atmosphere drives invasion of anthropogenic excess CO_2 into the oceans. However, this acidifies ocean waters and threatens life in the oceans. The reasons for CO_2 uptake on land are a bit less clear, although temperature increase at high latitudes increases forest growth whilst in the tropics increased atmospheric CO_2 levels may have contributed to observed increases in tropical forest carbon gains (Brienen et al., 2015). The future of this land carbon sink is uncertain, partially because not enough is known about sensitivity/vulnerability of land and ocean vegetation to rapid changes in climate, and partially because implementation of demographic processes like tree mortality are difficult to represent realistically in existing land vegetation models. Such processes may be major causes of feedbacks.

Large organic carbon pools, which potentially may be released to the atmosphere quite rapidly and thus feedback on climate, include forest vegetation, soil organic matter in frozen and unfrozen form, and peat. As an example, carbon stored in aboveground mass in forests is estimated to be on the order of 400–500 PgC (10^{15} g C). Another example is carbon stored in peat. Peatlands in the boreal and arctic zone are estimated to store on the order of 450 PgC (Gorham, 1991). To put this in perspective, the atmosphere currently holds approximately 800 PgC in form of CO_2. Another characteristic illustrating the importance of the carbon exchanged by the land biosphere and atmosphere is the seasonal cycle it causes in atmospheric CO_2 (Fig. 6). During northern hemisphere summer, land vegetation in the northern hemisphere takes up large amounts of carbon via photosynthesis of which a substantial part is released back during the rest of the year.

How large are these fluxes at the leaf and forest stand level? Leaf-level carbon assimilation rates are on the order of 10 $\mu mol\,m^{-2}\,s^{-1}$ or $\sim$3.5 $gC\,m^{-2}\,day^{-1}$. Assuming that photosynthesis occurs during 8 h day^{-1}, this corresponds to $\sim$350 $kgC\,ha^{-1}\,day^{-1}$. The process of photosynthetic conversion of light to metabolites, ATP (adenosine triphosphate) and NADPH (nicotinamide adenine dinucleotide phosphate), which are then used to fix carbon in the form of sugars and starch, is called gross primary productivity (GPP). The total amount of carbon fixed annually (i.e. GPP), has been estimated to be on the order of 100 to 150 PgC (1 Pg $= 10^{15}$ g) (e.g., Welp et al., 2011). A quantity related to GPP is net primary production (NPP) (e.g., Woodward, 2007). NPP is GPP minus the carbon lost due to respiration (oxidation of sugars) to produce the necessary energy for growth and maintenance of a plant. Thus, NPP is the uptake rate of carbon, which ends up as structural (wood) and nonstructural carbohydrates (sugars).

One way to estimate NPP is based on satellite information. The information used includes spatial maps of photosynthesising mass and of light absorbed by photosynthesising biomass. NPP is then estimated as

$$\mathrm{NPP} = \varepsilon \cdot I_{\mathrm{abs}} \cdot B \tag{1}$$

where ε is a light use efficiency (gC (mol photons)$^{-1}$), I_{abs} absorbed photosynthetically active radiation (mol photons (kg biomass)$^{-1}$day^{-1}), and B is photosynthesising biomass (kg m^{-2}) (Field et al., 1995). Photosynthetically active radiation refers to the solar radiation in the frequency band in which plant pigments are able to absorb solar radiation (which based on measurements has been defined as all solar radiation at earth's surface between 400 and 700 nm). The light use efficiency has been estimated using on-ground observations. The product of absorbed photosynthetic radiation and photosynthesising biomass (FAPAR) is estimated based on the so-called normalised difference vegetation index (NDVI). The definition of NDVI is motivated by the fact that pigments of plants absorb in the red (RED) frequency part of the solar spectrum but not in the near infrared (NIR) part of the spectrum. Thus, the relative difference of reflected light in the near infrared and the red, NDVI$\equiv \frac{\mathrm{NIR}-\mathrm{RED}}{\mathrm{NIR}+\mathrm{RED}}$ should be related to available pigments (chlorophyll) and the photosynthetically active radiation (PAR) they absorb.

The spatial pattern of annual mean net primary productivity estimated in this way (Fig. 8) reflects the main controls, precipitation, light, and temperature. Globally maxima are highly spatially correlated with precipitation, including the humid forest in the tropics (Amazon, Congo, Indonesia) as well as the North America East coast and Western Europe. Productivity decreases towards high latitudes where there is a strong seasonality in light and increasingly low temperatures.

Whilst NDVI and related remote sensing indices give an indication of greenness of the earth surface, which is related to chlorophyll spatial density, they are only indirectly related to the photosynthesis process in leaves. Another aspect of photosynthesis that can be measured from space is the so-called solar induced fluorescence (SIF). Solar energy captured by pigments in leaves (primarily chlorophyll) elevate electrons to higher energy levels. The electrons in

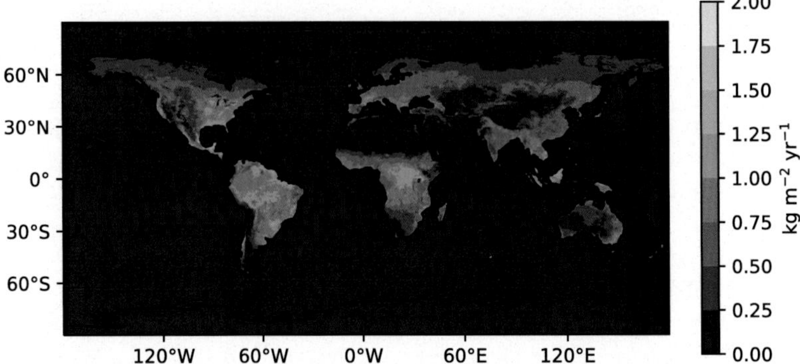

FIG. 8 Annual mean land vegetation net primary productivity (kg C m^{-2} yr^{-1}) predicted with the CASA model (Field et al., 1995) based on remote sensing data of absorbed photosynthetically active radiation using Eq. (1) (model used for the simulations provided to the authors by G. van der Werff).

an excited state can lose their energy in various ways; most importantly, they may be released causing effectively an electric current across a series of electron acceptors, which leads to the creation of high energy chemical compounds (ATP, adenine triphosphate). These are then used in the Calvin cycle to fix CO_2. Electrons in an excited state may also lose their energy transitioning back to the original energy level via radiation at slightly higher wavelengths around 700nm (for details, see Chapter 1 of this book). This phenomenon is called fluorescence (see Chapters 6 and 10 of this book). At most frequencies of the solar spectrum, the fluorescence signal is much smaller than solar radiation reflected from canopies. However, there are narrow wavelength windows where nearly all solar radiation is absorbed by oxygen in the earth's atmosphere, the oxygen bands. Thus in these windows reflected solar radiation is small enough that fluorescence from photosynthesis can be detected from space. Global SIF data permit to detect anomalies in photosynthesis, e.g., as a result of drought. Such data exist since approximately one decade and so-far are mainly used to determine how photosynthesis at large scale responds to stressors like heat and drought (e.g., Somkuti et al., 2020).

3 Future of the earth system—Modelling approaches, predictions of ongoing and future changes, and comparison with observations

In order to predict how earth's climate will evolve with increasing atmospheric CO_2 levels, highly sophisticated and increasingly realistic computer models of the coupled ocean atmosphere system, the so-called climate models, have been developed. Development of these began in the late 60s and early 70s of the last century (Manabe and Bryan, 1969). Starting from an initial atmosphere and ocean state, these models integrate forward in time the physics equations which describe how the so-called fluids like atmospheric air and ocean water, as well as ice of polar ice caps, evolve over time. The equations include the application of Newton's law of motion and the laws of thermodynamics to fluids. The equations are numerically solved at the corners of three-dimensional grids covering the earth surface and extending vertically into the atmosphere, consisting, e.g., of 1×1 degree longitude by latitude rectangles and $\sim$20 vertical layers both for oceans and atmosphere. For regular reports by the IPCC (United Nations International Panel on Climate Change), a set of fixed atmospheric greenhouse case scenarios are run regularly by approximately 20 climate models. IPCC reports are being prepared approximately every 5years. These simulation efforts are also known as Climate Modelling Intercomparison Projects (CMIPs) (e.g., Eyring et al., 2016).

A key element of these models is realistic representation of how solar radiation and long-wave radiation emitted from the earth are being absorbed and re-emitted by different molecules in the atmosphere. The process of how electromagnetic radiation interacts with gases in the atmosphere and thus to what extent and in what form it passes through the atmosphere is called 'radiative transfer'. Realistic representation of radiative transfer in the atmosphere is crucial for predicting the effects on climate of changes of greenhouse gas composition of the atmosphere due to fossil fuel burning.

One main purpose of these model simulations is to establish how earth's surface temperature will change for a doubling of CO_2 compared to preindustrial times, a parameter called the 'climate sensitivity' of the earth climate system (Charney et al., 1979; Sherwood et al., 2020). Important additional predictions of the models include how surface temperatures and the hydrological cycle, and thus precipitation, will change, as well as how sea levels will rise.

Climate models have been and continue to be constantly improved. One ongoing improvement is a steadily continuing increase of spatial resolution, which is made possible by increasingly powerful computers. Another improvement has been the incorporation of those components of the global carbon cycle, which operate on time-scales up to several centuries and millenia (i.e., neglecting processes operating on geological timescales). Incorporation of the global carbon cycle is important because CO_2 is the most important greenhouse gas and thus is central for prediction of the fate of CO_2 released to the atmosphere by fossil fuel burning and cement manufacture. It is also important because of the feedback between large organic carbon pools on land and climate. The carbon stored by these pools may potentially be released rapidly to the atmosphere and thus feedback on climate. Coupled climate and carbon cycle models are called Earth-System models (Foley et al., 1996).

Photosynthesis is at the heart of the global carbon cycle and thus an important component of Earth-System models. Realistic representation of photosynthesising organisms in these models is not only important for future predictions of climate, but, e.g., also to predict whether and which crops may be grown in various parts of the world in the future, particularly the tropics, which are predicted to host half of world's human population by 2050. A prerequisite for these predictions is determination and use of realistic sensitivities of vegetation to heat and drought stress.

In the sequel, we focus on and discuss how the land carbon cycle and specifically the functioning of vegetation is incorporated into these Earth-System models. Representation of ocean biota (phytoplankton, zooplankton) and primary productivity is done in many respects in a similar way but will not be covered in detail in this chapter. We will first explain how the carbon cycle on land is conceptualised in these models, focus on how photosynthesis is represented in these models, and then discuss observed and predicted changes of land vegetations over recent decades and the decades to come and highlight possible improvements of these models.

3.1 Carbon cycle representation in Earth-System models

The carbon cycle in land vegetation models is usually conceptualised as a suite of distinct carbon pools and a cascade of carbon fluxes from one pool to another and finally back to the atmosphere. Pools include inorganic carbon both in the atmosphere and oceans, biomass (or organic carbon) in living organisms (plants on land, biota in the oceans), dead organic matter like leaf litter, dead branches, dead roots, older decomposing organic matter in soils on land, and dead organisms both in the water columns and after sinking to the abyss at and buried in the ocean sediment.

Photosynthesis. Central for the carbon cycle component in these models is photosynthesis. It creates the initial flux of carbon from the atmosphere, or from dissolved carbon in the sea, into organic carbon pools, and thus is the starting point of the carbon cascade both on land and the oceans. Photosynthesis on land is probably the most complex process represented in these models. Its description in models includes two components, flux of CO_2 into the leaves via small valves, the stomata, and the biochemical process of carbon fixation, the so-called carboxylation reaction, in chloroplasts inside the leaves (Fig. 10). Stomata are located on the lower side of plant leaves. The length of stomata is on the order of 10–20 μm and its width 3–12 μm, and the spatial density on leaves is on the order of 100–1000 mm^{-2}. For each CO_2 molecule entering the leaf on the order of 300 to 500, H_2O molecules are lost to the atmosphere. As water is essential for keeping plant organs alive, opening and closing of these valves is regulated by plants. Two environmental variables are of particular importance for regulation of stomatal opening, soil water content, and the difference between saturated water vapour pressure inside the stomata and water vapour pressure in air. The larger this difference, the higher the loss of water via open stomata. Thus, plants regulate stomatal opening in response to environmental conditions, and therefore, the hydrological cycle plays an important role for carbon uptake. Because of its key role and because photosynthesis is the central theme of this book, the coupled leaf carbon uptake and carbon assimilation model used in Earth-System models will be explained in detail later.

Plant metabolism and growth. The photosynthesising part of land vegetation has until recently mostly been conceptualised as one homogeneous photosynthesising layer, or 'big leaf'. A few (~10–15) different plant functional types like tropical evergreen trees or temperate deciduous trees with somewhat different characteristics are usually distinguished. Carbon gained during photosynthesis is fixed as carbohydrates. These in turn are used as energy source to maintain the metabolism of the plant and for growth of new tissue. The loss of carbon used for metabolism is called autotrophic respiration. The remaining fixed carbon is distributed to the leaf, stem, and root carbon pools using allocation rules.

Dead organic carbon. Plants and trees shed their leaves, branches, and roots and eventually die. This dead material then follows two paths, both leading to release of CO_2 back to the atmosphere but on different time-scales (or equivalently turnover times). Partly the compounds are being oxidised when lying on the ground, releasing CO_2 directly back to the atmosphere, partly more recalcitrant compounds are infiltrated into the soil profile, where they decompose on a wide range of time-scales ranging from weeks to months, to hundreds and even thousands of years. As an example, lignins, which strengthen rigidity of cells, e.g., of tree trunks, and leaf waxes (Alkanes) which protect leaves, decompose very slowly (up to thousands of years 'decay time'). Decomposition of dead organic matter is strongly facilitated by soil organisms, such as invertebrates, bacteria, and fungi and called 'heterotrophic respiration'.

Altogether the land component of the carbon cycle in Earth-System models includes (i) a hydraulics component for predicting plant gas exchange regulation via soil water content and vapour pressure deficit, the difference between vapour pressure inside the stomata and in air, (ii) a carbon fixation 'engine' (photosynthesis), (iii) a plant growth model using an allocation scheme, and (iv) a dead organic carbon decomposition model (heterotrophic respiration) using a range of characteristic turnover times (Fig. 9). Importantly many of the modelled processes are directly sensitive to climate and give rise to a wide range of future predictions of how productivity and biomass will respond to climate.

3.1.1 Carbon assimilation at the leaf level

This process is represented in Earth-System models by combining models of two processes: provision of CO_2 via diffusion through stomata, and a model of carbon fixation capacity in the leaf dictated by light absorption and biochemical processes in the leaf. At the global scale, two types of photosynthesis are important:

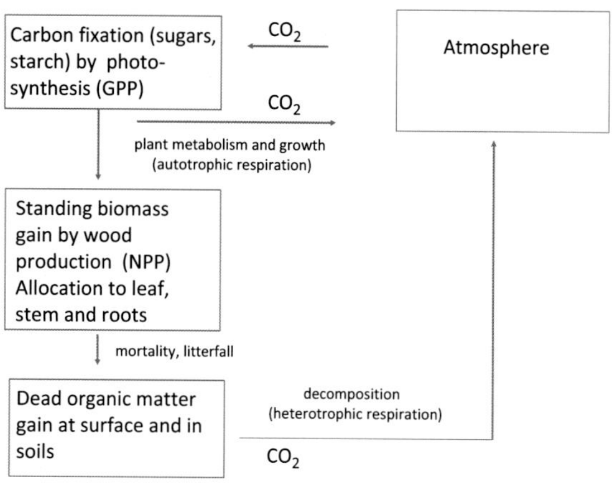

FIG. 9 Schematic of land vegetation representation in Earth-System models.

so-called C3 photosynthesis and C4 photosynthesis (see Chapter 3). Here, we discuss the C3 model used in Earth-System models, which has been formulated by Farquhar et al. (1980). The C4 model derivation follows similar steps. C3 is thought to account for ~90% of photosynthesis on land. Farquhar, von Caemmerer and Berry used the terminology 'supply' for the process of CO_2 provision and 'demand' for carbon fixation capacity. Following this conceptualisation, we first discuss the 'supply' component, followed by the 'demand' component and then explain how these two models are being combined to estimate carbon assimilation rates given atmospheric CO_2, relative humidity, wind speed, and leaf length.

3.1.2 *Carbon supply to leaf ('supply side')*

Carbon assimilation in chloroplasts via the Calvin cycle depends on CO_2 as one of two substrates. Thus, the supply rate of CO_2 is one limiting factor for CO_2 fixation (assimilation).

The flux of CO_2 into stomata and then to the carboxylating site is described as a diffusion process (also see Chapter 3). Diffusion smoothes out concentration differences. Diffusive transport is thus directed from high concentration towards low concentration. For the process of photosynthesis, diffusive CO_2 flux is described as $F = g \cdot \Delta c$, where g is a conductance in units of (mol m^{-2} s^{-1}) and Δc is CO_2 molar fraction difference (mol mol^{-1}) along the pathway (Cowan, 1977). Stomatal conductance differs for different compounds in the air, e.g., CO_2 and H_2O. The inverse of a conductance is a resistance. In most descriptions, two resistances in series along the pathway of air to the carboxylating site are taken into account: diffusive flux through the leaf boundary layer $F_b = \frac{1}{r_b} \cdot (c_a - c_s)$, a thin nonturbulent air layer directly adjacent to the leaf surface, followed by diffusive flux through stomata $F_s = \frac{1}{r_s} \cdot (c_s - c_i)$. Here c_a, c_s, and c_i are CO_2 molar fraction (μmol mol^{-1}) in the atmosphere, at the leaf surface and, inside stomata often assumed to be equal to CO_2 molar fraction at the carboxylation site (Fig. 10). Using the rule for combining resistances in series, the flux from the atmosphere to the carboxylating site can be described as

$$F_{CO_2} = \frac{1}{r_b + r_s}(c_a - c_i) = \frac{1}{\frac{1}{g_b} + \frac{1}{g_s}}(c_a - c_i)$$

The leaf boundary conductance is parameterized via a typical width L_w (m) of a leaf and wind speed W_s (m s^{-1}) as

$$g_b = c\sqrt{\frac{W_s}{L_w}}$$

where $c = 0.11$ for CO_2, is a constant (mol m^{-2} s$^{+1/2}$). Stomatal conductance is based on a parameterisation, which has been found by Ball et al. (1987) based on analysis of a large experimental dataset. When plotting stomatal conductance against a wide range of parameter combinations, they discovered that the data collapse on a line if they plot stomatal conductance against the ratio between CO_2 assimilation rate A (mol m^{-2} s^{-1}) multiplied by relative humidity h_s (−) and CO_2 molar fraction c_s at the leaf surface (Fig. 11). Thus, stomatal conductance can be expressed as

$$g_s = g_1 \cdot \frac{h_s}{c_s} \cdot A + g_0$$

where g_1, g_0 are constants. Based on these three equations, CO_2 assimilation rate A_{phys} can be predicted as a function of h_s, c_a, c_i, W_s, L_w, and if h_s, c_a, W_s, L_w are given as a function, $A_{phys} = A_{phys}(c_i)$. The stomatal conductance model has been extended by Leuning (1995) to $g_s = \frac{g_1 \cdot A}{(c_s - \Gamma_*)\left(1 + \frac{D}{D_0}\right)} + g_0$, where $D = (e_{sat} - e)$ is vapour pressure deficit with e_{sat} saturation vapour pressure and e actual vapour pressure, D_0 a reference vapour pressure deficit value, and Γ_* the CO_2

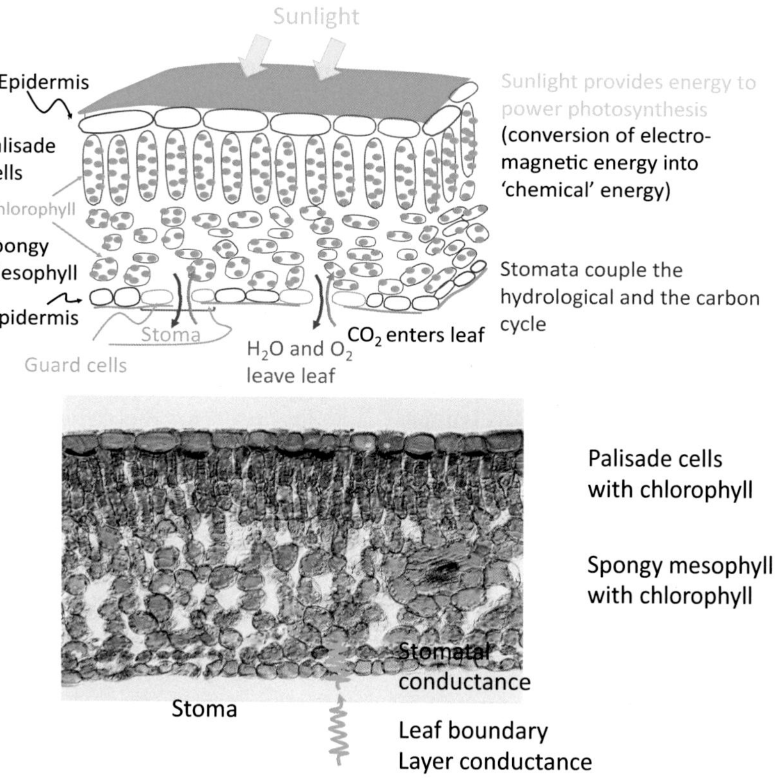

FIG. 10 Schematic and photograph of cross-section through a typical C3 leaf, with location of stomata and chlorophyll in the leaf.

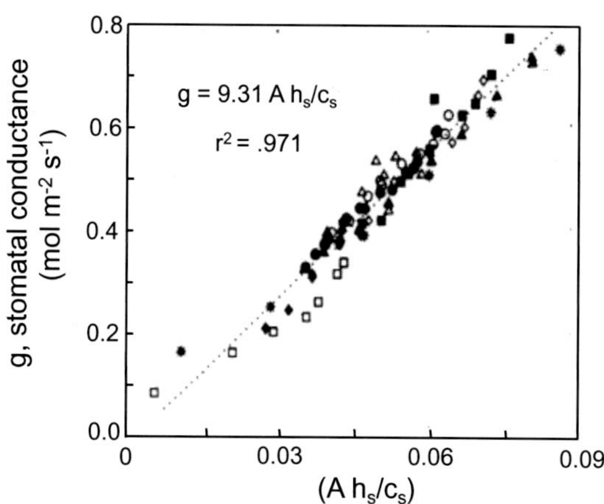

FIG. 11 A plot of stomatal conductance data versus $\frac{h_s}{c_s} \cdot A$, where A (μmol m^{-2} s^{-1}) is leaf CO_2 assimilation rate, h_s is relative humidity and c_s is CO_2 molar ratio (μmol mol^{-1}) at the leaf surface. *From Ball, J.T., et al., 1987. A model predicting stomatal conductance and its contribution to the control of photosynthesis under different environmental conditions. In: Biggens, J. (Ed.), Progress in Photosynthesis Research. Martinus Nijhoff, Netherlands, pp. 221–224.*

compensation point (see Section 3, Farquhar–von Caemmerer–Berry model).

More recently, stomatal conductance models have been developed, which take not only vapour pressure deficit but also soil water status into account (Sperry et al., 2017). To do so, they represent realistically water transport in plants and its dependence on stress on water in the plant xylem, and potential breakdown of hyraulics, as explained in more detail below.

3.1.3 *Carbon fixation capacity ('demand')*

The model for C3 photosynthesis carbon fixation capacity is based on (i) the main components of the carbon fixation process (see Chapter 3) and (ii) measurements of carbon assimilation rates versus leaf internal CO_2 concentrations, the so-called A–c_i curves. The carbon fixation cycle involves the two substrates RuBP (ribulose-1,5-bisphosphate) and CO_2, which catalysed by the enzyme rubisco produce a compound with three carbon atoms 3PGA (3-phosphoglycerate). In the next step, ATP (energy carrier) and NADPH (reducing power) are used to transform 3PGA into triose phosphate, a three carbon phosphate sugar. Next, triose phosphate is converted into sucrose, glucose, and starch, whilst phosphates are being recycled. The cycle closes itself by regeneration of the substrate RuBP using again ATP and NADPH as well as phosphates (Fig. 12).

The second piece of information for formulating a model of photosynthesis for use in Earth-System models is measurements of leaf carbon assimilation rates and their dependence on various parameters including, in particular, the CO_2 concentration inside stomata. Leaf carbon assimilation rate can be determined by enclosing a leaf in a purpose-built well-lit chamber through which a steady airflow passes. By measuring the CO_2 content of air before entering the chamber and after leaving the chamber, carbon uptake, or carbon assimilation rate, can be determined. With the same approach, at the same time, leaf H_2O loss is being measured. CO_2 and H_2O levels before and after entering the chamber are determined by measuring the absorption of an infrared laser beam. The devices used for these measurements are called Infrared Gas Analysers (IRGA).

One type of measurements done with such IRGAs is determination of how carbon assimilation rate A (μmol m^{-2} s^{-1}) varies with c_i, where c_i is CO_2 molar fraction (μmol (mol air)$^{-1}$) inside the stomata, so called 'A–c_i curves' (see Chapter 10). To measure these curves the light intensity in the chamber is chosen such that photosynthesis is a bit below, but close to, its maximum, and temperature in the chamber is fixed to 25°C chosen by convention as a reference point. The CO_2 level in the chamber is then raised slowly from 400 ppm (parts per million, i.e., number of CO_2 molecules per one million of air molecules) to higher CO_2 levels, and after a peak value, it is then steadily lowered. The CO_2 concentration c_i at the carboxylation site is determined using simultaneous measurement of leaf evaporation (and assuming that the ratio of stomatal conductance of H_2O and CO_2 is equal to the ratio of molecular diffusivities D

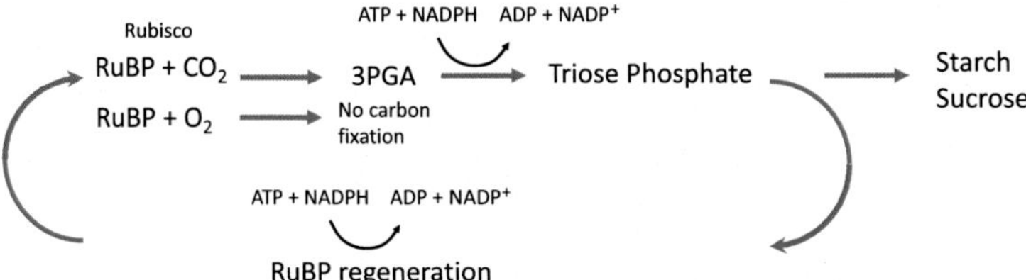

FIG. 12 Simplified schematic of CO_2 fixation by a plant which uses C3 photosynthesis for CO_2 fixation (Calvin Cycle).

of H_2O and CO_2 respectively, i.e., $\frac{g_{s,H_2O}}{g_{s,CO_2}} \approx \frac{D_{H_2O}}{D_{CO_2}}$, and that vapour pressure e (mol mol^{-1}) at the carboxylating site is at saturation ($e = e_{sat}$) (thus $\frac{A}{E} = \frac{g_{s,CO_2}(c_a - c_i)}{g_{s,H_2O}(e_{sat} - e_a)}$ where c_a and e_a is atmospheric CO_2 and H_2O molar fraction, respectively, which can be solved for c_i as $c_i = c_a - \frac{A}{E} \cdot \frac{D_{H_2O}}{D_{CO_2}} \cdot (e_{sat} - e_a)))$. By turning off the light source in the chamber, these analysers can also be used to measure leaf respiration under nonphotosynthesising conditions. This so-called 'dark respiration', R_d, is the result of the oxidation of hydrocarbons to produce energy carriers (ATP, adenosine triphosphate) and reducing power (NADPH) needed to sustain metabolism as well as for growth. Because of the way R_d is measured, it does not include carbon release caused by two processes occurring during the photosynthesis process under well-lit conditions: photorespiration (to be explained later on) and nitrate and sulphate reduction.

A–c_i curves measured by the approach just described typically look as in Fig. 13 for photosynthesis of one type of plants (primarily trees, so-called C3 photosynthesis).

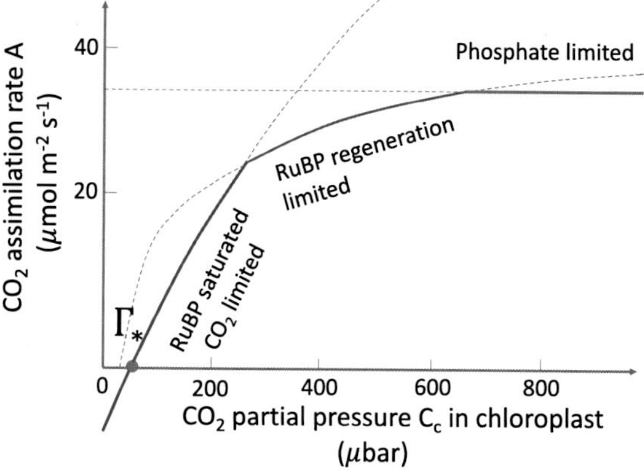

FIG. 13 Schematic of leaf CO_2 assimilation rate as a function of leaf internal CO_2 concentration under controlled conditions, a so-called 'A–c_i' curve. *Adapted from Von Caemmerer, S., 2000. Biochemical Models of Leaf Photosynthesis, Techniques in Plant Sciences No. 2, with permission from CSIRO Publishing, Australia.*

To derive the 'demand' component of the Farquhar–von Caemmerer–Berry model, we follow Von Caemmerer (2000). The derivation starts by expressing the carbon assimilation rate A in units of (moles m^{-2} s^{-1}) at the leaf level as the sum of carboxylation rate V_c (moles m^{-2} s^{-1}) of CO_2 by RuBP minus the fraction of CO_2 lost during oxygenation, which is known to be half the oxygenation rate V_o, and the so-called dark respiration R_d

$$A = V_c - 0.5 \cdot V_o - R_d = V_c \cdot (1 - 0.5 \cdot \phi) - R_d = \left(1 - \frac{\Gamma_*}{c_c}\right) \cdot V_c - R_d \tag{2}$$

where $\phi \equiv \frac{V_o}{V_c}$ is the ratio of oxygenation to carboxylation rate, and c_c is CO_2 molar fraction (e.g. in units (μmol mol^{-1})) at the carboxylating site in the chloroplast. The last identity in Eq. (2) holds as ϕ can be expressed as $= \frac{2\Gamma_*}{c_c}$. Γ_* is the 'CO_2 compensation point', defined as the CO_2 molar fraction at which carboxylation rate equals the photorespiration loss of CO_2 in the absence of dark respiration R_d. (The relationship is derived by expressing the carboxylation rate as $V_c = k \cdot c_c$, where k is the carboxylation reaction rate constant. Then, for $R_d = 0, A = k(c_c - \Gamma_*) = k \cdot c_c \cdot \left(1 - \frac{\Gamma_*}{c_c}\right) = (1 - 0.5 \cdot \phi) \cdot V_c = k \cdot c_c \cdot (1 - 0.5 \cdot \phi)$, and thus, $\phi = \frac{2\Gamma_*}{c_c}$).

The processes in the leaf involved in carbon fixation vary with the molar fraction c_c of CO_2 at the carboxylation site. CO_2 is fixed in the leaf by the enzyme rubisco (ribulose-1,5-bisphosphate carboxylase/oxygenase) from the substrates RuBP and CO_2. At comparably low levels of c_c, ample RuBP is available, and the enzyme reaction can be described by the Michaelis and Menten (1913) model of enzyme kinetics with limiting substrate CO_2. The Michaelis–Menten model conceptualises enzyme reactions as follows (Cermak, 2009)

$$\begin{array}{ccccc} & k_1 & & k_2 & \\ E+S & \leftrightarrows & E \cdot S & \leftrightarrows & E+P \\ & k_{-1} & & k_{-2} & \end{array}$$

In the first step, (i) the enzyme E forms a complex ES with substrates S (in our case, two substrates, RuBP and CO_2), it then (ii) catalyses the reaction, and (iii) it dissociates from product P. The k's are reaction constants. The model makes two assumptions, (a) $E \cdot S \approx \text{const}$ ('steady state kinetics') and (b) $k_{-2} \approx 0$. The production rate of product P can then be expressed as

$$\frac{d[P]}{dt} = \frac{V_{\max} \cdot [S]}{[S] + K_m}$$

where $V_{\max} \equiv k_2 \cdot [E_{\text{total}}]$ and E_{total} is the total amount of enzyme, and K_m the Michaelis–Menten constant. For small amounts of the substrate, here CO_2, the production rate increases linearly with $[S]$ with slope $\frac{V_{\max}}{K_m}$, whilst for large amounts, it asymptotes at a constant rate $V_{\max}$. As mentioned above, instead of using CO_2 as one of the substrates, rubisco can also use O_2, and thus, O_2 has the effect of inhibiting the carboxylation reaction. This is thought to be a relic of times during evolution when CO_2 levels in the atmosphere were much higher. The Michaelis–Menten model can be generalised to include such an inhibitor to

$$\frac{d[P]}{dt} = \frac{V_{\max} \cdot [S]}{[S] + K_m \cdot \left(1 + \frac{[I]}{K_i}\right)}$$

where $[I]$ is the molar concentration of the inhibitor and K_i is a measure of inhibitor efficiency.

For comparably low levels of c_c but ample availability of RuBP (RuBP saturated regime), this model is adapted for CO_2 carboxylation rate V_c: $V_c = \frac{V_{c,}\max \cdot c_c}{c_c + K_{CO_2} \cdot \left(1 + \frac{o_c}{K_o}\right)}$, where c_c and o_c are CO_2 and O_2 molar fractions at the carboxylating site in the chloroplast. Using Eq. (2), the RuBP saturated CO_2 assimilation rate, A_c, is

$$A_c = V_{c,}\max \cdot \left(\frac{c_c - \Gamma_*}{c_c + K_{CO_2} \cdot \left(1 + \frac{O_i}{K_o}\right)} \right) - R_d \ (\mu\text{mol}\,\text{m}^{-2}\,\text{s}^{-1}) \tag{3}$$

At higher c_c, RuBP regeneration may not be sufficiently rapid and thus RuBP starts to become limiting. RuBP regeneration requires energy provided by ATP and reducing power provided by NADPH. The NADPH consumption rate R to regenerate RuBP and the carboxylation rate V_c can be shown to be related as $R = (2 + 2 \cdot \phi) \cdot V_c$. Creation of NADPH from NADP+ needs two electrons; thus, electron transport rate J (μmol electrons $m^{-2}\,s^{-1}$) and carboxylation rate V_c are related by $V_c = \frac{J}{2^*(2+2\cdot\phi)}$. Using Eq. (2) and the relation $\phi = \frac{2\Gamma_*}{c_c}$ yields the RuBP regeneration limited CO_2 assimilation rate A_j (μmol $m^{-2}\,s^{-1}$)

$$A_j = \frac{1}{4} J \cdot \left(\frac{c_c - \Gamma_*}{c_c + 2 \cdot \Gamma_*} \right) - R_d \tag{4}$$

The electron transport rate in turn is the result of energy gain by photon capture by pigments (chlorophyll) of photosystem II. Absorbed irradiation I_{abs} (μmol photons $m^{-2}\,s^{-1}$) by photosystem II is given by $I_{\text{abs}} = \frac{1}{2} \cdot f_{\text{abs}} \cdot (1 - f) \cdot I$, where I (μmol photons $m^{-2}\,s^{-1}$) is incoming solar radiation, f_{abs} is fraction of absorbed radiation, f is a correction factor for nonuniformity of the spectrum of solar radiation, and the factor 1/2 takes into account the assumption that photosystem II absorbs half of incoming radiation whilst photosystem I absorbs the other half. An empirical relationship between electron transport rate and absorbed solar radiation is usually used

$$J = \frac{1}{1.4} (I_{\text{abs}} + J_{\max}) \cdot \left(1 - \sqrt{1 - 2.8 \frac{I_{\text{abs}} \cdot J_{\max}}{(I_{\text{abs}} + J_{\max})^2}} \right)$$

where $J_{\max}$ is an experimentally determined maximum electron transport rate, which is a function of temperature. In Eqs. (3), (4), it is usually assumed that the CO_2 concentration at the carboxylation site, c_c, is equal to the CO_2 concentration c_i inside the stomata and thus c_c is replaced by c_i.

At yet higher CO_2 molar fractions at the carboxylation site, CO_2 assimilation rate is limited if triose phosphate is not sufficiently rapidly removed to produce sugars and starch, and thus, phosphate becomes limiting for RuBP regeneration. This 'Phosphate limited assimilation rate' (μmol m^{-2} s^{-1}) regime can be described as (Amthor, 1984; Foley et al., 1996)

$$A_s = \frac{1}{2} V_{c,\max} - R_d \tag{5}$$

The realised assimilation capacity (or demand) is the minimum of these three regimes

$$A_{\text{biochem}} = \min\{A_c, A_j, A_s\} = f(c_c)$$

which is a function of c_c.

To estimate CO_2 assimilation rate, the demand and supply components are solved numerically for the CO_2 concentration c_c at the carboxylation site at which demand equals supply

$$A_{\text{biochem}}(c_c) = A_{\text{phys}}(c_c)$$

Mathematically, a root, $c_{c,\text{root}}$, of the function $h(c_c) = A_{\text{biochem}}(c_c) - A_{\text{phys}}(c_c)$ is determined. The CO_2 assimilation rate estimate is then given by $A_{\text{biochem}}(c_{c,\text{root}})$ (or equivalently $A_{\text{phys}}(c_{c,\text{root}})$).

Various components of this model depend strongly on environmental conditions like temperature, atmospheric CO_2, and light and are thus important determinants for future predictions of land vegetation productivity and functioning in a warming world (Fig. 14, Bernacchi et al., 2001). Generally, photosynthesis rates decrease above 30°C.

3.2 Observed and predicted future land vegetation changes

The primary purpose of the land vegetation component in Earth-System models is to predict to what extent, where and for how long into the future, land vegetation and soils will take up carbon from the atmosphere and thus will moderate the carbon problem, or whether instead land carbon pools including soils will start to loose carbon at some point and thus make the problem more difficult to solve.

A prediction of most Earth-System models is that land vegetation is currently and will continue to be a sink for atmospheric carbon during large parts of the 21st century (e.g., Cox, 2019, Fig. 15). One reason for this predicted continued carbon uptake is the specific formulation of the Farquhar–von Caemmerer–Berry photosynthesis model. According to this model, an increase in atmospheric CO_2 will lead to an increase in predicted photosynthesis rates, unless the model operates in the phosphate limited regime of photosynthesis (Fig. 13). This effect is known as 'CO_2 fertilisation' (Lloyd, 1999). Such carbon sink predictions tend to be reduced in models that include also cycles of growth co-limiting elements like the nitrogen and phosphorous cycles. Another prediction of these models is that stomatal opening will tend to be downregulated in a higher CO_2 world as a consequence of the Ball–Berry and similar models of stomatal control. This has not only implications for CO_2 but also on release of water vapour to the atmosphere via trees and thus on precipitation downwind of forest vegetation, or, i.e., the hydrological cycle.

Remote sensing observations of leaf area index (LAI) defined as total leaf area per ground area, (m^2 m^{-2}) derived in essence from NDVI seem to be consistent with a CO_2 fertilisation effect. Studies analysing these data report a positive trend in leaf area (Fig. 16), a so-called greening trend. Because satellite missions are of limited duration and thus data from several missions need to be combined, trend analysis of satellite data is a difficult task and the uncertainty of estimated trends high.

However, as is illustrated in Fig. 9, net gains by photosynthesis are not the only determinant of carbon store changes in land vegetation. It is

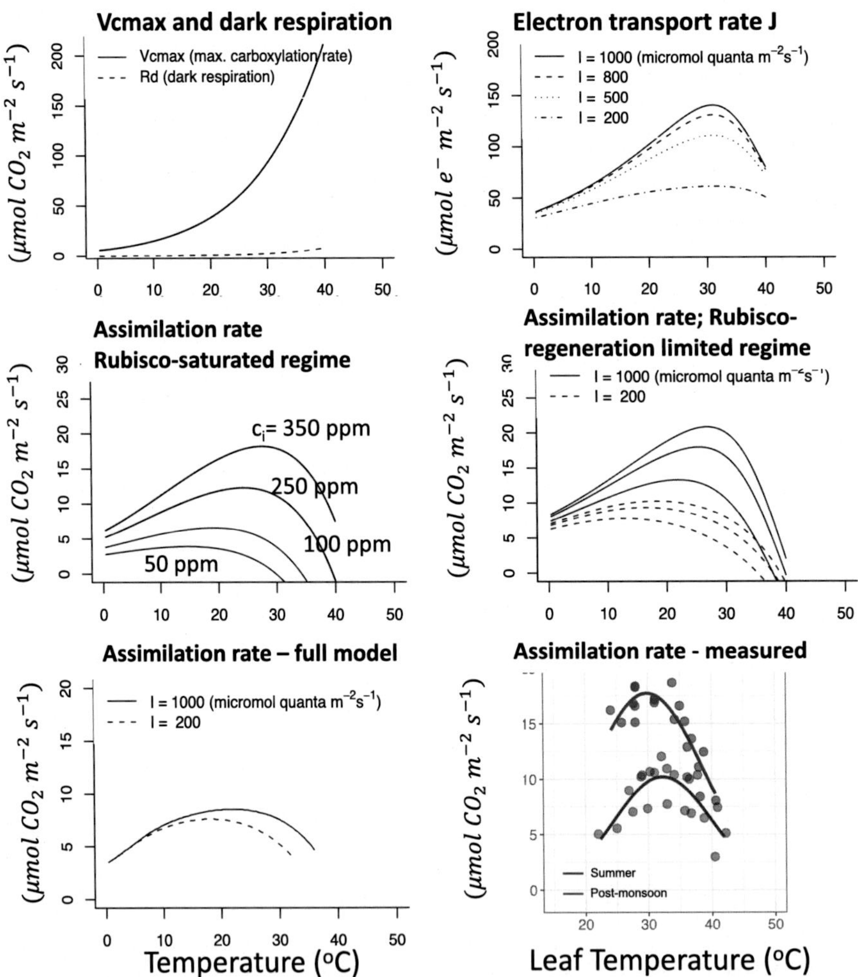

FIG. 14 Dependence of components of the Farquhar–Berry–von Caemmerer photosynthesis model on environmental variables (temperature, light, ambient CO_2) and model predicted and observed assimilation rates as a function of temperature (observed assimilation rates courtesy R. Tiwari).

rather changes in the balance between gains (NPP) and losses (mortality and heterotrophic respiration). In most simple terms

$$\frac{dB}{dt} = G - \mu \cdot B = G - \frac{1}{\tau} \cdot B$$

where B is carbon stocks in living and dead organic matter, e.g., in units of ($\mathrm{tC\,ha^{-1}}$), t is time (year), G is biomass gain ($\mathrm{tC\,ha^{-1}\,year^{-1}}$), and μ ($\mathrm{year^{-1}}$) is mortality. The inverse of mortality μ is called turnover time τ (year). At steady state, $B = \tau \cdot G$. Stocks as well as changes of stocks, thus,

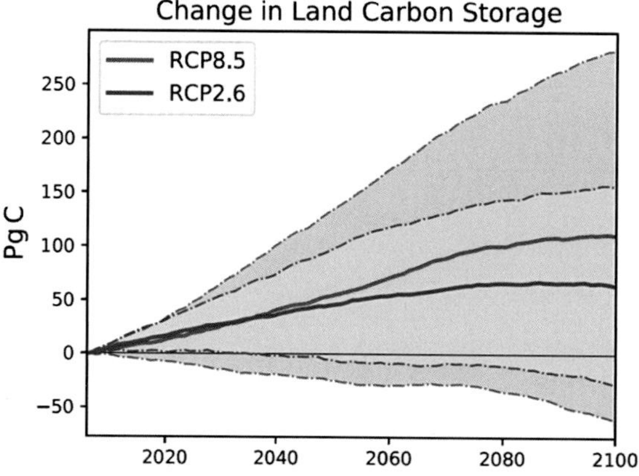

FIG. 15 Earth-System model predictions of land CO_2 uptake during this century for a low and a high anthropogenic CO_2 emissions scenario. *From Cox, P., 2019. Emergent constraints on carbon-cycle climate feedbacks. Curr. Clim. Change Rep. 5, 275–281.*

depend both on changes in turnover times (or tree mortality rates) as well as gains. Corresponding to the various organic carbon pools in these models, there is a spectrum of turnover times (or decay time-scales) as mentioned earlier.

Another source of information of how land plants and specifically trees have responded over recent decades to changes in environmental conditions is data from regularly repeated censuses at fixed forest plots (typically of 1 ha size). Censused information includes tree diameter, tree species, tree birth, and tree mortality. Typically trees with diameter equal or larger than 10 cm are being censused. Such data exist for the Amazon covering approximately the past 40 years (Fig. 17). Amazon humid forests are well suited to establish how tree growth and mortality has changed over recent decades because they have not been disturbed by human activity for a long time. Time-series of biomass gain and loss based on these forest inventory data reveal two interesting features: first humid Amazon forests have taken up carbon over past decades although the net carbon uptake (or 'sink') is decreasing over time. Second, tree mortality is increasing over time.

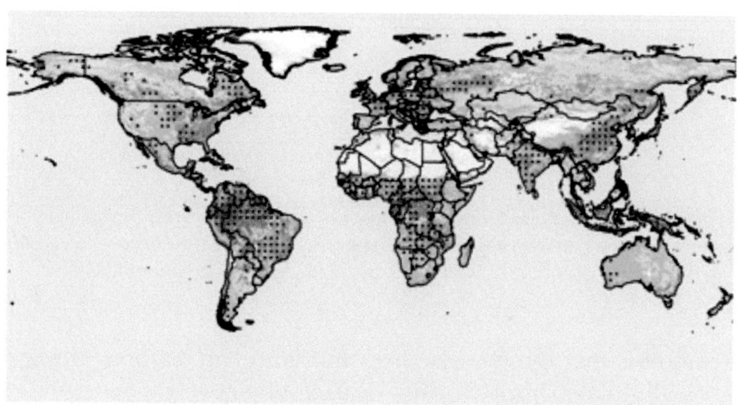

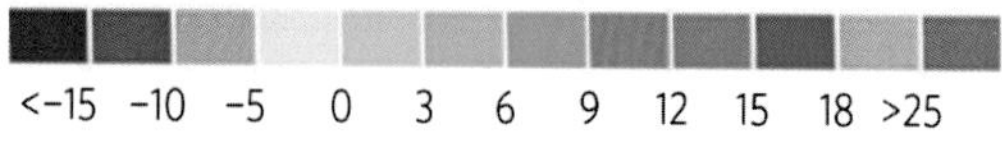

FIG. 16 Time trend of leaf area index (LAI) based on GLOBMAP LAI. *From Zhu, Z., Piao, S., Myneni, R.B., Huang, M., Zeng, Z., Canadell, J.G., Ciais, P., Sitch, S., Friedlingstein, P., Arneth, A. Cao, C., Cheng, L., Kato, E., Koven, C., Li, Y., Lian, X., Liu, Y., Liu, R., Mao, J., Pan, Y., Peng, S.., Peñuelas, J., Poulter, B., Pugh, T.A.M., Stocker, B.D., Viovy, N., Wang, X., Wang, Y., Xiao, Z., Yang, H., Zaehle, S., Zeng, N., 2016. Greening of the earth and its drivers. Nat. Clim. Chang. 6, 791–796.*

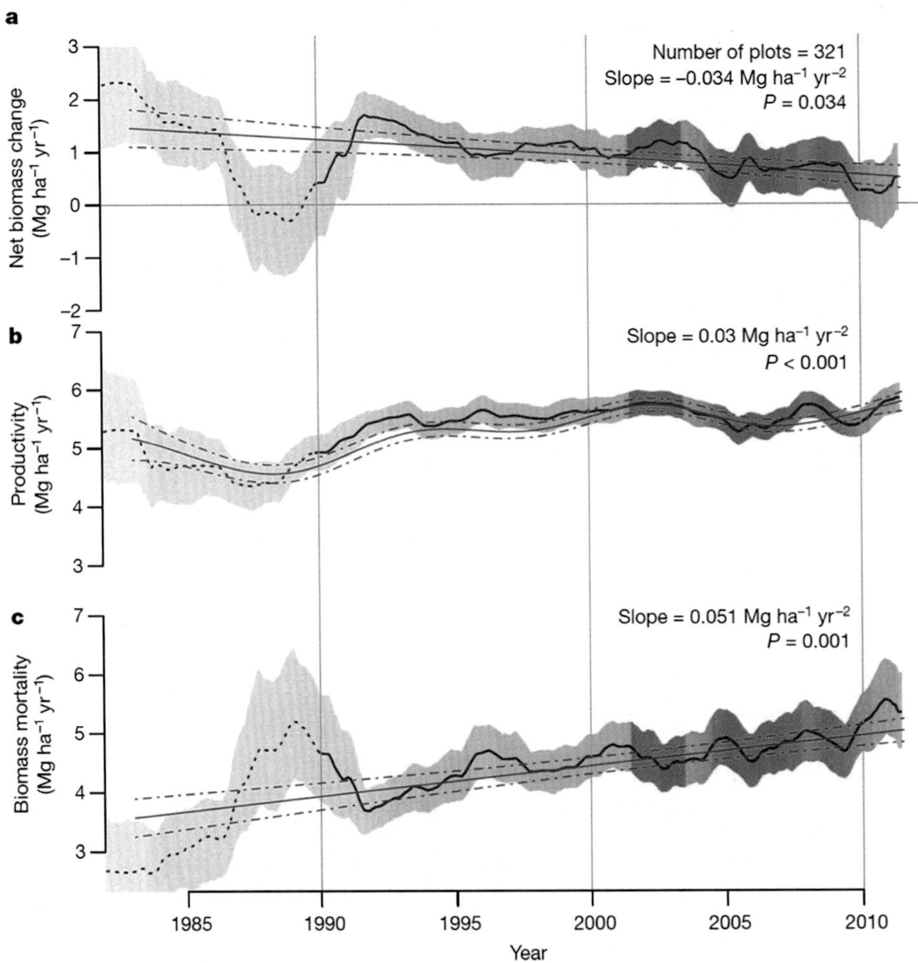

FIG. 17 Net biomass gain, productivity and biomass mortality of trees from a forest census network of approximately 250 1 ha plots distributed widely over the Amazon basin. *From Brienen, R., Phillips, O., Feldpausch, T., Gloor, M., Galbraith, D., et al., 2015. Long-term decline of the Amazon carbon sink. Nature 519, 344–348. doi:10.1038/nature14283.*

A possible interpretation is that the increase in atmospheric CO_2 has stimulated growth causing net carbon uptake at an early stage of the stimulus. More rapid growth however will eventually—with a lag time on the order of the life-time of a tree—lead to earlier exposure of trees to increased mortality risk and thus will cause an increasing mortality trend. Eventually, a new steady state will be reached, and the opposing effect of trends in productivity and mortality on carbon storage will then offset each other.

Earth-System models, which represent land vegetation as a photosynthesising spatially homogeneously distributed medium, are not able to capture such effects. Proper representation of these effects needs inclusion of tree population dynamics: birth, growth, competition, and death (Manusch et al., 2012). Representation of population dynamics is also necessary for mechanism-based prediction of species specific effects of drought and heat. Predictions are particularly important in the increasingly hot tropics where temperatures are reaching levels, which may already affect functioning of plants adversely. Efforts to include population dynamics have been pioneered by Moorcroft et al. (2001); nonetheless, spatial scaling from forest plot (1 ha) to climate model grid cell (e.g. 100 km × 100 km) remains a challenge and the problem has not yet been fully resolved.

Whether natural land vegetation in a new steady state will hold more carbon compared to preindustrial times depends on the relationship between enhanced growth and mortality risk (or, i.e., turnover time) as explained earlier. Insight into this relationship can be gained from tree ring data and forest inventory data. A recent analysis of these data demonstrates that there is a nearly universal trade-off between early growth rate and longevity both across species and within species (Fig. 18). This universality suggests that increased carbon uptake by land vegetation by a growth stimulation due to CO_2 will be offset by an increase in mortality due to shortening of tree lifespans (i.e., reduction in forest turn-over time, τ) after reaching a new steady state. Thus, the current land carbon sink component caused by CO_2 fertilisation is likely a transient phenomenon unlike what Earth-System models predict. Currently, the architecture of Earth-System models is not very well suited to properly reproduce this trade-off.

As mentioned earlier, a second prediction of Earth-System models is that stomatal conductance will be downregulated in a high CO_2 world. As CO_2 in the atmosphere increases, plants need to open stomata to a lesser degree in order to capture the same amount of carbon and thus also less water is being transpired. Thus, one may expect that the ratio of carbon gained to water lost is higher in an elevated CO_2 world, or that so-called water use efficiency of plants has increased.

An indication whether this is indeed the case comes from a budget of atmospheric $^{13}CO_2$ (Fig. 7). As discussed earlier, fossil fuel burning will lead to increasingly isotopically lighter atmospheric CO_2 (or, i.e., increasingly negative $\delta^{13}CO_2$) as is indeed observed. However, the decreasing trend cannot be fully explained by fossil fuel emissions alone. The trend is not

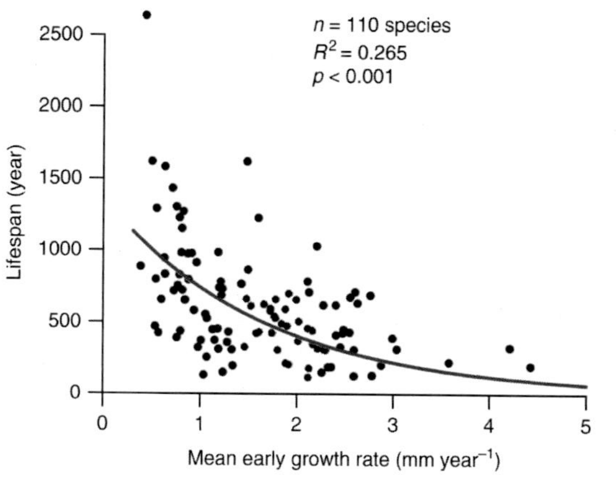

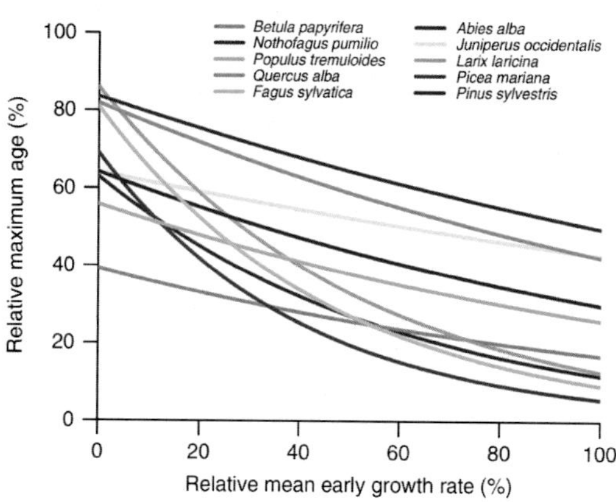

FIG. 18 Growth-Longevity trade-off across species and within a species inferred from tree ring data (Brienen et al., 2020). *From Brienen, R.J.W., Caldwell, L., Duchesne, L., Voelker, S., Barichivich, J., Baliva, M., Ceccantini, G., Di Filippo, A., Helama, S., Locosselli, G.M., Lopez, L., Piovesan, G., Schöngart, J., Villalba, R., Gloor, E., 2020. Forest carbon sink neutralized by pervasive growth-lifespan trade-offs. Nat. Commun. 11, 4241.*

sufficiently steep. The trend can be explained, however, if plant ^{13}C discrimination $\Delta \equiv \delta^{13}C_{atm} - \delta^{13}C_{plant}$ increases over time.

Based on (i) the equations for assimilation $A = k(c_i - \Gamma_*) - R_d$, $A' = k'(c_i - \Gamma_*') - R_d'$ using $V_{CO2} = kc_c$ and an apostrophe (′) designates quantities analogue to $^{12}CO_2$ but for $^{13}CO_2$, (ii) that at steady state assimilation rate equals CO_2 flux through stomata $A = F_{CO2} = g \cdot (c_a - c_i)$, $A' = g' \cdot (c_a' - c_i')$,and that $R_{plant} = \frac{A'}{A}$. Farquhar et al. (1982) derived the following model of plant ^{13}C discrimination

$$\Delta = a + (b-a)\frac{c_i}{c_a} - \frac{f\Gamma_* + \frac{e}{k}R_d}{c_a}$$

where a (=4.4‰) is the fractionation caused by slower diffusion of $^{13}CO_2$ compared with $^{12}CO_2$ through stomata, b (=30‰) is fractionation during carboxylation caused by discrimination of rubisco against $^{13}CO_2$ inside the leaf, f (=12‰) is the discrimination due to photorespiration, and e discrimination during dark respiration and thought to be small. The first two terms of the equation for Δ, $a + (b-a)\frac{c_i}{c_a}$, can be interpreted as follows: if stomata are wide open, $\frac{c_i}{c_a}$ is close to one, and thus, Δ approaches b, or, i.e., all fractionation is predominantly due to fractionation of the carboxylation reaction. If, on the other extreme, stomata are nearly closed, $\frac{c_i}{c_a}$ decreases and Δ approaches a, or, i.e., fractionation is dominated by fractionation during diffusion through stomata. Because, in this case, nearly all carbon inside the leaf has been fixed, carboxylation fractionation is offset by enrichment of ^{13}C inside the stomata caused by carboxylation fractionation (a process known as Rayleigh distillation). In reality, these two extremes are not reached in nature.

As mentioned above, the decreasing trend in atmospheric $\delta^{13}CO_2$ cannot be explained by the Suess effect alone, which overpredicts the decrease in atmospheric $\delta^{13}CO_2$. A possible process that reduces the decreasing trend of atmospheric $\delta^{13}CO_2$ is an increase of discrimination of plants against atmospheric $^{13}CO_2$. Assuming that it is indeed an increasing trend over time in plant discrimination against $^{13}CO_2$ which explains the missing component of the atmospheric $\delta^{13}CO_2$ trend then based on the model for discrimination of Farquhar et al. (1982), the ratio $\frac{c_i}{c_a}$ can be estimated. It turned out that the data imply that $\frac{c_i}{c_a}$ has remained approximately constant over time (Keeling et al., 2017). Now water use efficiency WUE is defined as $\text{WUE} \equiv \frac{A}{E}$, where E is evapotranspiration and can be expressed as $\text{WUE} = \frac{g_{s,CO_2}(c_a - c_i)}{g_{s,H_2O}(e_i - e_a)} \approx \frac{1}{1.6}c_a\left(\frac{1 - \frac{c_i}{c_a}}{e_i - e_a}\right)$. Since atmospheric CO_2, c_a, has increased rapidly over recent decades, and $\frac{c_i}{c_a}$ has remained constant, WUE is likely to have increased over recent decades. Interestingly, and consistent with this result, the density of stomata has decreased compared to preindustrial times (Woodward, 1987).

Earth-System Model Climate projections suggest that extreme climate events like severe floods and droughts will increase in the future. This will be in addition to steadily increasing temperatures which will increase plant vapour pressure deficit. For future carbon storage and forest composition, it is necessary to represent realistically in models how these events will affect trees, specifically which tree species are robust and which ones threatened under such conditions, and thus eventually how vegetation will change. Earth-System models are likely not yet able to predict how vegetation will shift over the coming decades. One reason is lack of knowledge and data about the sensitivity of vegetation to climate stress.

One important predictor of tree stress and potentially mortality is tree hydraulic failure. Over the past decade or so (McDowell et al., 2008; Anderegg et al., 2015), it has become clear that the combination of low soil water content and vapour pressure deficit can severely damage trees and may lead to tree death. Water is transported by trees from the soil to the canopy as thin water filaments in conducting vessels (xylem). The drier the soil and the stronger the vapour pressure deficit, the stronger the stress, or negative pressure, exerted on these filaments. Water under negative pressure, according to the laws of physics, should change its aggregate state to water vapour, a transition which plants are usually able to prevent. However, if stress exceeds species specific thresholds, water filaments embolise and rupture. As a consequence, water conductivity is impaired preventing water from reaching the canopy, which may lead to tissue death and ultimately tree death due to desiccation. Increasingly data of tree hydraulic failure thresholds for a wide range of species have become available, which provide a basis to represent such processes in Earth-System models, and developments are underway to incorporate tree hydraulics into models.

Vegetation will also be exposed to increasing temperatures, particularly in tropical regions, where temperatures are already high. This may possibly affect tree performance negatively, although existing results are not fully conclusive how temperature extremes will affect vegetation. Various methods have and are being applied to determine how heat affects vegetation, including tree performance, particularly in the tropics. Methods for trees include comparisons of forest censuses, tree ring based growth data, forest stand scale flux measurements and SIF data with climate fields (e.g., Goulden et al., 2004), controlled greenhouse heating experiments, as well as approaches probing the temperature sensitivity of photosynthesis at the leaf level.

As photosynthesis is a central theme of this book we discuss an approach that has recently been applied to get some insight into temperature sensitivity of photosynthesis of natural vegetation. The methodology assesses limits of photosystem II (PSII) by exposing leaves in the dark to saturating light pulses (PAM, pulse amplitude modulation) and measuring the yield of chlorophyll fluorescence in responses to these pulses (Sastry and Barua, 2017; Schreiber and Berry, 1977; Tiwari et al., 2021). From the fluorescence signal, the efficiency of PSII photochemistry, or quantum yield, can be estimated (see Chapter 10). PSII quantum yield or ϕPSII as a function of leaf temperature measured using this methodology tends to follow a sigmoidal curve. The decrease at high temperatures has been interpreted as a decrease in functioning of photosystem II. Once very low levels of quantum efficiency have been reached photosystem II may not recover to normal functioning when temperatures are reduced to normal levels. The temperature corresponding to a 50% ϕPSII decrease is referred to as T_{50} and is a frequently used metric of leaf thermotolerance. Sastry and Barua (2017) have compiled T50 data for tropical trees and compared them with expected future air temperatures (Fig. 19). A slightly more realistic comparison would compare leaf temperatures with air temperatures. Leaf temperatures tend to be higher than air temperatures during the day as a result of the heating by solar radiation. Thus, the comparison of Sastry and Barua (2017) is rather conservative. The comparison suggests that leaf photosynthesis will be affected by future temperature levels. The exact consequences of this result on future tropical vegetation is not straightforward; nonetheless, it clearly raises concerns.

In summary, many aspects of the carbon cycle necessary to represent land vegetation in Earth-System models are well understood and quite ingenious models, particularly of photosynthesis, have successfully been developed and been included in land vegetation representation. It has also been possible to include successfully comparably simple predictors of vegetation type. Nonetheless it seems that key trade-offs are likely not yet properly represented, as they need to include realistic population dynamic mechanisms. Similarly sensitivities and thresholds of vegetation functioning caused by heat and drought stress, particularly in the tropics,

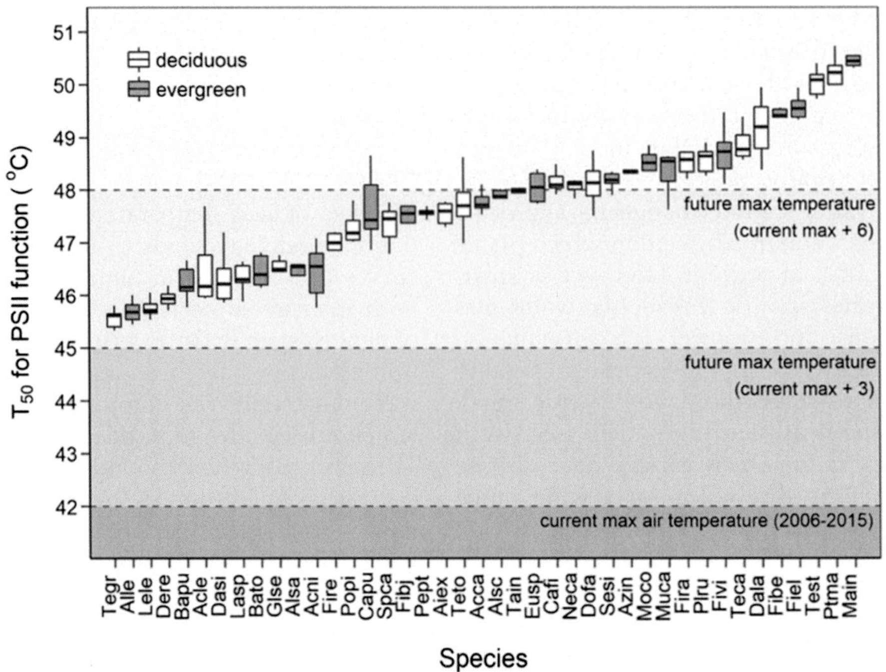

FIG. 19 Comparison of leaf thermal tolerance T_{50} for 41 tropical tree species and expected air temperatures in 3 and 6 degree warmer environment. *From Sastry, A., Barua, D., 2017. Leaf thermotolerance in tropical trees from a seasonally dry climate varies along the slow-fast resource acquisition spectrum. Sci. Rep. 7, 11246.*

are not fully understood yet and thus not yet realistically represented in Earth-System models. Efforts to make progress along these lines are currently underway.

Finally, given that photosynthesis captures CO_2 and that plants have substantially contributed to the gradual decrease of atmospheric CO_2 from 45 million to 20 million years ago, it is natural to ask to what extent photosynthesis may play a role in reducing the anthropogenic perturbation of atmospheric CO_2 caused by fossil fuel burning. On time-scales relevant to humans one possible role played by photosynthesis is via its link to carbon uptake and storage in land vegetation. Amongst land vegetation pools, forests store the most carbon and thus carbon can be sequestered most effectively via afforestation. As afforestation needs land, an indirect role of photosynthesis to increase carbon sequestration is by increasing crop productivity/yield, e.g., by increasing its efficiency (Evans, 2013). This would free up land to grow forests. Crop productivity has indeed increased substantially over the past decades.

Another second possible role for photosynthesis is via stimulation of growth of oceanic photo-synthesising micro-organisms by adding nutrients. As a result, the flux of dead organic matter to the deep sea is expected to increase and a larger amount of carbon be buried in sediments.

To evaluate the first of these mechanisms, several factors need to be considered. First, how much carbon per area can be hold by such forests, and second, what fraction of land which can potentially be afforested can realistically be used for this purpose, given pressure from other land uses like agriculture (or wood production). Existing studies suggest that maximally on the order of 50–100 PgC can be sequestered in forests (Lewis et al., 2019), roughly 5 to 8 years worth of fossil fuel emissions at the emission rate of the year 2021, i.e., not that large an amount. It also needs to be recognised that there is no guarantee that the use of these forests as carbon stores will really be uphold in the future. Also currently (the year 2021) globally forest area is shrinking (by approximately an area of 50 million ha in the decade from 2010 to 2020, roughly two times the area of the United Kingdom, although the decrease rate has slowed down from 150 million ha in 1980 to 1990).

The main approach considered and experimentally explored with regards to ocean productivity stimulation is iron fertilisation (release) in the Southern Ocean. Iron is an essential trace element for growth of ocean micro-organisms and the Southern Ocean is depleted in iron thus it is an ideal place to test the methodology. Model simulations suggest that, similarly to forest sequestration, possibly on the order of 50–100 PgC could be sequestered in this way. However, there are various side effects that may at least partially offset the greenhouse gas balance of such an approach. For example, it may lead to anoxic zones and in turn denitrification and release of N_2O, a very strong greenhouse gas, to the atmosphere; thus, it is not clear whether ocean productivity fertilisation is a safe option (e.g., Williamson et al., 2012). Altogether photosynthesis can thus contribute to moderate the carbon problem humanity is currently facing, but it is not a panacea.

References

Amthor, J.S., 1984. The role of maintenance respiration in plant growth. Plant Cell Environ. 7, 561–569.

Anderegg, W.R.L., Flint, A., Huang, C.-Y., Flint, L., Berry, J.-A., Davis, F.W., Sperry, J.S., Field, C.B., 2015. Tree mortality predicted from drought-induced vascular damage. Nat. Geosci. 8, 367–371.

Andres, R.J., Fielding, D.J., Marland, G., Boden, T.A., Kumar, N., 1999. Carbon dioxide emissions from fossil-fuel use, 1751–1950. Tellus 51B, 759–765.

Ball, J.T., et al., 1987. A model predicting stomatal conductance and its contribution to the control of photosynthesis under different environmental conditions. In: Biggens, J. (Ed.), Progress in Photosynthesis Research. Martinus Nijhoff, Netherlands, pp. 221–224.

Bernacchi, C.J., Singsaas, E.L., Pimentel, C., Portis Jr., A.R., Long, S.P., 2001. Improved temperature response functions for models of Rubisco-limited photosynthesis. Plant Cell Environ. 24, 253–259.

Berner, R.A., 1990. Atmospheric carbon dioxide levels over phanerozoic time. Science 249, 1382–1386.

Brienen, R., Phillips, O., Feldpausch, T., Gloor, M., Galbraith, D., et al., 2015. Long-term decline of the Amazon carbon sink. Nature 519, 344–348. https://doi.org/10.1038/nature14283.

Brienen, R.J.W., Caldwell, L., Duchesne, L., Voelker, S., Barichivich, J., Baliva, M., Ceccantini, G., Di Filippo, A., Helama, S., Locosselli, G.M., Lopez, L., Piovesan, G., Schöngart, J., Villalba, R., Gloor, E., 2020. Forest carbon sink neutralized by pervasive growth-lifespan trade-offs. Nat. Commun. 11, 4241.

Cermak, N., 2009. Fundamentals of Enzyme Kinetics: Michaelis-Menten and Deviations. Lecture Notes https://www.coursehero.com/file/13158116/383final-cermak-enzymekinetics-20090312/.

Charney, J.G., Arakawa, A., Baker, D.J., Bolin, B., Dickinson, R.E., Goody, R.M., et al., 1979. Carbon Dioxide and Climate: A Scientific Assessment. National Academy of Sciences, Washington, DC.

Cloud, P., 1972. A working model of the primitive earth. Am. J. Sci. 272, 537–548.

Cowan, I.R., 1977. Stomatal behaviour and environment. Adv. Bot. Res. 4, 117–228.

Cox, P., 2019. Emergent constraints on carbon-cycle climate feedbacks. Curr. Clim. Change Rep. 5, 275–281.

Demoulin, C.F., Lara, Y.J., Cornet, L., Francois, C., Baurain, D., Wilmotte, A., Javaux, E.J., 2019. Cyanobacteria evolution: insight from the fossil record. Free Radic. Biol. Med. 140, 206–223.

Evans, J.R., 2013. Improving photosynthesis. Plant Physiol. 162, 1780–1793.

Eyring, V., Bony, S., Meehl, G.A., Senior, C.A., Stevens, B., Stouffer, R.J., Taylor, K.E., 2016. Overview of the coupled model intercomparison project phase 6 (CMIP6) experimental design and organization. Geosci. Model Dev. 9, 1937–1958. https://doi.org/10.5194/gmd-9-1937-2016.

Farquhar, G.D., von Caemmerer, S., Berry, J.A., 1980. A biogeochemical model of photosynthesis in leaves of C3 species. Planta 149, 78–90.

Farquhar, G.D., O'Leary, M.H., Berry, J.A., 1982. On the relationship between carbon isotope discrimination and the intercellular carbon dioxide concentration in leaves. Aust. J. Plant Physiol. 9, 121–137.

Farquhar, J., Savarino, J., Airieaua, S., Thiemens, M.H., 2001. Observation of wavelength-sensitive mass-independent sulfur isotope effects during SO_2 photolysis: implications for the early atmosphere. J. Geophys. Res. 106 (E12), 32829–32839.

Field, C.B., Randerson, J.T., Malmstroem, C., 1995. Global net primary production: combining ecology and remote sensing. Remote Sens. Environ. 51, 74–88.

Foley, J.A., Prentice, I.C., Ramankutty, N., Levis, S., Pollard, D., Sitch, S., Haxeltine, A., 1996. An integrated biosphere model of land surface processes, terrestrial carbon balance, and vegetation dynamics. Glob. Biogeochem. Cycles 10, 603–628.

Friedlingstein, P., et al., 2019. Global carbon budget 2019. Earth Syst. Sci. Data 11, 1783–1838.

Gorham, E., 1991. Northern peatlands: role in the carbon cycle and probable responses to climatic warming. Ecol. Appl. 1 (2), 182–195. https://doi.org/10.2307/1941811.

Goulden, M.L., Miller, S.D., da Rocha, H.R., Menton, M.C., den Freitas, H.C., Figueira, A.M.E.S., de Sousa, C.A.D., 2004. Diel and seasonal patterns of tropical forest CO_2 exchange. Ecol. Appl. 14 (4), S42–S54. https://doi.org/10.1890/02-6008.

Graven, H., Allison, C.E., Etheridge, D.M., Hammer, S., Keeling, R.F., Levin, I., Meijer, H.A.J., Rubino, M., Tans, P.P., Trudinger, C.M., Vaughn, B.H., White, J.W.C., 2017. Compiled records of carbon isotopes in atmospheric for historical simulations in CMIP6. Geosci. Model Dev. 10, 4405–4417. https://doi.org/10.5194/gmd-10-4405-2017.

Holland, H.D., 2006. The oxygenation of the atmosphere and oceans. Philos. Trans. R. Soc. B 361, 903–915. https://doi.org/10.1098/rstb.2006.1838.

Keeling, R., Shertz, S.R., 1992. Seasonal and interannual variations in atmospheric oxygen and implications for the global carbon cycle. Nature 358, 727–732.

Isley, A., Abbott, D., 1999. Plume-related mafic volcanism and the deposition of banded iron formation. J. Geophys. Res. 104 (B7), 15461–15477.

Keeling, R.F., Graven, H.D., Welp, L.R., Resplandy, L., Bia, J., Piper, S.C., Sun, Y., Bollenbacher, A., Meijer, H.A.J., 2017. Atmospheric evidence for a global secular increase in carbon isotopic discrimination of land photosynthesis. Proc. Natl. Acad. Sci. USA 114 (39), 10361–10366.

Leuning, R., 1995. A critical appraisal of a combined stomatal-photosynthesis model for C_3 plants. Plant Cell Environ. 18, 339–355.

Lewis, S., Wheeler, C.E., Mitchard, E., 2019. Regenerate natural forests to store carbon. Nature 568, 25–28.

Lloyd, J., 1999. The CO_2 dependence of photosynthesis, plant growth responses to elevated CO_2 concentrations and their interaction with soil nutrient status, II. Temperate and boreal forest productivity and the combined effects of increasing CO_2 concentrations and increased nitrogen deposition at a global scale. Funct. Ecol. 13, 439–459.

Manabe, S., Bryan, K., 1969. Climate calculation with a combined ocean-atmosphere model. J. Atmos. Sci. 26 (4), 786–789.

Manusch, C., Bugmann, H., Heiri, C., Wolf, A., 2012. Tree mortality in dynamic vegetation models—a key feature for accurately simulating forest properties. Ecol. Model. 243, 101–111. https://doi.org/10.1016/j.ecolmodel.2012.06.008.

McDowell, N., Pockman, W.T., Allen, C.D., Breshears, D.D., Cobb, N., Kolb, T., Plaut, J., Sperry, J., West, A., Williams, D.G., Yepez, E.A., 2008. Mechanisms of plant survival and mortality during drought: why do some plants survive while others succumb to drought? New Phytol. 178, 719–739.

Michaelis, L., Menten, M.L., *1913*. Die Kinetik der Invertinwirkung. Biochem. Z. *49*, *333–369*.

Moorcroft, P., Hurtt, G., Pacala, S., 2001. A method for scaling vegetation dynamics: the ecosystem demography model (ED). Ecol. Monogr. 71 (4), 557–586.

Pagani, M., Caldeira, K., Berner, R., Beerling, D.J., 2009. The role of terrestrial plants in limiting atmospheric CO_2 decline over the past 24 million years. Nature 460, 85–88.

Randerson, J.T., Masiello, C.A., Still, C.J., Rahn, T., Poorter, H., Field, C.B., 2006. Is carbon within the global terrestrial biosphere becoming more oxidized? Implications for trends in atmospheric O2. Glob. Chang. Biol. 12, 260–271. https://doi.org/10.1111/j.1365-2486.2006.01099.x.

Sastry, A., Barua, D., 2017. Leaf thermotolerance in tropical trees from a seasonally dry climate varies along the slow-fast resource acquisition spectrum. Sci. Rep. 7, 11246.

Schreiber, U., Berry, J.A., 1977. Heat-induced changes of chlorophyll fluorescence in intact leaves correlated with damage of the photosynthetic apparatus. Planta 136, 233–238.

Sherwood, S., Webb, M.J., Annan, J.D., Armour, K.C., Forster, P.M., Hargreaves, J.C., Hegerl, G., Klein, S.A., Marvel, K.D., Watanabe, M., Andrews, T., Bretherton, C.-S., Foster, G.L., Hausfather, Z., von der Heydt, A.S., Knutti, R., Mauritsen, T., Norris, J.R., Proistosescu, C., Rugenstein, M., Schmidt, G.A., Tokarska, K.B., Zelinka, M.D., 2020. An assessment of earth's climate sensitivity using multiple lines of evidence. Rev. Geophys. https://doi.org/10.1029/2019RG000678.

Somkuti, P., Bösch, H., Feng, L., Palmer, P.I., Parker, R.J., Quaife, T., 2020. A new space-borne perspective of crop productivity variations over the US Corn Belt. Agric. For. Meteorol. 281, 107815.

Sperry, J.S., Venturas, M.D., Anderegg, W.R.L., Mencuccini, M., Mackay, D.S., Wang, Y., Love, D.M., 2017. Predicting stomatal responses to the environment from the optimization of photosynthetic gain and hydraulic cost. Plant Cell Environ. 40, 816–830.

Tiwari, R., Gloor, E., Foyer, C.H., da Cruz, W.J.A., Marimon, B.S., Marimon-Junior, B., Reis, S.M., de Souza, I.A., Slot, M., Winter, K., Krause, H.G., Ashley, D., Fauset, S., Galbraith, D., 2021. Photosynthetic quantum efficiency in South-Eastern Amazonian trees may be already affected by climate change. Plant Cell Environ. 44, 2428–2439.

Von Caemmerer, S., 2000. Biochemical Models of Leaf Photosynthesis, Techniques in Plant Sciences No. 2. CSIRO Publishing, Australia.

Welp, L.R., Keeling, R.F., Meijer, H.A.J., Bollenbacher, A.F., Piper, S.C., Yoshimura, K., Francey, R.J., Allison, C.E., Wahlen, M., 2011. Interannual variability in the oxygen isotopes of atmospheric CO_2 driven by El Niño. Nature 477, 579–582.

Williamson, P., Wallace, D.W.R., Law, C.S., Boyd, P.W., Collos, Y., Croot, P., Denman, K., Riebesell, U., Takeda, S., Vivian, C., 2012. Ocean fertilization for geoengineering: a review of effectiveness, environmental impacts and emerging governance. Process. Saf. Environ. Prot. 90, 475–488.

Woodward, I.F., 1987. Stomatal numbers are sensitive to increases in CO_2 from pre-industrial levels. Nature 327, 617–618.

Woodward, I.F., 2007. Global primary production. Curr. Biol. 17 (8), R269–R273.

Xiong, J., Bauer, C.E., 2002. Complex evolution of photosynthesis. Annu. Rev. Plant Biol. 53, 503–521.

Further reading

Doughty, C.E., Goulden, M.L., 2008. Are tropical forests near a high temperature threshold? J. Geophys. Res. 113, G00B07. https://doi.org/10.1029/2007JG000632.

Meroni, M., Rossini, M., Guanter, L., Alonso, L., Rascher, U., Colombo, R., Moreno, J., 2009. Remote sensing of solar-induced chlorophyll fluorescence: review of methods and applications. Remote Sens. Environ. 113, 2037–2051.

Ruban, A., 2013. Photosynthetic Membrane. John Wiley & Sons, West Sussex, United Kingdom.

Zhu, Z., Piao, S., Myneni, R.B., Huang, M., Zeng, Z., Canadell, J.G., Ciais, P., Sitch, S., Friedlingstein, P., Arneth, A., Cao, C., Cheng, L., Kato, E., Koven, C., Li, Y., Xu, L., Liu, Y., Liu, R., Mao, J., Pan, Y., Peng, S., Peñuelas, J., Poulter, B., Pugh, T.A.M., Stocker, B.D., Viovy, N., Wang, X., Wang, Y., Xiao, Z., Yang, H., Zaehle, S., Zeng, N., 2016. Greening of the earth and its drivers. Nat. Clim. Chang. 6, 791–796.

Index

Note: Page numbers followed by *f* indicate figures, *t* indicate tables, and *b* indicate boxes.

D

E

Printed in the United States
by Baker & Taylor Publisher Services